Alcohol and Aldehyde Metabolizing Systems

Volume II

Johnson Research Foundation Colloquia

Energy-Linked Functions of Mitochondria
Edited by Britton Chance
1963

Rapid Mixing and Sampling Techniques in Biochemistry
Edited by Britton Chance, Quentin H. Gibson, Rudolph H. Eisenhardt, K. Karl Lonberg-Holm
1964

Control of Energy Metabolism
Edited by Britton Chance, Ronald W. Estabrook, John R. Williamson
1965

Hemes and Hemoproteins
Edited by Britton Chance, Ronald W. Estabrook, Takashi Yonetani
1966

Probes of Structure and Function of Macromolecules and Membranes
Volume I Probes and Membrane Function
Edited by Britton Chance, Chuan-pu Lee, J. Kent Blasie
1971

Probes of Structure and Function of Macromolecules and Membranes
Volume II Probes of Enzymes and Hemoproteins
Edited by Britton Chance, Takashi Yonetani, Albert S. Mildvan
1971

Biological and Biochemical Oscillators
Edited by Britton Chance, E. Kendall Pye, Amal K. Ghosh, Benno Hess
1973

Alcohol and Aldehyde Metabolizing Systems
Edited by Ronald G. Thurman, Takashi Yonetani, John R. Williamson, Britton Chance
1974

Alcohol and Aldehyde Metabolizing Systems
Volume II Enzymology and Subcellular Organelles
Edited by Ronald G. Thurman, John R. Williamson, Henry R. Drott, Britton Chance
1977

Alcohol and Aldehyde Metabolizing Systems
Volume III Intermediary Metabolism and Neurochemistry
Edited by Ronald G. Thurman, John R. Williamson, Henry R. Drott, Britton Chance
1977

Papers contributed to Second International Symposium on Alcohol and Aldehyde Metabolizing Systems. Held in Philadelphia, Pennsylvania October 15-17, 1976.

Alcohol and Aldehyde Metabolizing Systems

Volume II

ENZYMOLOGY AND SUBCELLULAR ORGANELLES

Edited by

Ronald G. Thurman

Department of Pharmacology
University of North Carolina
Chapel Hill, North Carolina

John R. Williamson **Henry R. Drott**
Britton Chance

The Johnson Research Foundation
University of Pennsylvania
Philadelphia, Pennsylvania

Academic Press, Inc. New York San Francisco London 1977
A Subsidiary of Harcourt Brace Jovanovich, Publishers

ACADEMIC PRESS, INC.
111 Fifth Avenue, New York, New York 10003

United Kingdom Edition published by
ACADEMIC PRESS, INC. (LONDON) LTD.
24/28 Oval Road, London NW1

Library of Congress Cataloging in Publication Data

Main entry under title:

Alcohol and aldehyde metabolizing systems.

Vol. 1 contains papers presented at the 1st International Symposium on Alcohol and Aldehyde Metabolizing Systems, Stockholm, 1973; v. 2–3, papers from the 2d, University of Pennsylvania, 1976.
Includes bibliographical references.
1. Alcohol metabolism—Congresses. 2. Alcohol dehydrogenase—Congresses. 3. Catalase—Congresses. I. Thurman, Ronald G., ed. II. International Symposium on Alcohol and Aldehyde Metabolizing Systems, 1st, Stockholm, 1973. III. International Symposium on Alcohol and Aldehyde Metabolizing Systems, University of Pennsylvania, 1976. IV. Series: Pennsylvania. University. Eldridge Reeves Johnson Foundation for Medical Physics. Colloquia.
[DNLM: 1. Alcohols—Metabolism—Congresses. 2. Aldehydes—Metabolism—Congresses. W3 IN915S 1973a / QV82 I 61 1973a]
QP801.A3 I 62 1973 599'.01'33 73-5316
ISBN 0-12-691402-8 (v. 2)

PRINTED IN THE UNITED STATES OF AMERICA

CONTENTS

PEROXISOMES AND MICROSOMES

LIST OF PARTICIPANTS

Numbers in parentheses indicate the pages on which authors' contributions begin.

Abe, H. (393), First Department of Internal Medicine, Osaka University Medical School, Fukushima-ku, Osaka 553, Japan

Barry, A. (103), Department of Biochemistry, Rutgers Medical School, Piscataway, New Jersey 08854

Baumbach, G. (419), The Toxicology Center, Department of Pharmacology, University of Iowa, Iowa City, Iowa 52242

Bohlken, D. P. (63), Department of Biochemistry, University of Iowa, Iowa City, Iowa 52242

Boveris, A. (261, 441), Johnson Research Foundation, University of Pennsylvania, Philadelphia, Pennsylvania 19174

Brändén, C. I. (17), Department of Agricultural Chemistry I, Agricultural College of Sweden, Uppsala F, Sweden

Brentzel, H. James (373), Department of Biochemistry and Biophysics, University of Pennsylvania, Philadelphia, Pennsylvania 19174

Cancilla, P. (419), The Toxicology Center, Department of Pharmacology, University of Iowa, Iowa City, Iowa 52242

Chacos, N. (351), Department of Biochemistry, University of Texas Health Science Center, Dallas, Texas 75235

Chance, Britton (213, 261, 361, 393), Johnson Research Foundation, University of Pennsylvania, Philadelphia, Pennsylvania 19174

Chou, Ta-hsu (33), Department of Physiology, Medical School, Vanderbilt University, Nashville, Tennessee 37232

Cohen, Gerald (403), Department of Neurology, Mount Sinai School of Medicine, New York 10029

Collins, A. C. (223), Institute for Behavioral Genetics, University of Colorado, Boulder, Colorado 80309

Coon, Minor J. (307), Department of Biological Chemistry, University of Michigan Medical School, Ann Arbor, Michigan 48109

Creighton, Donald J. (53), Institute for Cancer Research, Philadelphia, Pennsylvania 19111

Deitrich, Richard A. (223), Department of Pharmacology, University of Colorado, Denver, Colorado

Drott, Henry R. (87, 119), Department of Biochemistry and Biophysics, University of Pennsylvania, Philadelphia, Pennsylvania 19174

Dubied, A. (63), Medizinisch-chemisches Institut der Universität Bern, CH-3012 Berne, Switzerland

Dworschack, R. T. (43), Department of Biochemistry, University of Iowa, Iowa City, Iowa 52242

Edson, C. R. (195), Center for Alcohol Studies, Rutgers University, New Brunswick, New Jersey 08903

Estabrook, Ronald W. (351), Department of Biochemistry, University of Texas Health Science Center, Dallas, Texas 75235

Greenfield, N. J. (195), Center for Alcohol Studies, Rutgers University, New Brunswick, New Jersey 08903

Hackenbrock, C. R. (235), Department of Cell Biology, University of Texas Health Science Center, Dallas, Texas 75235

Hagihara, B. (393), First Department of Internal Medicine, Osaka University Medical Center, Fukushima-ku, Osaka 553, Japan

Harvey, Richard A. (103, 119), Department of Biochemistry, Rutgers Medical School, Piscataway, New Jersey 08854

Hayreh, M. M. (419), The Toxicology Center, Department of Pharmacology, University of Iowa, Iowa City, Iowa 52242

Hayreh, S. S. (419), The Toxicology Center, Department of Pharmacology, University of Iowa, Iowa City, Iowa 52242

Hill, Edward J. (33), Department of Physiology, Medical School, Vanderbilt University, Nashville, Tennessee 37232

Höchli, M. (235), Department of Cell Biology, University of Texas Health Science Center, Dallas, Texas 75235

Jörnvall, Hans E. (145), Department of Chemistry I, Karolinska Institutet, S-104 01 Stockholm 60, Sweden

Jones, A. L. (291), Cell Biology Section—151 E, VA Hospital, San Francisco, California 94121

Kamada, T. (393), First Department of Internal Medicine, Osaka University Medical Center, Fukushima-ku, Osaka 553, Japan

Klinman, Judith (53), Institute for Cancer Research, Philadelphia, Pennsylvania 19111

Koch, O. R. (441), Centro de Patología Experimental, II Cátedra de Patología e Instituto de Química Biológia, Facultad de Medicina, Universidad de Buenos Aires, J.E. Uriburu 950, Buenos Aires, Argentina

Lazarow, Paul B. (275), Department of Biochemical Cytology, The Rockefeller University, New York 10021

Lee, In-Young (213), Johnson Research Foundation, University of Pennsylvania, Philadelphia, Pennsylvania 19174

Lennon, Michael B. (123), Department of Biochemistry, School of Medicine, Temple University, Philadelphia, Pennsylvania 19140

LeQuire, Virgil (33), Department of Physiology, Medical School, Vanderbilt University, Nashville, Tennessee 37232

Levin, W. (323), Hoffmann–La Roche, Inc., Nutley, New Jersey 07110

Lieber, Charles S. (341), Laboratory of Liver Disease, Nutrition, and Alcoholism, Bronx VA Hospital and Mount Sinai School of Medicine (CUNY), Bronx, New York 10468

Lindberg, P. (203), Institute of Zoophysiology, Box 560, S-751 22, Uppsala, Sweden

Lu, A. Y. (323), Hoffmann–La Roche, Inc., Nutley, New Jersey 07110

McMartin, Kenneth E. (419, 429), The Toxicology Center, Department of Pharmacology, University of Iowa, Iowa City, Iowa 52242

Makar, Adeeb B. (413, 419, 429), The Toxicology Center, Department of Pharmacology, University of Iowa, Iowa City, Iowa 52242

Malin, Edyth L. (137), Department of Biochemistry and Biophysics, University of Pennsylvania, Philadelphia, Pennsylvania 19174

Marchner, Hans (203), Institute of Zoophysiology, Box 560, S-751 22, Uppsala, Sweden

Martin-Amat, Gladys (419, 429), The Toxicology Center, Department of Pharmacology, University of Iowa, Iowa City, Iowa 52242

Mildvan, Albert S. (109, 119), Institute for Cancer Research, Philadelphia, Pennsylvania 19111

Miwa, G. T. (323), Hoffmann–La Roche, Inc., Nutley, New Jersey 07110

Mope, L. (157), Department of Biochemistry and Biophysics, University of Pennsylvania, Philadelphia, Pennsylvania 19174

Ohnishi, K. (341), Laboratory of Liver Disease, Nutrition, and Alcoholism, Bronx VA Hospital and Mount Sinai School of Medicine (CUNY), Bronx, New York 10468

Oshino, N. (261), Johnson Research Foundation, University of Pennsylvania, Philadelphia, Pennsylvania 19174

Park, Charles (33), Department of Physiology, Medical School, Vanderbilt University, Nashville, Tennessee 37232

Park, Jane H. (33), Department of Physiology, Medical School, Vanderbilt University, Nashville, Tennessee 37232

Petersen, Dennis R. (223), Institute for Behavioral Genetics, University of Colorado, Boulder, Colorado 80309

Pietruszko, Regina (79, 195), Center for Alcohol Studies, Rutgers University, New Brunswick, New Jersey 08903

Pinson, Richard (33), Department of Physiology, Medical School, Vanderbilt University, Nashville, Tennessee 37232

Plapp, Bryce V. (43, 63), Department of Biochemistry, University of Iowa, Iowa City, Iowa 52242

Ris, Margret M. (185), Medizinisch-chemisches Institut der Universität Bern, CH-3012 Berne, Switzerland

Roelofs, Robert (33), Department of Physiology, Medical School, Vanderbilt University, Nashville, Tennessee 37232

Rossmann, Michael G. (1), Department of Biological Science, Purdue University, Lafayette, Indiana 47907

Ryzewski, C. N. (79), Center for Alcohol Studies, Rutgers University, New Brunswick, New Jersey 08903

Sanny, Charles G. (167), Department of Biochemistry, Purdue University, W. Lafayette, Indiana 47907

Sato, Nobuhiro (393), First Department of Internal Medicine, Osaka University Medical School, Fukushima-ku, Osaka 553, Japan

Schmucker, D. L. (291), Cell Biology Section—151 E, VA Hospital, San Francisco, California 94121

Schulman, Martin P. (361, 381), Forensic Medicine, Karolinska Institutet, Stockholm, Sweden

Spring-Mills, Elinor J. (291), Cell Biology Section—151 E, VA Hospital, San Francisco, California 94121

Stoppani, A. O. M. (441), Centro de Patología Experimental, II Cátedra de Patología e Instituto de Química Biológica, Facultad de Medicina, Universidad de Buenos Aires, J.E. Uriburu 950, Buenos Aires, Argentina

Suhadolnik, Robert J. (123), Department of Biochemistry, School of Medicine, Temple University, Philadelphia, Pennsylvania 19140

Tank, A. W. (175), Department of Biochemistry, Purdue University, W. Lafayette, Indiana 47907

Tephly, Thomas R. (413, 419, 429), The Toxicology Center, Department of Pharmacology, University of Iowa, Iowa City, Iowa 52242

Thomas, P. E. (323), Hoffmann–La Roche, Inc., Nutley, New Jersey 07110

Thurman, Ronald G. (335, 373), Department of Pharmacology, University of North Carolina School of Medicine, Chapel Hill, North Carolina 27514

Tottmar, Olof (203), Institute of Zoophysiology, Box 560, S-751 22, Uppsala, Sweden

Vatsis, Kostas P. (307, 335, 361, 381), Department of Biological Chemistry, University of Michigan Medical School, Ann Arbor, Michigan 48109

Venteicher, Robert F. (157), Department of Biochemistry and Biophysics, University of Pennsylvania, Philadelphia, Pennsylvania 19174

von Wartburg, Jean-Pierre (63, 185), Medizinisch-chemisches Institut der Universität Bern, CH-3012 Berne, Switzerland

Weiner, Henry (167, 175), Department of Biochemistry, Purdue University, W. Lafayette, Indiana 47907

Welsh, Katherine M. (53), Institute for Cancer Research, Philadelphia, Pennsylvania 19111

Wermuth, B. (185), Medizinisch-chemisches Institut der Universität Bern, CH-3012 Berne, Switzerland

Werringloer, J. (351), Department of Biochemistry, University of Texas Health Science Center, Dallas, Texas 75235

Winer, A. D. (71), Department of Biochemistry, University of Kentucky, Lexington, Kentucky 40506

Yonetani, T. (157), Department of Biochemistry and Biophysics, University of Pennsylvania, Philadelphia, Pennsylvania 19174

Young, J. Maitland (109, 119, 137), Institute for Cancer Research, Philadelphia, Pennsylvania 19111

PREFACE

The papers in this book represent about one-half of the work presented at the Second International Symposium on Alcohol and Aldehyde Metabolizing Systems which was held on the campus of the University of Pennsylvania in October 1976. The other half of the contributions to the symposium will be found in Volume III of "Alcohol and Aldehyde Metabolizing Systems." The purpose of this meeting was to bring together experts in the field from a wide variety of backgrounds in an attempt to gain some clarity and insight into the problems of alcohol and aldehyde metabolism. There are over 10,000 annual publications relating to alcohol, and hundreds of these deal with alcohol and aldehyde metabolism. The Second International Symposium attempted to focus on this literature, to bring new findings into perspective, and to try to clarify controversial issues ranging from the molecular structure of enzymes on the one hand, to the clinical actions of alcohol in man on the other.

The papers in this and the companion volume are of two types. In each of the four major topics covered in the symposium (enzymology, subcellular organelles, intermediary metabolism, and neurochemistry), experts were invited to present material covering general principles in depth, but not necessarily related to alcohol or aldehydes. These lectures were followed by presentations of work relating to alcohol and aldehydes. Volume III contains papers in intermediary metabolism and neurochemistry. In this volume, papers dealing with general concepts of enzymology are followed by material dealing with the effects of alcohol and aldehydes on various dehydrogenases. The same general pattern is followed in the subsequent section on subcellular organelles.

I wish to thank Ms. Julie Revsin and Mr. David Prager for their dedicated assistance in the organization of the symposium and the manuscripts, respectively.

Ronald G. Thurman
Chapel Hill, N.C.

A COMPARISON OF THE ENZYMICALLY CATALYZED OXIDATION OF GLYCERALDEHYDE-3-PHOSPHATE AND LACTATE

M.G. Rossmann

Purdue University

Glyceraldehyde-3-phosphate dehydrogenase and lactate dehydrogenase have a similarity to alcohol dehydrogenase in the nature of their substrates, products and co-factor. However, these enzymes do not require any metal such as zinc for catalysis. The negatively charged substrates are bound to their respective enzymes by providing a positively charged environment. The reactive carbon atom (C_1 for glyceraldehyde-3-phosphate and C_2 for lactate) is then brought between the nicotinamide group of the NAD co-factor and a base (histidine in both cases). The relative position of these groups must be precise. Small spatial changes, as in different lactate dehydrogenase isozymes, cause large differences in rate constants. During the binding of co-factor and (subsequently) substrate to these enzymes there are conformational changes which orient and position the reactive groups. Chemical modification of the amino acids involved in these processes reduces or destroys activity.

I. INTRODUCTION

The scientific method consists of the unification of diverse facts in terms of the fewest possible concepts. This mode of reasoning leads to the expectation that reactions as similar as those characterizing the enzymic oxidation of alcohol, lactate and glyceraldehyde-3-phosphate (Fig. 1) might have a common basis. The similarity of some general properties of the corresponding enzymes (Table 1) is equally provocative. Waley (1969), for instance, speculated on the characteristics of dehydrogenases as well as kinases in terms of their common evolutionary origin (Fig. 2). Fortunately the three-dimensional structures of the four dehydrogenases listed in Table 1 were determined a few years ago giving some ground

Reaction catalyzed by alcohol dehydrogenase (ADH) is:

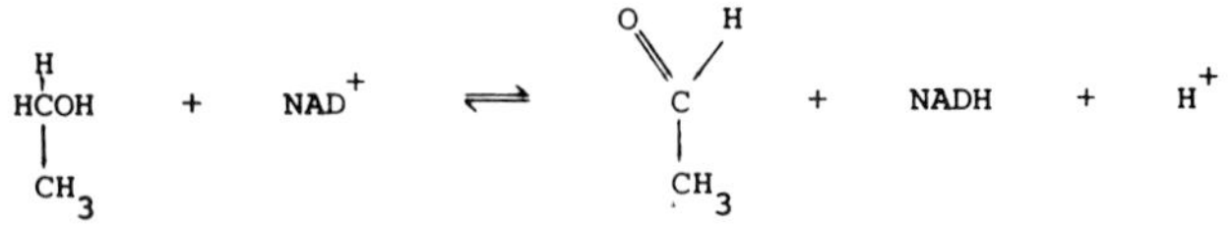

$$H\text{-}HCOH\text{-}CH_3 + NAD^+ \rightleftharpoons (O{=})C(H)\text{-}CH_3 + NADH + H^+$$

Reaction catalyzed by lactate dehydrogenase (LDH) is:

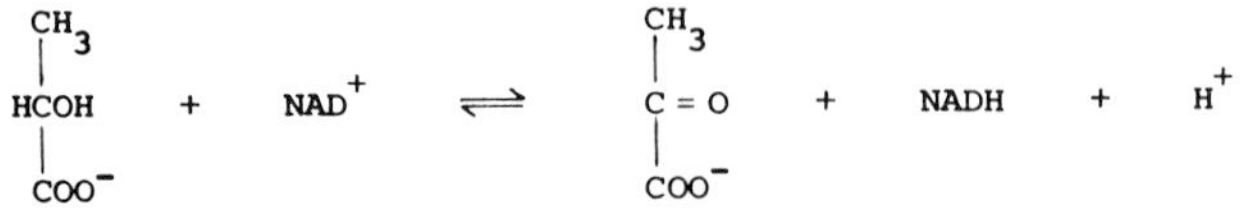

$$CH_3\text{-}HCOH\text{-}COO^- + NAD^+ \rightleftharpoons CH_3\text{-}C{=}O\text{-}COO^- + NADH + H^+$$

Reactions catalyzed by glyceraldehyde-3-phosphate dehydrogenase (GAPDH) are:

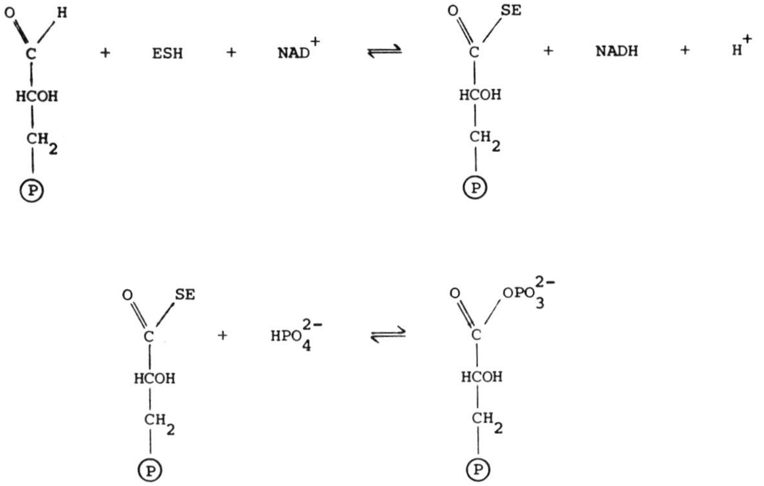

$$(O{=})C(H)\text{-}HCOH\text{-}CH_2\text{-}Ⓟ + ESH + NAD^+ \rightleftharpoons (O{=})C(SE)\text{-}HCOH\text{-}CH_2\text{-}Ⓟ + NADH + H^+ \quad (1)$$

$$(O{=})C(SE)\text{-}HCOH\text{-}CH_2\text{-}Ⓟ + HPO_4^{2-} \rightleftharpoons (O{=})C(OPO_3^{2-})\text{-}HCOH\text{-}CH_2\text{-}Ⓟ \quad (2)$$

Fig. 1. Reactions Catalyzed by Three Dehydrogenases Involved in Glycolysis.

for such hypotheses. Some detailed analyses of their mode of catalysis is now emerging, in part as a consequence of the amino acid sequence determination of a variety of different isozymes and species. Hence it has now become possible to examine enzymes related to alcohol metabolism so as to discern some basic principles.

TABLE 1
Comparison of Basic Properties of Dehydrogenases Catalyzing Reactions Shown in Fig. 1

	MW per subunit	*Coenzyme*	*Specificity of Nicotinamide Ring*	*Essential Metals*	*Number of Subunits*
Alcohol dehydrogenase	40,000	NAD+	A	Zn	2 (liver) 4 (yeast)
Lactate dehydrogenase	35,000	NAD+	A	None	4
Glyceraldehyde-3-Phosphate Dehydrogenase	35,000	NAD+	B	None	4
s-Malate Dehydrogenase	35,000	NAD+	A	None	2

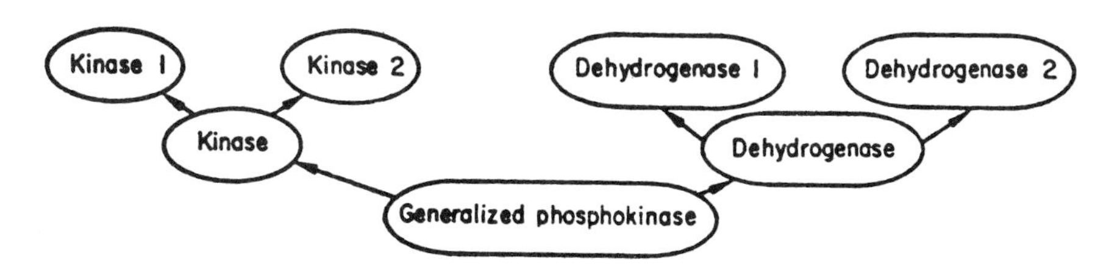

Fig. 2. General Scheme for the Evolution of Enzymes According to Waley (1969).

II. RESULTS AND DISCUSSION

A. Structural Similarities among Dehydrogenases

The subunit structures of lactate dehydrogenase (LDH), soluble malate dehydrogenase (sMDH), liver alcohol dehydrogenase (LADH) and glyceraldehyde-3-phosphate dehydrogenase (GAPDH) have a dinucleotide binding domain in common (Rossman *et al.*, 1974, 1975; Ohlsson *et al.*, 1974). The position of this domain within each polypeptide chain varies (Fig. 3), and its structure (Fig. 4) is subject to smaller modifications. For example, about 100 of the 150 residues of the

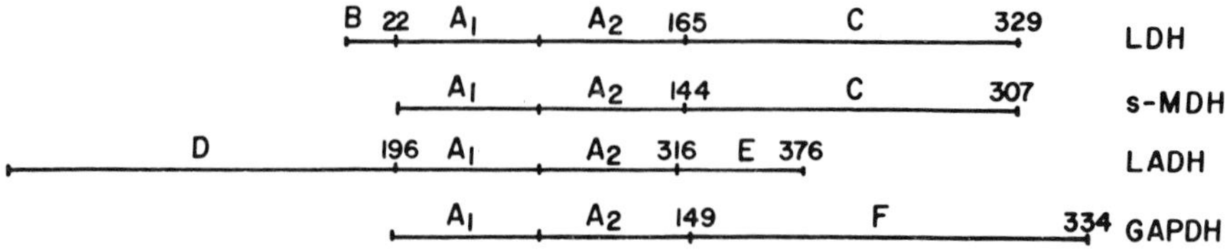

B 22 A1 A2 165 C 329 LDH
A1 A2 144 C 307 s-MDH
D 196 A1 A2 316 E 376 LADH
A1 A2 149 F 334 GAPDH

Figure 3. Comparison of Domain Structures Within Subunits of Four Dehydrogenases.

Lactate dehydrogenase (LDH) consists of an arm (B), a dinucleotide binding domain (A_1 A_2), and a catalytic domain (C). Soluble malate dehydrogenase (sMDH) has a similar structure as LDH but is missing the amino terminal arm. Glyceraldehyde-3-phosphate dehydrogenase (GAPDH) has a similar dinucleotide binding domain but a different catalytic domain (F). In liver alcohol dehydrogenase, the dinucleotide binding domain follows the catalytic domain (D).

αIF
βF
βE
αE
COOH
αD
βD
NH2
βA
βB
αB
βC
αC

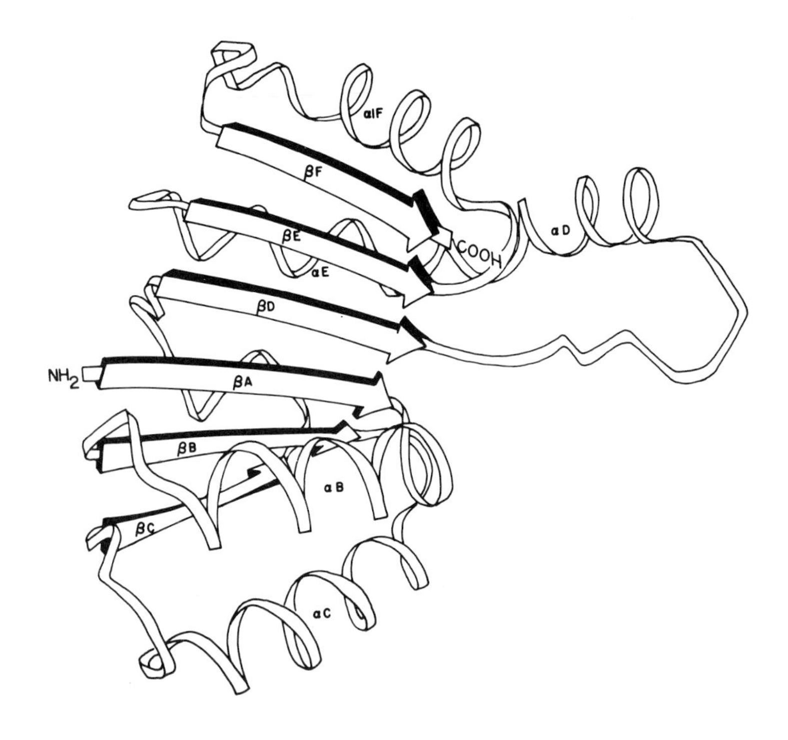

Fig. 4. Diagrammatic representation of the NAD binding domain as found in LDH.

LDH and GAPDH domains are structurally equivalent (Rossmann & Argos, 1976a). The dinucleotide binding domain is probably the expression of a very ancient gene which has been duplicated and fused to other genes. The essential nature of glycolysis to even the most primitive prokaryotes suggests that the NAD binding fold found in dehydrogenases must have remained substantially unchanged for most of biological evolution on Earth. The remainder of the polypeptide chain is the catalytic domain involved in providing residues for substrate binding, substrate specificity and catalysis. Thus, these domains are mostly different among those dehydrogenases which have been studied.

The dinucleotide binding domain in dehydrogenases consists of a β-pleated sheet made of six parallel strands connected by four helices, two on either side of the sheet. The first three strands of the sheet (A, B, C) are associated with the binding of AMP and the last three strands (D, E, F) with nicotinamide mononucleotide binding. Similar structures have been found in kinases (Steitz *et al.*, 1976; Schulz *et al.*, 1974; Schulz & Schirmer, 1974; Rossman & Argos, 1976b) and other nucleotide binding proteins as predicted by Waley (Fig. 2). The cofactors invariably bind to the carboxy end of the sheet, and their conformation in relationship to the protein is closely conserved among the dehydrogenases. The similarity of these domains permits the alignment of their primary structures (Fig. 5) when the amino acid sequence alone does not itself contain sufficient information to detect homology. Figure 5 shows examples of residues which are completely conserved among known dehydrogenases (Gly 28, Gly 33, Asp 53 and Gly 99 using the LDH numbering scheme), as well as other residues where only the chemical character is conserved. Such conservation is essential to either the function of the domain or the polypeptide folding. Wootton (1974) has used the known characteristics of the NAD binding domain to predict the position of two such domains in glutamate dehydrogenase.

B. The Active Center in Dehydrogenases

The conformation of the coenzyme as it is bound to LDH is shown in Fig. 6 (Rossmann *et al.*, 1971; Adams *et al.*, 1973a). Very similar conformations are observed for the other dehydrogenases discussed here. The adenosine binds into a hydrophobic pocket aided by the hydrogen bond between 02' Asp 53, a bond which is likely to be present in all NAD-linked dehydrogenases. Arginine 101 is probably the trigger which produces a large conformational change during the binding of coenzyme

	← βA →		← βB →
	2	2	3 4
	2	9	3 3
Dogfish LDH	N K I T V V G	C B A V	Ⓖ M A D A I S V L M K
			1 2
	1	8	2 2
Lobster GAPDH	S K I G I D G	F G R I	Ⓖ R L V L R A A L S C
	1	2	2 2
	9	0	0 1
	3	0	4 4
Horse LADH	S T C A V F G	L G G V	Ⓖ L S V I M G C K A A
	2	2	2 2
	4	5	5 6
	5	2	6 6
Bovine GluDH	K T F A V Q G	F G N V	Ⓖ L H S M R Y L H R F
		1	2 2 3
	9	6	0 2 1
Pig AK	K I I F V V G	G P G S	Ⓖ G T Q C E K I V Q K
			1 1 2
Clostridium MP	1	6	0 7 6
Flavodoxin	M K I V Y	W S G T	Ⓖ E L I A K G I I E S
	6	7	7 8
	0	6	9 9
Carboxypeptidase	P A I W I D	N A T	Ⓖ V W F A K K F T E N
	2	6	6
	6	2	4
Subtilisin	V K V A V I	N S H	Ⓖ T H V A

Fig. 5. Alignment of Amino Acid Sequences in Dehydrogenases and Other Proteins with Similar Folds. Shown is only the small portion of the dinucleotide binding domain represented by βA and αB. Note the hydrophobic character of the residues in the β-sheet and the conservation of glycines corresponding to residues 28 and 33 in LDH.

that brings arginine 109 close to the substrate site. The presence of this positive charge reduces the strength of binding of NAD^+ due to charge repulsion. A closer look at the substrate binding site is given in Fig. 7. An anion binding site (Adams *et al.*, 1973b), created by the positive charges on arginines 171 and 109, binds L-lactate in an orientation suitable for the transfer of a proton to histidine 195 and a hydride to the C4 of the nicotinamide moiety of the coenzyme. Thus, D-lactate must bind non-productively.

The general arrangement depicted in Fig. 7 for LDH is identical to that in s-malate dehydrogenase (Hill *et al.*, 1972; Webb *et al.*, 1973), similar to that in liver alcohol dehydrogenase (Eklund *et al.*, 1974, 1976) and glyceraldehyde-3-phosphate dehydrogenase (Fig. 8; Buehner *et al.*, 1974;

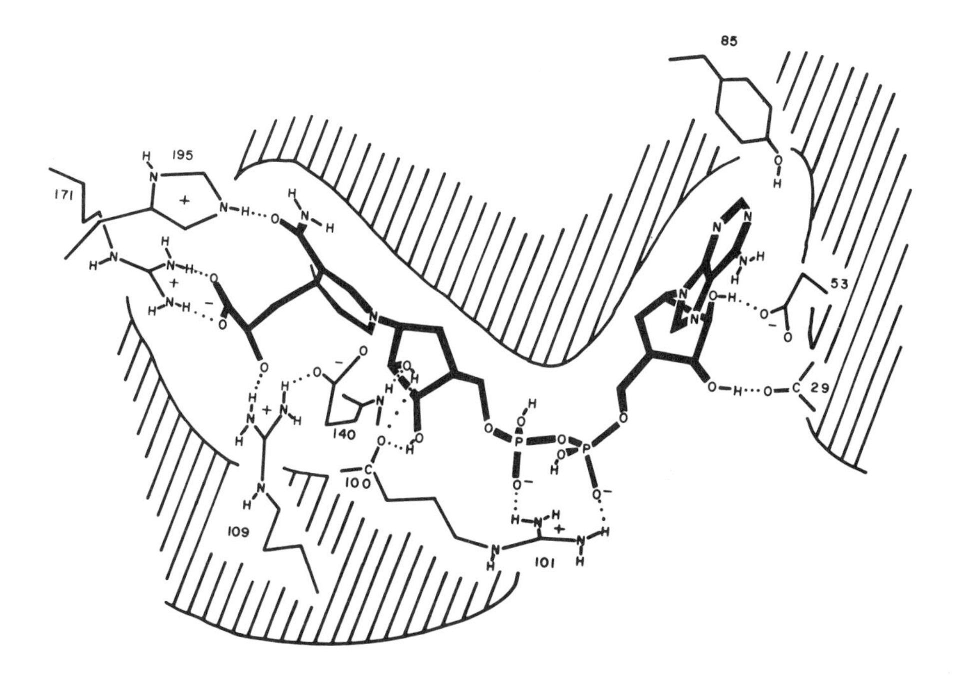

Fig. 6. NAD as Bound to Lactate Dehydrogenase.

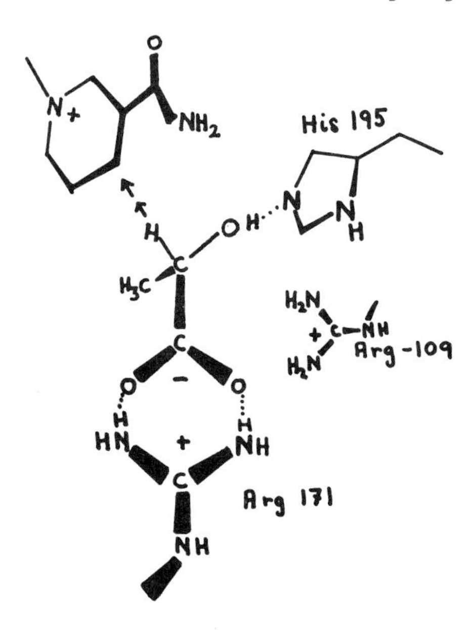

Fig. 7. Substrate Binding in LDH, Showing the Oxidation of Lactate by Removal of a Hydride to the Nicotinamide and proton to the imidazole group.

Moras *et al.*, 1975). The substrate position in GAPDH is also an anion binding site between histidine 176, the nicotinamide ring, and the positively charged arginine 231 (Fig. 8). One

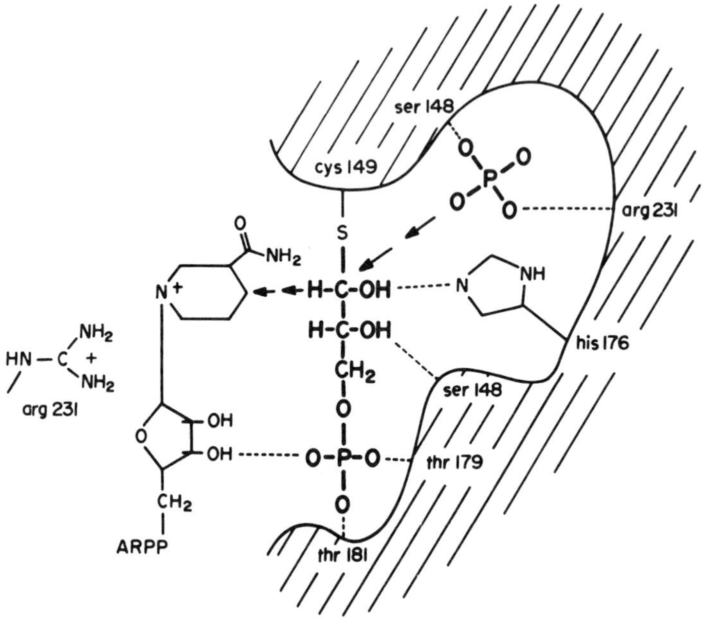

Fig. 8. The GAPDH Active Center Must Provide for the Catalysis of the Redox Step and the Phosphorylation.

difference, however, occurs in the orientation of the nicotinamide which has been rotated by 180°, exposing its B face to the substrate in GAPDH (see Table 1). The precise orientation of the nicotinamide ring in LDH might be associated with the formation of a temporary hydrogen bond between the carboxyamide group and the carbonyl group of residue 139, while in GAPDH there is a hydrogen bond between the carboxyamide and asparagine 313.

During the oxidation of glyceraldehyde-3-phosphate, GAPDH forms a thioester bond with the substrate at cysteine 149. This is subsequently broken during the phosphorylation of the C1 carbon atom. That process requires a binding site for an inorganic phosphate at a position suitable for a nucleophilic attack (Fig. 8). The occurrence of this site was demonstrated by the binding of citrate to the holoenzyme which displaces sulfate ions from both the substrate phosphate and inorganic phosphate sites in the crystal (Fig. 9; Olsen *et al.*, 1976).

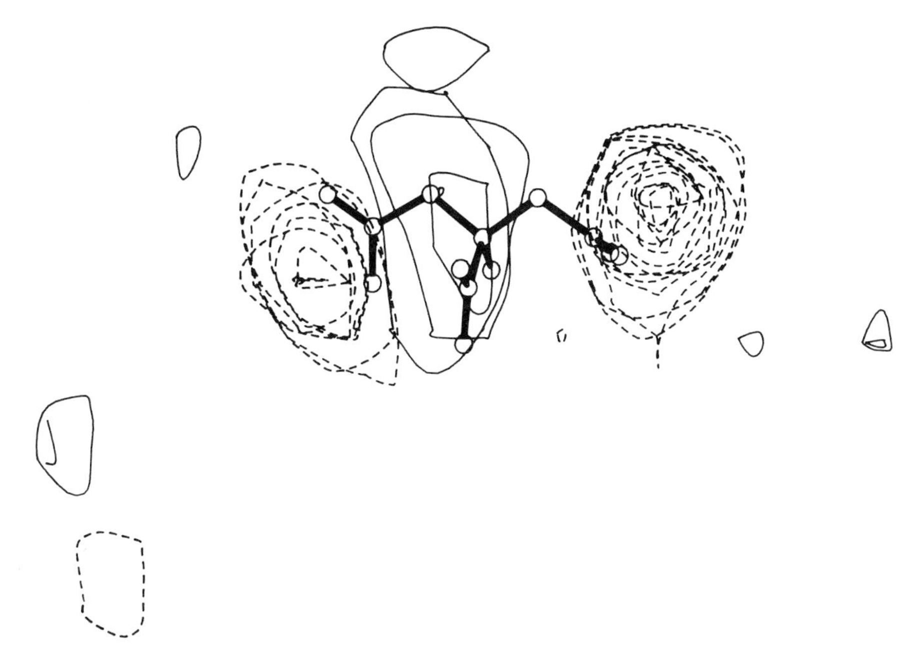

Fig. 9. The Binding of a Citrate Ion to Holo-GAPDH at pH 6.2 Bridges the Gap between the Substrate Phosphate Site and the Inorganic Phosphate Site. Both sites were occupied by SO_4^{2-} in the crystals grown from ammonium sulfate solution.

C. Specialization

The above considerations have established some basic properties of dehydrogenases. However, it is also necessary to provide for their adaptation to specialized functions. For instance, there are two principal types of polypeptide chains in LDH molecules in higher vertebrates. These are the H chain (synthesized primarily in heart muscle) and the M chain (synthesized primarily in skeletal muscle). The properties of H_4 and M_4 LDH are suitable for their aerobic and anaerobic physiological environments, respectively. The amino acid sequences of dogfish M_4 (Taylor, 1976), pig H_4 and M_4 (H.H. Kiltz, private communication) and chicken H_4 and M_4 (H. J. Torff, private communication) LDH have been determined. About one-third of all amino acids are altered between M and H chains (Table 2). However, the conservation of residues at the coenzyme and substrate binding site is far greater (Table

TABLE 2
Conservation of Amino Acids between Different LDH Species and Isozymes

a) Percent Changes in External Residues (112 residues in all)

		1	2	3	4	5
1.	Dogfish M_4		32	29	36	37
2.	Pig M_4			25	37	38
3.	Chicken M_4				34	36
4.	Pig H_4					20
5.	Chicken H_4					

b) Percent Changes in Internal Residues (109 residues in all)

		1	2	3	4	5
1.	Dogfish M_4		20	20	23	23
2.	Pig M_4			14	17	17
3.	Chicken M_4				18	17
4.	Pig H_4					6
5.	Chicken H_4					

3). The most radical change is that of alanine with glutamine at position 31 close to the NAD phosphate site. This slightly alters the position of the nicotinamide moiety with respect to the substrate (Eventoff *et al.*, 1976), giving rise to the specialized properties of H_4 and M_4 LDH.

Knowledge of some LDH amino acid sequences can also give information as to the antigenic binding properties of this molecule. It has been shown that antiserum against the M_4 isozyme of one species will cross-react with the M_4 isozyme from several species but not with the H_4 isozyme of the same species (Nisselbaum & Bodansky, 1963; Markert & Apella, 1963; Pesce *et al.*, 1964). Thus, antigenic sites should be identifiable by surface residues which distinguish M and H chains (Table 4).

GAPDH differs from LDH in that it shows cooperative

TABLE 3

Lactate Dehydrogenase Active Center Residues

	Number		Dogfish M_4	Pig M_4	Chicken M_4	Pig H_4	Chicken H_4
Adenine	27		Val	Val	?	Val	?
	52		Val	Val	Val	Val	Val
	53		Asp	Asp	Asx	Asp	Asx
	54		Val	Val	Val	Val	Val
	85		Tyr	Tyr	Tyr	Tyr	Tyr
	96	*	Ile	Ile	Val	Val	Val
	98		Ala	Ala	Ala	Ala	Ala
	119	*	Ile	Ile	Ile	Val	Val
	123		Ile	Ile	Ile	Ile	Ile
Adenine Ribose	28		Gly	Gly	?	Gly	?
	30		Asx	Gly	?	Gly	?
	53		Asp	Asp	Asx	Asp	Asx
	55	*	Met	Met	Val	Leu	Leu
	58		Lys	Lys	Lys	Lys	Lys
Pyrophos- phate	31	*	Ala	Ala	?	Gln	?
	58		Lys	Lys	Lys	Lys	Lys
	99		Gly	Gly	Gly	Gly	Gly
	101		Arg	Arg	Arg	Arg	Arg
	245		Tyr	Tyr	Tyr	Tyr	Tyr
Nicotinamide Ribose	32		Val	Val	?	Val	?
	97		Thr	Thr	Thr	Thr	Thr
	100	*	Ala	Ala	Ala	Val	Val
	139		Ser	Ser	?	Ser	Ser
Nicotinamide	32		Val	Val	?	Val	?
	138		Val	Val	?	Val	Val
	140		Asn	Asn	?	Asn	Asn
	167		Leu	Leu	Leu	Leu	Leu
	246		Thr	Thr	Thr	Thr	Thr
	250		Ile	Ile	Ile	Ile	Ile
Substrate	109		Arg	Arg	Arg	Arg	Arg
	171		Arg	Arg	Arg	Arg	Arg
	195		His	His	His	His	His

Note: The asterisk (*) denotes residues which are not completely conserved.

TABLE 4
Possible Antigenic Sites on the Surface of the Lactate Dehydrogenase Molecule

Site	*Residue Number*	*in M*	*in H*
A	5	Asp	Glu
	213	His	Glx
	214	Pro	Leu
	304	Lys	Asn
B	80	Ser	Ala
	81	Gly	Asn
C	100	Ala	Val
	119	Ile	Val
	126	Asn	Gln
	330B	deletion	Asp
D	177	Gly	Ala
	301	Asx	Ser
E	282	Lys	Gln
	309	Pro	Asp
	312	Glu	Val

phenomena (occurrences in one subunit alter the properties of neighboring subunits). Such properties are often vital in the control of enzymes or proteins. An excellent example occurs in the positive cooperativity of hemoglobin to the binding of oxygen, a phenomenon which is now largely understood at the molecular level (Perutz, 1970). Discussion of such properties centers around the occurrence of asymmetry in quaternary structure. Bode *et al.* (1975) showed that the trifluoroacetonylated (TFA) GAPDH derivative is bound asymmetrically on addition of the substrate analogue. X-ray crystallographic studies (Garavito & Rossmann, unpublished results) demonstrate the asymmetry of the TFA adduct when bound to the coenzyme (Fig. 10).

III. ACKNOWLEDGEMENTS

I have been most fortunate to have the collaboration of many able scientists in pursuing the studies described here. Among these I mention only my recent co-workers Drs. W. Eventoff and J.L. White who have participated in the study of

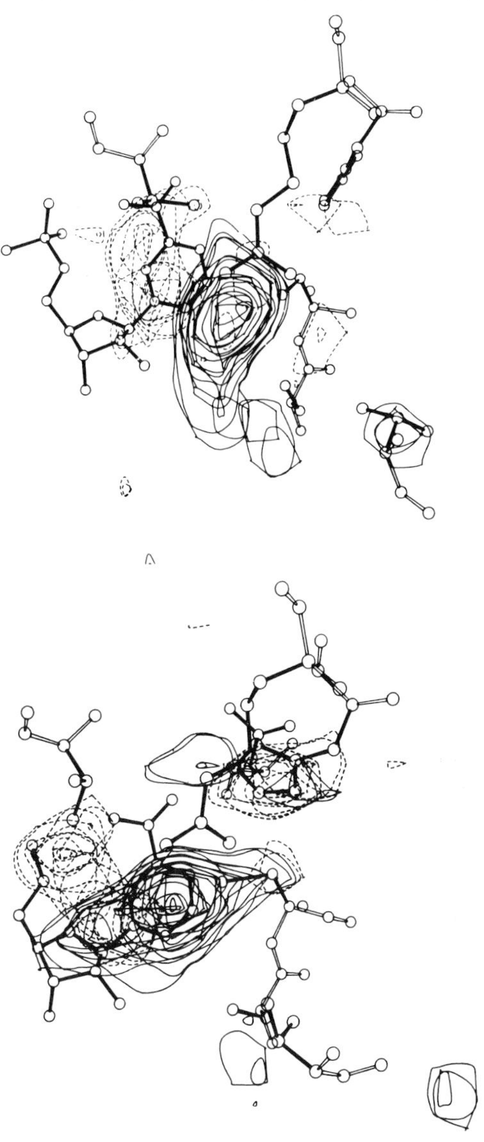

Fig. 10. Conformation of the Trifluoroacetonyl adduct to GAPDH in (a) the "red" and "yellow" subunits and (b) the "green" and "blue" subunits (for more details, see Buehner, et al., 1974).

LDH, Mr. R.M. Garavito and Dr. K.W. Olsen who have investigated GAPDH, and Dr. G.C. Ford who has been a constant aid in providing suitable computer programs for the many intricate crystallographic operations. Finally, I would like to thank Sharon Wilder for her help in the preparation of this manuscript. The work was supported by the National Institutes of Health (grant # GM 10704) and the National Science Foundation (grant # BMS74-23537).

IV. REFERENCES

(1). Adams, M.J., Buehner, M., Chandrasekhar, K., Ford, G.C., Hackert, M.L., Liljas, A., Rossmann, M.G., Smiley, I.E., Allison, W.S., Everse, J., Kaplan, N.O., and Taylor, S. S., Proc. Natl. Acad. Sci. U.S.A. 70, 1968-1972 (1973a).

(2). Adams, M.J., Liljas, A., and Rossman, M.G., J. Mol. Biol. 76, 519-531 (1973b).

(3). Bode, J., Blumenstein, M., and Raftery, M.A., Biochemistry 14, 1153-1160 (1975).

(4). Buehner, M., Ford, G.C., Moras, D., Olsen, K.W., and Rossmann, M.G., J. Mol. Biol. 90, 25-49 (1974).

(5). Eklund, H., Nordström, B., Zeppezauer, E., Söderlund, G., Ohlsson, I., Boiwe, T., and Brändén, C.I., FEBS Lett. 44, 200-204 (1974).

(6). Eklund, H., Nordström, B., Zeppezauer, E., Söderlund, G., Ohlsson, I., Boiwe, T., Soderberg, B.O., Tapia, O., Brändén, C.I., and Åkeson, Å., J. Mol. Biol. 102, 27-59 (1976).

(7). Eventoff, W., Rossmann, M.G., Taylor, S.S., Torff, H.J., Meyer, H., Keil, W., and Kiltz, H.H., Proc. Natl. Acad. Sci. U.S.A., submitted for publication (1976).

(8). Hill, E., Tsernoglou, D., Webb, L., and Banaszak, L.J., J. Mol. Biol. 72, 577-591 (1972).

(9). Markert, C.L., and Apella, E., Ann. N.Y. Acad. Sci. 103, 915-929 (1963).

(10). Moras, D., Olsen, K.W., Sabesan, M.N., Buehner, M., Ford, G.C., and Rossmann, M.G., J. Biol. Chem. 250, 9137-9162 (1975).

(11). Nisselbaum, J.S., and Bodansky, O., J. Biol. Chem. 238, 969-974 (1973).

(12). Ohlsson, I., Nordström, B., and Brändên, C.I., J. Mol. Biol. 89, 339-354 (1974).

(13). Olsen, K.W., Garavito, R.M., Sabesan, M.N., and Rossmann, M.G., J. Mol. Biol., in press (1976).

(14). Perutz, M.F., Nature (London) 228, 726-739 (1970).

(15). Pesce, A., Fondy, T.P., Stolzenbach, F.E., Castillo, F., and Kaplan, N.O., J. Biol. Chem. 239, 1753-1761 (1964).

(16). Rossmann, M.G., Adams, M.J., Buehner, M., Ford, G.C., Hackert, M.L., Lentz, P.J., Jr., McPherson, A., Jr., Schevitz, R.W., and Smiley, I.E., Cold Spring Harbor Symp. Quant. Biol. 36, 179-191 (1971).

(17). Rossmann, M.G., and Argos, P., J. Mol. Biol. 105, 75-96 (1976a).
(18). Rossmann, M.G., and Argos, P., J. Mol. Biol., in press (1976b).
(19). Rossmann, M.G., Liljas, A., Brändén, C.I., and Banaszak, L.J., in "The Enzymes" (P.D. Boyer, Ed.), 3rd ed., Vol. 11, pp. 61-102. Academic Press, New York, 1975.
(20). Rossmann, M.G., Moras, D., and Olsen, K.W., Nature (London) 250, 194-199 (1974).
(21). Schulz, G.E., Elzinga, M., Marx, F., and Schirmer, R.H., Nature (London) 250, 120-123 (1974).
(22). Schulz, G.E., and Schirmer, R.H., Nature (London) 250, 142-144 (1974).
(23). Steitz, T.A., Fletterick, R.J., Anderson, W.F., and Anderson, C.M., J. Mol. Biol. 104, 197-222 (1976).
(24). Taylor, S.S., J. Biol. Chem., submitted for publication (1976).
(25). Waley, S.G., Comp. Biochem. Physiol. 30, 1-11 (1969).
(26). Webb, L.E., Hill, E.J., and Banaszak, L.J., Biochemistry 12, 5101-5109 (1973).
(27). Wootton, J.C., Nature (London) 252, 542-546 (1974).

MECHANISM OF ACTION OF LIVER ALCOHOL DEHYDROGENASE

C.-I. Brändén

Agricultural College of Sweden

The recent X-ray structure determination of a triclinic holoenzyme LADH *complex to 4.5 Å resolution shows that no gross structural changes occur in the active site of the enzyme as a result of the coenzyme induced conformational change. The nicotinamide moiety of the coenzyme is positioned in the active site by a hydrogen bond from the carboxamide group to the side chain of Thr 178. The distance from the catalytic zinc atom to the center of the nicotinamide ring is 4.5 Å and approximately the same to the C4 atom.*

A model of the active site has been built based on this position of the nicotinamide ring and the detailed structure of the apoenzyme previously determined to 2.4 Å resolution. Using this model we have, in collaboration with Dr. H. Dutler, correlated his extensive and systematic kinetic data on alkyl-substituted cyclohexanol derivatives with the steric requirements of the active site. By assuming that the oxygen atom of the substrate binds to zinc and that the hydrogen to be transfered points towards C4 of the nicotinamide we find that methyl substitutions in those four positions which decrease or abolish the activity seriously overlap regions of enzyme or coenzyme in our model. In contrast, space is available for substitutions in the reamining six positions. The corresponding methylsubstituted compounds are all active. Using this model we can also correlate the kinetic properties of 3α and 3β - hydroxy steroids both in the 5α and 5β configurations with the steric requirements of the substrate binding pocket. Here we predict that the basis for the difference in activity towards steroids for the E and S chains is that substitution of Asp 115 in the E-chain for serine or a deletion in the S-chain changes the position of Leu 116 in the structure by breaking a salt bridge between Asp 115 and Arg 120. Leu 116 overlaps the 18 CH_3 *group of the 3β -hydroxy steroids in the E-chain. Similar correlations have also been made for the inhibitor strength of different pyrazole derivatives, by assuming that the nitrogen atoms of the pyrazole ring bind to the zinc atom and C4 of the coenzyme. The excellent correlation obtained with kinetic data in solution shows that this model can be used to design substrate and inhibitor molecules of desired properties.*

Based on the X-ray structure and these model building studies we also discuss possible mechanisms of action. Two mechanisms are considered, electrophilic catalysis by a four-coordinated zinc atom and a general acid-base catalysis by a pentacoordinate zinc complex.

I. INTRODUCTION

During the catalytic action the horse liver alcohol dehydrogenase molecule is subject to conformational changes (1). As a consequence different states of the enzyme form different types of crystals (2). Thus apoenzyme and binary complexes with inhibitors such as ADP-ribose, imidazole or 1,10-phenanthroline give orthorhombic crystals with a strict two-fold symmetry that relates the two chemically identical subunits of the molecule, whereas binary complexes with reduced coenzyme and some ternary complexes such as enzyme-NAD^+-pyrazole or enzyme-NADH-isobutyramide give triclinic crystals with the whole molecule in the asymmetric unit (3). Since the binding of ADP-ribose and NADH give rise to different crystal forms with somewhat different protein conformations, the nicotinamide moiety of the coenzyme is essential for this isomerization reaction to occur. In addition, it has recently been shown (E. Zeppezauer to be published) that the catalytic zinc atom or its ligands play an essential role in this process. Thus, when one of the zinc ligands, Cys 46, is carboxymethylated (4), orthorhombic binary complexes with NADH are formed with no conformational changes of the protein. Furthermore, when imidazole is coordinated to the catalytic zinc atom, ternary complexes with reduced coenzyme form orthorhombic crystals and not triclinic such as the enzyme-NADH-isobutyramide complex. In these crystals small local conformational changes can be observed in the vicinity of the catalytic site.

The results to be described here are based on the following X-ray studies: independent structure determinations of the orthorhombic apoenzyme to 2.4 Å resolution (5) and a triclinic enzyme-NADH-dimethylsulfoxide complex to 4.5 Å resolution (H. Eklund and C.-I. Branden to be published), ligand binding studies by difference Fourier techniques of the orthorhombic complexes enzyme-ADPribose (6), enzyme-NADH-imidazole and enzyme-tetrahydroNAD to 2.9 Å resolution (E. Zeppezauer to be published) as well as enzyme-imidazole-pyridine adenine dinucleotide and enzyme-imidazole-3-iodopyridine adenine dinucleotide to 4.5 Å resolution (7).

II. RESULTS AND DISCUSSION

A. Tertiary Structure of the Horse Liver Alcohol Dehydrogenase Molecule.

Fig. 1 shows a stereodiagram of one molecule of LADH.

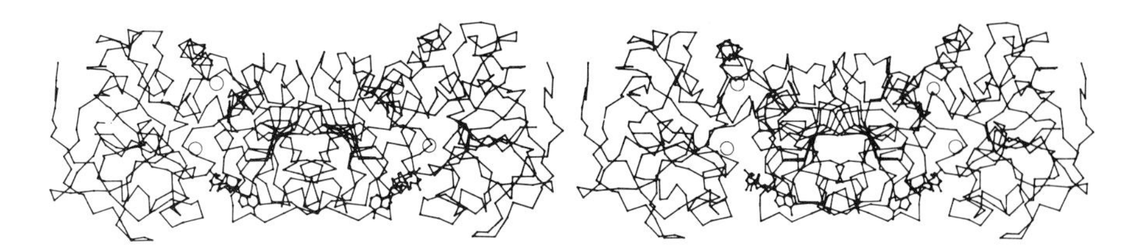

Fig. 1. Stereodiagram of One Molecule of Liver Alcohol Dehydrogenase (LADH).

A detailed description of this structure including a number of different stereodiagrams has been published recently (5,8). The 374 amino acids of each subunit are arranged in two clearly separated domains. One of these binds the coenzyme and has a structure very similar to corresponding domains in other dehydrogenases and in some kinases (9,10). The second domain which is larger and comprises residues 1-173 and 319-374 binds the two zinc atoms of the subunit.

The two subunits of the dimer are linked together mainly by interactions within the coenzyme binding domains which form a core in the middle of the molecule. The catalytic domains are at the two ends of the molecule and the catalytic sites are in the junctions between these domains and the core as is schematically illustrated in Fig. 2a.

The catalytic zinc atom is situated in the middle of the subunit close to the junction between the two domains. The domains are oriented relative to each other in such a way that the catalytic zinc atom is accessible to the solvent in some regions of the interface. These accessible regions can be described as one crevice in which the nicotinamide moity of the coenzyme is bound and one deep pocket perpendicular to this crevice where the substrate binds. Both the crevice and the pocket are thus in the interface region and are lined by sidechains from both domains.

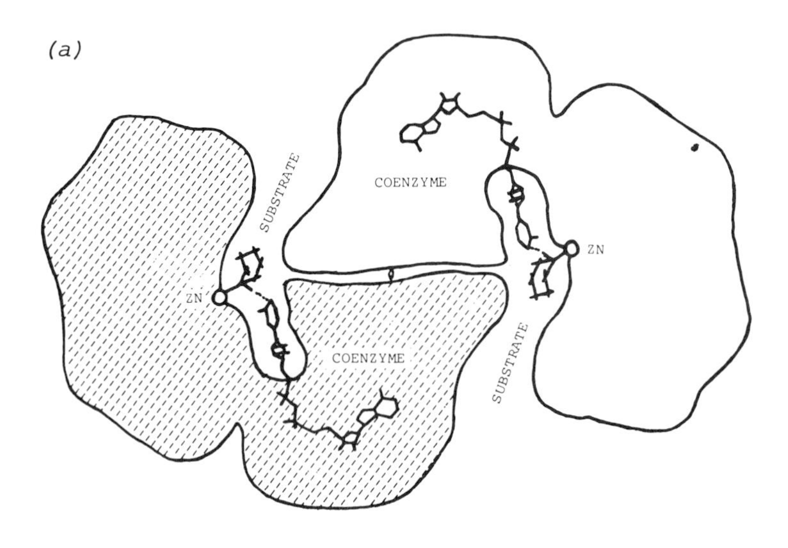

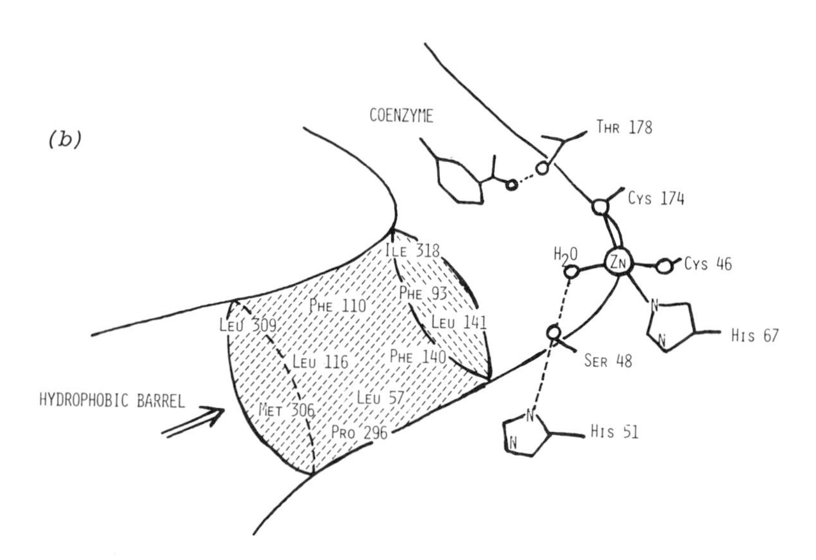

Fig. 2. (a) Schematic Diagram of a Section through the Dimeric LADH *Molecule with Bound Coenzyme Projected onto this Section. The substrate molecule, cyclohexanol, has been positioned by modelbuilding assuming direct binding to zinc.*

(b) Schematic Diagram of the Active Site of LADH *as observed from X-ray studies.*

Three protein ligands anchor the catalytic zinc atom firmly to the bottom of the substrate binding pocket, two sulphur atoms from Cys 46 and Cys 174 and one nitrogen atom from His 67. A water molecule or hydroxyl ion, depending on pH, completes a tetrahedral coordination in the apoenzyme structure.

The substrate binding pocket can be divided into three parts (see Fig. 2b); a hydrophilic bottom where the catalytic action occurs, a large hydrophobic barrel through which the substrate must enter and a rim where both polar and nonpolar residues are present. The bottom part contains the following hydrophilic groups: 1) The zinc atom and its coordination sphere including the water molecule which projects into the pocket. 2) The nicotinamide moity of the coenzyme and the hydroxyl group of Thr 178. 3) A general acid-base system consisting of the hydroxyl group of Ser 48 and the imidazole group of His 51. The oxygen atom of the zinc bound water molecule forms a hydrogen bridge to the serine oxygen which in turn is hydrogen bonded to the imidazole nitrogen.

The hydrophobic barrel provides a completely non-polar lining of the pocket by the sidechains of residues Leu 57, Phe 93, Phe 110, Leu 116, Phe 140, Leu 141, Pro 296 and Ile 318 from one subunit and Met 306 and Leu 309 from the other subunit. Hydrophobic regions of large substrates such as cyclohexanol, steroids and ω-hydroxy fatty acids would be expected to bind to some of these residues.

A comparison of the electron density to 4.5 Å resolution of the triclinic crystals which reflects the enzyme conformation that binds substrates in a productive mode, with that of the orthorhombic apoenzyme crystals to 2.4 Å resolution from which this detailed description has been derived shows that no gross conformational differences are present except possibly for some of the residues close to the subunit interface. These would effect the properties of the rim of the substrate binding pocket but they are rather far from the catalytic zinc atom. No extra residues are thus brought into the active site by the coenzyme induced conformational change.

As a measure of the similarities and differences of these two conformations I have listed distances in Table 1 between the four zinc atoms of the molecule. The largest difference is between the two "structural" zinc atoms where the distances differ by 2.3 Å.

TABLE 1
Distances in A units between the Four Zinc Atoms of the LADH *molecule in the Orthorhombic and Triclinic Conformations*

Zn - Zn	*orthorh.*	*tricl.*	*Zn - Zn*	*orthorh.*	*tricl.*
C1 - S1	20.2	19.7	C2 - S2	20.2	18.3
C1 - S2	42.7	43.5	C2 - S1	42.7	43.3
C1 - C2	46.2	44.9	S1 - S2	43.9	46.2

C denotes the catalytic zinc atom and S the other zinc atom of the subunit. 1 and 2 refer to the two different subunits of the molecule.

B. Coenzyme Binding

By comparing the electron density maps of the orthorhombic complex with tetrahydroNAD to 2.9 Å resolution with those of the triclinic complex with NADH it was found that within the limits of resolution the conformations of the dinucleotides were the same. The mode of binding and the conformation of the AMP portion were similar to that described earlier for the ADP-ribose complex (5,6). The nicotinamide moiety is positioned in the active site with the A-side of the ring facing the zinc atom. The carboxamide group is hydrogen bonded to the side chain of Thr 178. The distance from the zinc atom to the center of the nicotinamide ring is 4.5 Å and approximately the same of somewhat shorter to the C4 atom of this ring.

The coenzyme molecule is thus positioned in the crevice of the NAD-binding domain by a number of relatively weak contacts which extend over the whole coenzyme molecule. These contacts are more frequent in the AMP-part which is thus firmly anchored to the enzyme in agreement with binding studies of coenzyme fragments (8). Since there are no other apparent contacts with the nicotinamide moiety than the hydrogen bond between Thr 178 and the carboxamide group there is some degree of flexibility for the nicotinamide to move during catalytic action.

In order to deduce the possible importance of this hydrogen bond for proper positioning of the nicotinamide moiety we have studied the binding of coenzyme analogues where this group is either removed, pyridine-adenine-dinucleotide, $PyAD^+$, or substituted with an iodine atom, 3-I $PyAD^+$. In both these complexes the binding of the pyridine mononucleotide part is the same, but it is very different from that of the NMN-part of the coenzyme (7). The pyridine ring is positioned on the surface of the molecule quite far from the active site, the distance from zinc to the center of the pyridine ring being 13 Å. The AMP-parts of these analogues, on the other hand, bind in a similar manner as the corresponding part of the coenzyme. The hydrogen bond between the carboxamide group and Thr 178 is thus quite essential for positioning the nicotinamide moiety of the coenzyme in the active site.

C. Substrate Binding

By model building experiments we have been able, in collaboration with Dr. H. Dutler, to correlate the geometry of the substrate binding pocket described here with kinetic results in solution of systematically varied alkylsubstituted cyclohexanone derivatives and bridged carbocyclic ketones of stable conformations as substrate molecules (11). These kinetic studies (12) were designed to obtain a map of the free space available for substrate molecules at the active site using the "diamond-lattice" concept of Prelog (13). In our model building we assumed that the oxygen atom of the alcohol binds directly to zinc and that the H-atom to be transfered is 2.5 Å away from and points towards the C4 atom of the nicotinamide ring. Using these assumptions we built a cyclohexanol molecule into the active site and studied the steric effects of all possible substitutions of methyl groups. We found that in all compounds where the activity was decreased or abolished as measured kinetically (12), the substituted methyl group on the cyclohexanol skeleton overlapped sterically with the enzyme or coenzyme. This is schematically illustrated in Fig. 3a.

In particular three of the four possible methyl substitutions on the two carbon atoms adjacent to the hydroxyl group of the substrate overlapped groups adjacent to the active site zinc atom. These derivatives were inactive for which there would be no steric reason if the substrate was bound indirectly to zinc through a water molecule as has recently been suggested (14). For the remaining substitutions which did not affect the kinetic parameters of the derivatives as substrates we found that space was available for the corresponding substitution in the substrate binding pocket.

Using this position of the cyclohexanol ring we could also correlate the fact that adamantanol is inactive as a substrate (15) with the presence of a CH_2 group of this rigid

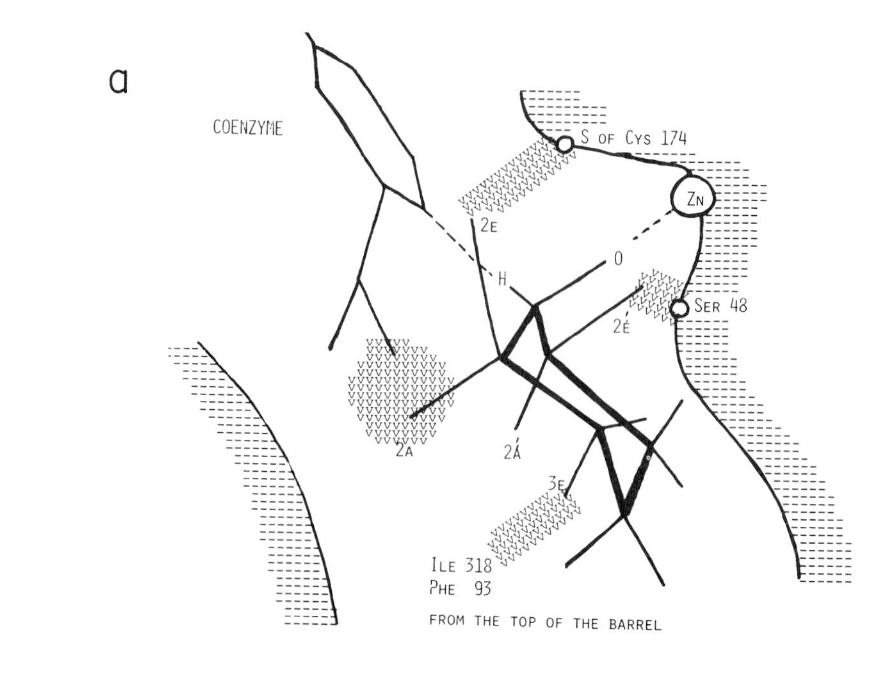

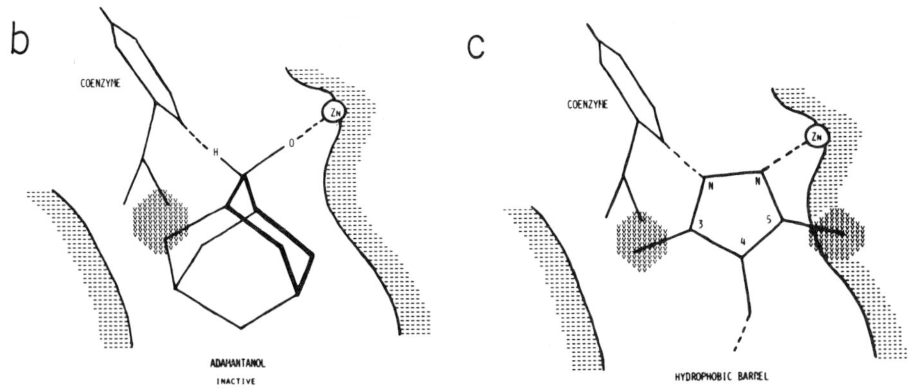

Fig. 3. Schematic Diagram of Different Molecules Built Into the Substrate Binding Pocket of LADH. *Light shades represent protein, darker shades represent steric overlap between atoms of the protein or coenzyme and atoms from the added molecule.*

(a) Methylsubstituted Cyclohexanol. Substitution in position 2e overlaps Cys 174, in 2'e Ser 48, in 2a the carboxamide group of the coenzyme and in 3e Ile 318 and Phe 93.

(b) Adamantanol. The CH_2 *group corresponding to substitution in position 2a of cyclohexanol overlaps the carboxamide group of the coenzyme.*

(c) Methylsubstituted Pyrazole. Substitution in position 3 overlaps the carboxamide group, in position 5 Ser 48 whereas substitution in position 4 is directed into the hydrophobic barrel of the pocket.

molecule in one of the sterically hindered substituted positions of cyclohexanol, 2a (see Fig. 3b). Bicyclo (3.2.1)-octan-2-ol, which lacks this CH_2 group but otherwise is identical to adamantanol, is active as a substrate (16) in complete agreement with our X-ray model.

We have also used these results to study possible binding of 3-hydroxy steroid molecules in the active site. The A-ring of these molecules can be considered as a partly substituted cyclohexanol ring. The steroid molecules are sufficiently large to fill up the hydrophobic barrel when the 3-hydroxyl group is bound to zinc.

It has been shown kinetically that the enzymes from horse (isozyme SS and ES but not EE), rat and human liver catalyze the oxidation of 3β-hydroxy groups of both 5α and 5β steroids (17,18,19). Corresponding 3α-hydroxysteroids were not active as substrates. The X-ray model is based on the horse EE isozyme. However, since only six amino acid differences have been observed between the E and the S chain (20) it can be safely assumed that the structure of the S-subunit is similar to our model except in the vicinity of these differences.

Our model building studies (see Figs. 4a - d) show that 3β-hydroxy-5β-cholanoic acid fits very nicely in the substrate binding pocket except at one particular point. The methyl group at position 18 of the steroid overlaps sterically with the side chain of Leu 116 close to the outer rim of the hydrophobic barrel. This interaction would thus prevent the steroid to bind in the assumed productive orientation in the E-subunit. However, in the S-subunit the situation could be quite different. Leu 116 is part of a loop on the outside of the molecule with no organized secondary structure. There is a salt bridge in the E-subunit from Asp 115 to Arg 120 which might contribute significantly to the conformation of this loop and thus to the observed position of Leu 116. This salt bridge cannot occur either in the S-subunit or in the rat enzyme since Asp 115 is substituted by serine (or a deletion) and asparagine, respectively. It is thus quite possible that the residues in the vicinity of 115 have a somewhat different structure in these enzymes and thus no steric overlap for our assumed orientation of the steroid molecule.

It is apparent from Fig. 4b that 3α-hydroxy-5β-steroids cannot bind in a productive mode. The pocket is sufficiently narrow to create impossible overlaps both at the junction between rings A and B on one side of the pocket and for the whole of ring D on the other side.

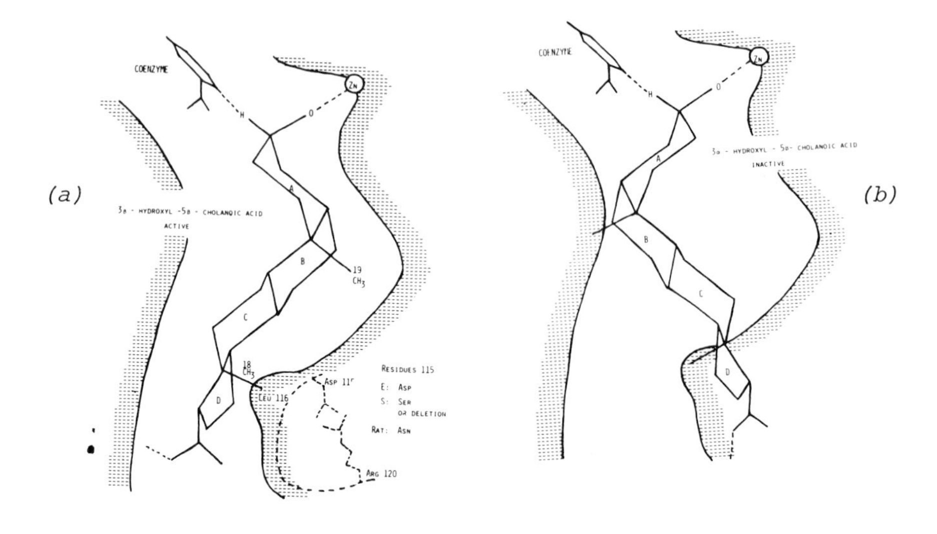

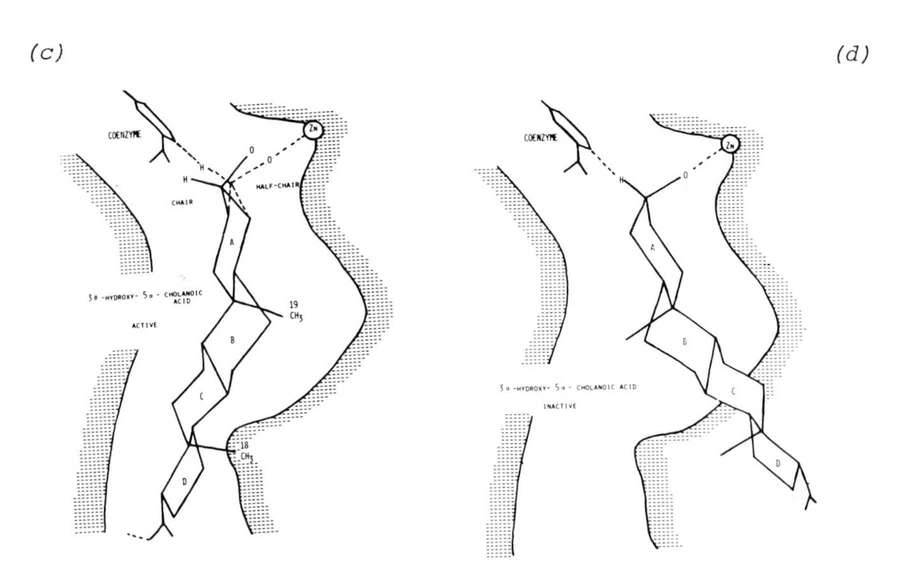

Fig. 4. Schematic Diagram of Steroid Molecules Built Into the Substrate Binding Pocket of LADH.

(a) 3β-hydroxyl-5β-cholanoic acid.
(b) 3α-hydroxyl-5β-cholanoic acid.
(c) 3β-hydroxyl-5α-cholanoic acid.
(d) 3α-hydroxyl-5α-cholanoic acid.

The pocket is in fact so narrow that it cannot accomodate a 3β-hydroxy-5α-steroid in this assumed orientation if ring A has its normal chair form. Rings C and D would then overlap with the enzyme on one side of the pocket. However, by changing ring A to a half-chair conformation, which would not be very unfavorable energetically, the steroid fits nicely into the pocket (Fig. 4c) except for the same overlap between CH_3 in position 18 and Leu 116. Considering the fact that the complete steroid ring structure is contained within the pocket it is remarkable that the shape of the pocket fits these 3β-hydroxy-steroid molecules so nicely. In agreement with the kinetic data the 3α-hydroxyl-5α-steroid cannot bind in this orientation, as is evident from Fig. 4d. It should be realized that both 3α-hydroxysteroids could bind in the pocket in a nonproductive manner.

Using this model we can also examine the binding of different pyrazole derivatives which are powerful inhibitors in the presence of NAD^+. It has been suggested (21) that one nitrogen atom of the pyrazole ring binds to zinc and the other to C4 of the nicotinamide ring. Fig. 3c shows schematically the result of positioning a pyrazole ring in this manner in our model. One nitrogen atom superimposes the oxygen atom of the cyclohexanol molecule and the other nitrogen atom superimposes the hydrogen to be transfered.

It is obvious that substitutions in 3- and 5- positions give steric overlap with enzyme and coenzyme which is consistent with the much larger dissociation constants for these derivatives (22). The K_i is 0.22 μM for pyrazole compared to 7100 μM for 3,5-dimethyl pyrazole. Alkyl-substitutions in the 4-position, on the other hand, are not only sterically possible but would be expected to contribute to tighter binding since they are positioned in the hydrophobic barrel. This is also the case. The K_i for 4-methylpyrazole is 0.08 μM and 0.001 μM for 4-pentylpyrazole (23). The introduction of charged groups on these alkyl chains would decrease the binding strength when they are inside the barrel but would lead to further enhancement of binding and specificity if they are suitably introduced to form salt bridges with charged groups at the rim of the pocket.

This excellent agreement between X-ray structure and kinetic data in solution for the cyclohexanol derivatives, steroids and pyrazole derivatives indicates that the geometry of the active site described here is relevant for deductions about the catalytic action of the enzyme and that the substrate in all probability binds directly to the zinc atom. Furthermore, it is obvious that the X-ray structure can be used not only to obtain correlations with existing kinetic data but also to design substrate or inhibitor molecules of desired properties. Work in this direction is now in progress.

D. Mechanism of Action

The X-ray structure determination shows that the only groups in the enzyme that may be responsible for binding and polarization of the reactive part of the substrate are the zinc atom, the zinc bound water molecule, the hydroxyl group of Ser 48 and, indirectly, the imidazole group of His 51 through its hydrogen bond to Ser 48. Previously suggested mechanisms based on direct participation of lysine (24), histidine (25), cysteine (26) or a combination of these (28) are thus highly unlikely.

We have suggested a mechanism (8,27) based on electrophilic catalysis mediated by the active site zinc atom. In this mechanism alcohol binds directly to zinc as the negatively charged alcoholate ion (see Fig. 5a). Formation of this ion is facilitated by the previous step in the reaction, NAD^+ binding, which converts the water molecule bound to zinc to a hydroxyl ion. It is known from direct determination of proton release on binding of NAD^+ (29) and the pH dependence of this binding (30) that the apoenzyme has a functional group with a pK_a about 9.6 that is perturbed to a pK_a of about 7.6 in the binary enzyme-NAD^+ complex. Our crystallographic studies indicate (8) that this is the ionization of the zinc bound water molecule. This hydroxyl ion can combine with the hydrogen atom of the hydroxyl group of alcohol to produce water and alcoholate ion.

The hydrogen bond system of Ser 48 and His 51 might be involved both in the proton release and in substrate binding and polarization. The water molecule is in the hydrophobic substrate binding pocket, whereas the εN atom of His 51 points into the solution at the surface of the molecule. It might thus be energetically more favorable for proton release to occur through this system of hydrogen bonds to the polar surface of the molecule than directly into the pocket. Preliminary theoretical calculations (31) have shown that such proton transfer is greatly facilitated in the presence of a local electric field such as that caused by the charged NAD^+.

The assumption that the substrate is directly coordinated to zinc is supported not only by the model building studies reported here, but also by spectroscopic evidence (32,33)

(a)

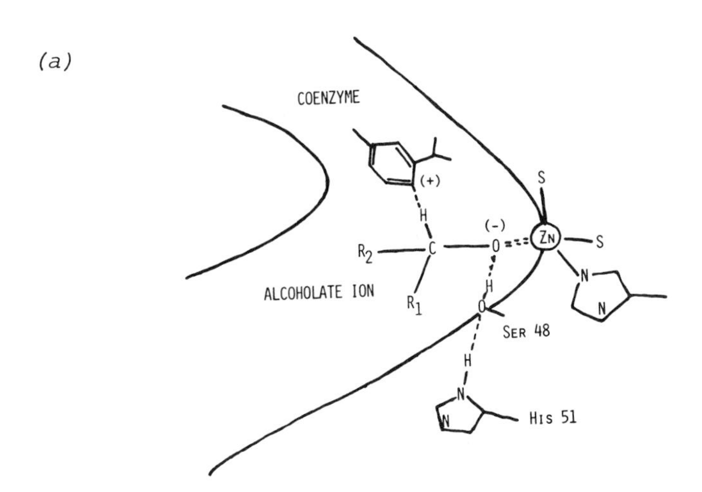

(b)

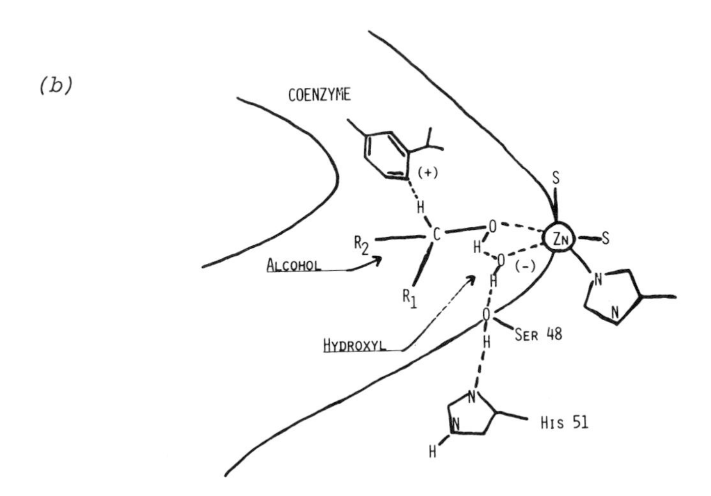

Fig. 5. Schematic Diatram of Productive LADH - NAD$^+$ - *Alcohol Complex suggested by (a) Branden et al. (8) and (b) Dworschack and Plapp (35).*

using chromophoric aldehydes as substrates which suggest that the carbonyl oxygen atom of the reaction intermediate is coordinated to zinc. Sloan *et al.* (14) have suggested that the substrate binds indirectly to zinc *via* a water molecule on the basis of NMR investigations of a cobalt-substituted LADH. From these studies they calculate distances from substrate or inhibitor protons to the cobalt atoms of the molecule. However, their distance calculations do not agree with the X-ray model. The sum of their distances from one of these inhibitor protons to the catalytic and structural cobalt atoms is 15.2 Å. According to the X-ray model this distance should be approximately 20 Å independent of direct or indirect binding to cobalt. The cause of this discrepancy might be that their assumption that the cobalt atom occupies the site of the catalytic atom in the hybrid enzyme, $LADHCo_2Zn_2$ is incorrect, as was actually recently demonstrated (34).

A modification of the mechanism suggested by us has recently been proposed (35) in order to explain some recent structure-reactivity studies (35,36,37). These studies indicate that no charge develops on the substrate during catalysis and hence that release of the proton of the alcohol and the hydride transfer occur simultaneously. A general acid-base catalysis would be consistent with these results. In the modified mechanism (35) the hydroxyl ion remains bound to zinc while the neutral alcohol molecule adds as a fifth ligand to zinc. The suggested productive LADH-NAD^+-alcohol complex is schematically illustrated in Fig. 5b. The zinc bound hydroxide ion is postulated to be the general base catalyst for the oxidation of alcohols, and the zinc bound water molecule as the general acid catalyst for the reduction of aldehydes. Hopefully, a decisive answer to this point will be obtained by X-ray studies now in progress of suitably chosen triclinic enzyme-coenzyme substrate complexes which might be crystallized because of favorable equilibrium constants.

III. REFERENCES

(1). Shore, J.D., Gutfreund, H., and Yates, D., J. Biol. Chem. 250, 5276 (1975).
(2). Brändén, C.-I., Arch. Biochem. Biophys. 112, 215 (1965).
(3). Zeppezauer, E., Söderberg, B.-O., Brändén, C.-I., Åkeson, Å., and Theorell, H., Acta Chem. Scand. 21, 1099 (1967).
(4). Zeppezauer, E., Jörnvall, H., and Ohlsson, I., Eur. J. Biochem. 58, 95 (1975).

(5). Eklund, H., Nordström, B., Zeppezauer, E., Söderlund, G., Ohlsson, I., Boiwe, T., Söderberg, B.-O., Tapia, O., Brändén, C.-I., and Åkeson, Å., J. Mol. Biol. 102, 27 (1976).
(6). Abdallah, M., Biellmann, J.-F., Nordström, B., and Brändén, C.-I., Eur. J. Biochem. 50, 475 (1975).
(7). Samama, J.-P., Biellmann, J.-F., Zeppezauer, E., and Brändén, C.-I., Eur. J. Biochem., in press.
(8). Brändén, C.-I., Jörnvall, H., Eklund, H., and Furugren, B., "The Enzymes" (P.D. Boyer, Ed.), 3rd edn., Vol. 11, p. 93. Academic Press, New York, 1975.
(9). Ohlsson, I., Nordström, B., and Brändén, C.-I., J. Mol. Biol. 89, 339 (1974).
(10). Rossmann, M.G., Liljas, A., Brändén, C.-I., and Banaszak, L.J., "The Enzymes" (P.D. Boyer, Ed.), 3rd edn., Vol. 11, p. 62. Academic Press, New York, 1975.
(11). Dutler, H., and Brändén, C.-I., in press.
(12). Dutler, H., in "Structure-Activity Relationships in Chemoreception" (G. Benz, Ed.), p. 65. Information Retrieval, London, 1976.
(13). Prelog, V., Pure and Applied Chemistry 9, 119 (1964).
(14). Sloan, D.L., Young, J.M., and Mildvan, A.S., Biochemistry 14, 1998 (1975).
(15). Graves, J.M.H., Clark, A., and Ringold, H.J., Biochemistry 4, 2655 (1965).
(16). Jones, J.B., and Beck, J.F., in "Techniques of Chemistry," Vol. 10, Part 1. Wiley & Sons, New York, in press.
(17). Waller, G., Theorell, H., and Sjövall, J., Arch.Biochem. Biophys. 111, 671 (1965).
(18). Reynier, M., Theorell, H., and Sjövall, J., Acta Chem. Scand. 23, 1130 (1969).
(19). Cronholm, T., Larsen, C. Sjövall, J., Theorell, H., and Åkeson, Å., Acta Chem. Scand. B29, 571 (1975).
(20). Jörnvall, H., Eur. J. Biochem. 16, 41 (1970).
(21). Theorell, H., and Yonetani, T., Biochem. Z. 338, 537 (1963).
(22). Theorell, H., Yonetani, T., and Sjöberg, B., Acta Chem. Scand. 23, 255 (1969).
(23). Dahlbom, R., Tolf, B.R., Åkeson, Å., Lundquist, G., and Theorell, H., Biochem. Biophys. Res. Comm. 57, 549 (1974).
(24). Kosower, E.M., Biochem. Biophys. Acta 56, 474 (1962).
(25). Ringold, H.J., Nature 210, 535 (1966).
(26). Wang, J.H., Science 161, 328 (1968).
(27). Eklund, H., Nordström, B., Zeppezauer, E., Söderlund, G., Ohlsson, I, Boiwe, T., and Brändén, C.-I., FEBS Lett. 44, 200 (1974).
(28). Rubin, B.R., and Whitehead, E.P., Nature 196, 658 (1962).

(29). Shore, J.D., Gutfreund, H., Brooks, R.L., Santiago, D., and Santiago, P., Biochemistry 13, 4185 (1974).
(30). Taniguchi, S., Theorell, H., and Åkeson, Å., Acta Chem. Scand. 21, 1903 (1967).
(31). Tapia, O., Sussmann, F., Poulain, E., Chem. Phys. Lett. 33, 65 (1975).
(32). Dunn, M.F., Biellmann, J.F., and Bruylant, G., Biochemistry 14, 3176 (1975).
(33). McFarland, J.T., Chu, Y.-H., and Jacobs, J.W., Biochemistry 13, 65 (1974).
(34). Sytkowski, A.J., and Vallee, B.L., Proc. Natl. Acad. Sci. USA 73, 344 (1976).
(35). Dworschack, R.T., and Plapp, B.V., in press.
(36). J.P. Klinman, Biochemistry 15, 2018 (1976).
(37). Blackwell, L.F., and Hardmann, M.J., Eur. J. Biochem. 55, 611 (1975).

EXTRAPOLATION OF MECHANISM STUDIES ON GLYCERALDEHYDE-3-PHOSPHATE DEHYDROGENASE TO THE TREATMENT OF AVIAN MUSCULAR DYSTROPHY

Jane H. Park, Edward J. Hill, [+]Ta-hsu Chou, Richard Pinson
Charles R. Park, Virgil LeQuire, and Robert Roelofs

Vanderbilt University Medical School

This paper presents the rationale and results of our work on the use of penicillamine in the treatment of hereditary avian muscular dystrophy. We have shown that penicillamine, a cysteine analogue with a reduced sulfhydryl group, delays the onset of symptoms and alleviates the debilitating aspects of the disease (1). In avian dystrophy, deterioration of the muscle fibers is evidenced in the 2nd month by an inability of the birds to rise after falling on their backs and by a progressive rigidity of the wings. Penicillamine produced three major improvements: (a) better righting ability when birds were placed on their backs; (b) greater wing flexibility; (c) and suppression of plasma creatine phosphokinase activity. The beneficial effects of penicillamine on muscle function and biochemistry are discussed in terms of the mechanism of drug action.

I. INTRODUCTION

This paper describes how the results of our theoretical studies on the mechanism of action of glyceraldehyde-3-phosphate dehydrogenase were extrapolated to the practical problem of the treatment of muscular dystrophy. It is hoped that this presentation may demonstrate a manner in which the experimental findings in this basic science workshop can be utilized in considering disease processes. First, the rationale for the penicillamine treatment program will be outlined. The beneficial results of this program will then be presented. Lastly, the mechanism of drug action will be briefly discussed.

[+]*Recipient of Andrew Mellon Teacher-Scientist Award*

A. Rationale for Penicillamine Treatment

Glyceraldehyde-3-phosphate dehydrogenase contains the most reactive cysteine residue of all the glycolytic enzymes and is extremely sensitive to inactivation by inhibitors or oxidants (2). In human Duchenne dystrophy, this dehydrogenase has been shown to lose 70% of its activity compared to normal muscle (3). This inactivation is greater than that of 15 other enzymes of the glycolytic pathway and the Krebs cycle (3). Using an animal model, we have observed a 75% decrease in specific activity of this enzyme in the pectoral muscles of the dystrophic chicken (4). This dehydrogenase is more abundant in white (fast) fibers where glycolysis is the main source of ATP during contraction. The fact that white fibers are most severely affected by the disease than red (5,6) suggested the possible involvement of the glycolytic pathway due to oxidation of the essential sulfhydryl group in the active site. We therefore considered the relationship of the catalytic activities of the dehydrogenase to the degenerative process.

When the active site cysteine residues of glyceraldehyde 3-phosphate dehydrogenase (ESH) are in the reduced state, the substrate is converted in the presence of NAD and P_i to the high energy compound, 1,3-diphosphoglyceric acid (2).

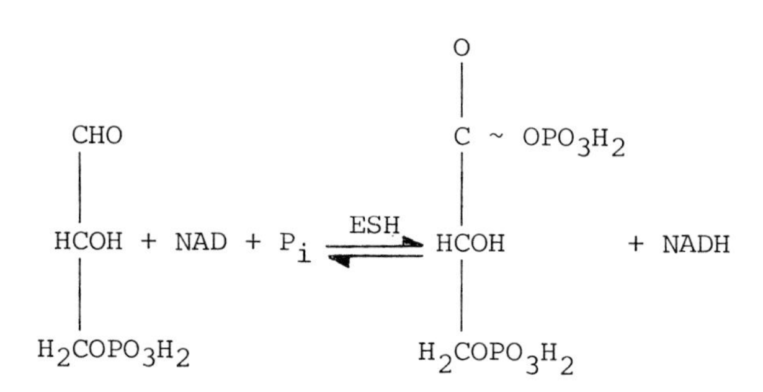

DEHYDROGENASE REACTION

This high energy phosphate is transferred to ADP to give ATP required for contraction and relaxation.

When the active site sulfhydryl groups are oxidized with oxygen (7) or chemically with H_2O_2 or iodosobenzoate (8), the dehydrogenase is converted into an acyl phosphatase. Converted phosphatase activity promotes the hydrolysis of 1,3-diphosphoglyceric acid to 3-phosphoglyceric acid with the loss of the high energy phosphate bond as shown below.

$$
\begin{array}{l}
\overset{O}{\overset{\|}{C}} \sim OPO_3H_2 \\
| \\
HCOH \\
| \\
H_2COPO_3H_2
\end{array}
+ H_2O \xrightarrow{E_{(NAD)}}
\begin{array}{l}
\overset{O}{\overset{\|}{C}}OH \\
| \\
HC\ OH \\
| \\
H_2COPO_3H_2
\end{array}
+ P_i
$$

ACYL PHOSPHATASE REACTION

The enzyme is represented as $E_{(NAD)}$ to indicate the loss of -SH groups and the requirement for the coenzyme (7). The oxidation of the sulfhydryl groups is not a simple conversion to a disulfide bridge but rather the formation of a sulfenic acid as shown by Ehring and Colowick (8) and Benitez and Allison (9). This interconversion of activities is reversible as the dehydrogenase activity is restored with sulfhydryl compounds such as glutathione, cysteine, or penicillamine[1] (7).

$$
\text{Dehydrogenase} \underset{\substack{\text{glutathione}\\ \text{cysteine}\\ \text{penicillamine}}}{\overset{O_2,\ H_2O_2}{\rightleftharpoons}} \text{Phosphatase}
$$

Thus sulfhydryl compounds protect the dehydrogenase and thereby promote the formation of 1,3-diphosphoglyceric acid which supplies high energy phosphate for ATP and subsequently phosphocreatine.

The rationale for using penicillamine, a sulfhydryl compound with reducing properties, was to protect the thiol enzymes of the glycolytic cycle and also important enzymes in anabolic pathways. In dystrophic muscle, there is an imbalance of the anabolic and catabolic processes. During progressive muscle degeneration, catabolic activity begins to predominate. Catabolic or hydrolytic reactions are often promoted by very stable proteins with disulfide bonds (S-S) such as the proteases, chymotrypsin, pepsin, elastin, carboxypeptidase A (10,11), RNases (12), DNases (13), glycosidases (14), phospholipases (15), and acyl phosphatase (16). On the other hand, anabolic and glycolytic pathways are characterized by more labile enzymes with reduced sulfhydryl groups (SH). These sulfhydryl enzymes catalyze protein, DNA, RNA, and fatty acid synthesis (17-20) and ATP production via oxidative phosphorylation or glycolysis (21,22). For maximal activity, sulfhydryl

[1] J.H. Park and R. Patnode, unpublished observations.

compounds or metal chelators are usually required in the assay systems to preserve essential thiol groups and might, therefore, be necessary for *in vivo* catalysis.

For a treatment program with this rationale, penicillamine appeared to be the most suitable drug. Unlike cysteine or glutathione, it is not readily autooxidizable and therefore persists in the serum in the reduced (-SH) form shown below (23).

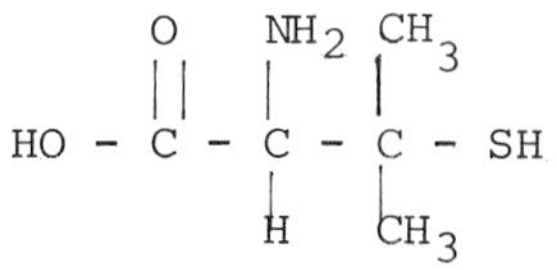

$$HO-\overset{\overset{\displaystyle O}{\|}}{C}-\underset{\underset{\displaystyle H}{|}}{\overset{\overset{\displaystyle NH_2}{|}}{C}}-\underset{\underset{\displaystyle CH_3}{|}}{\overset{\overset{\displaystyle CH_3}{|}}{C}}-SH$$

Moreover, the side effects and toxicity of the drug are well known (24,25) from the studies on Wilson's Disease and cystinuria. Thus we began a treatment program using the dystrophic chicken as the animal model.

B. Beneficial Results of Penicillamine Treatment

Genetic muscular dystrophy in the chicken has been considered as a good experimental model for inherited Duchenne dystrophy in humans (5). During the course of both avian and human dystrophy, there is a progressive destruction of muscle fibers with necrosis, phagocytosis, and replacement of muscle with adipose and collagenous connective tissue. The degeneration is accompanied by a rise in serum levels of muscle enzymes such as, creatine phosphokinase (CPK). The onset of symptoms in dystrophic chickens occurs in the second month after hatching. Chickens are unable to right themselves when placed on their backs, and their wings become excessively stiff and eventually cannot be elevated beyond a horizontal plane. The chicken model was selected for our penicillamine studies for the following reasons: (a) Muscle function tests are readily performed as a measure of the progress of the disease. (b) Pectoral and thigh muscles provide a large mass of tissue for enzymatic and biochemical analyses. (c) There is a natural anatomical separation of muscles into the white and red fiber types.

The results of a sample experiment on penicillamine treatment are briefly described in this section. Dystrophic chicks, which are obtained from Dr. Louis Pierro, University of Connecticut, were divided into 2 groups, an untreated and penicillamine-treated group. Normal white leghorn chicks were purchased locally and similarly divided into untreated and treated groups. There were 7 untreated and 8 treated dystrophic chicks as well as 9 untreated and 10 treated normal birds in the experimental groups. The experiment was designed as a modified double-blind study.

Penicillamine treatment was begun on the 9th day after hatching. The drug was prepared for administration from capsules of D-penicillamine (Cuprimine) obtained from Merck, Sharp and Dohme. Dosages and procedures for administering penicillamine have been detailed elsewhere (1).

Three methods are presented for evaluating the effects of penicillamine treatment: (1) righting ability, (2) wing apposition, and (3) CPK levels in the plasma. Since the general conclusions were the same for both the males and females, only the actual data for the roosters will be presented (1).

1. Righting ability

The righting ability was tested as follows: The bird was placed on its back and, if a bird could not rise, it was allowed to flap until exhausted. Two righting trials were done consecutively, followed by the measurements of body weight and wing apposition. Then two more trials for righting were performed. The total number of successful trials were recorded for each chicken. Righting tests and wing apposition measurements were carried out twice a week, but for simplification, only selected days are shown in the Tables.

TABLE 1

Condition of bird	*Average Number of Successful Rightings*		
	Day 25	*Day 50*	*Day 100*
Normal	4.0	4.0	4.0
Normal Treated	4.0	4.0	4.0
Dystrophic	2.2	0	0
Dystrophic Treated	4.0	3.2	3.0

Normal chickens could always right themselves immediately when placed on their backs on a flat surface. Dystrophic birds show some difficulty in rising at about 30 days of age. At approximately 50 days, striking effects of penicillamine treatment were observed. The untreated males were unable to right whereas penicillamine-treated males could right themselves on the average more than 70% of the time. The data all were statistically analyzed. Using Student's t test, penicillamine-treated males showed significantly better righting during the last 2½ months of the experiment. P values of less than 0.05 were observed for measurements on 17 out of 21 days (1).

2. *Wing Apposition*

Wing apposition was measured by raising the wings vertically until the styloid processes of the radii touched. If contractures prevented apposition, the distance between the styloid processes was measured in inches. Typical measurements are shown below.

TABLE 2

Condition of Bird	*Average Wing Apposition (inches)*		
	Day 25	*Day 50*	*Day 100*
Normal	0	0	0
Normal treated	0	0	0
Dystrophic	0	3.2	5.2
Dystrophic treated	0	2.2	2.0

During the entire experiment, the two wings of normal chickens could be raised vertically without any resistance until they touched each other. For the first month the measurements of the distance between the raised wings of the dystrophic chicks were essentially zero indicating perfect flexibility. Subsequently, untreated males showed the greatest rigidity, with an average of 5.2 inches between the wings by the third month. Penicillamine treatment partially alleviated the stiffness as evidenced by the final distance of 2.0 inches in the treated males. The treatment produced a statistically significant improvement in wing flexibility during the last month on 8 out of 9 testing days, as indicated by P values less than 0.05.

3. *CPK Levels*

The plasma CPK activity of every chicken was analyzed on the 105th day (Table III).

TABLE III

Condition of Birds	CPK Activity ± S.E.	
	IU/liter	
Normal	1,800	240
Normal, treated	1,900	100
Dystrophic	25,500	5,000
Dystrophic, treated	7,800	1,700

Untreated dystrophic males had high CPK values of approximately 25,000 IU/liter, and penicillamine suppressed the CPK levels to about 8,000 IU/liter. Statistical analysis of the treatment showed a P value of 0.03 when comparing the treated and untreated dystrophic males. This suppression did not result from an inhibition of plasma CPK by the drug since 50 mM penicillamine had no inhibitory effect on the CPK assay system. There was no significant difference in the plasma CPK activity of untreated and penicillamine-treated normal chickens.

III. DISCUSSION OF MECHANISM OF DRUG ACTION

Penicillamine imporved the dystrophic chickens as evidenced by righting ability and wing flexibility. The improved righting was not related to body weight but correlated with greater flexibility of the wings. Improved muscle function in righting tests is compatible with the hypothesis that penicillamine protects critical sulfhydryl enzymes by maintaining a suitable intracellular redox state. Sulfhydryl compounds, such as penicillamine and N-acetyl-cysteine, penetrate cells and increase protein sulfhydryl levels (26). The catalytic site of glyceraldehyde-3-phosphate dehydrogenase has an exceptionally reactive sulfhydryl group required for high energy phosphate production (2). Recent experiments in our laboratory have shown that penicillamine treatment of dystrophic chickens protects glyceraldehyde-3-phosphate dehydrogenase against inactivation (27). A number of other glycolytic enzymes, including CPK, also require sulfhydryl groups for maximal activity (28). Penicillamine *in vivo* could protect these enzymes, either by disulfide interchange or by chelation of divalent metal ions that cause oxidation of sulfhydryl groups. In dystrophic chickens, the white muscle fibers depend primarily on the glycolytic pathway for ATP and are more severely affected than red muscle fibers (5,6). These facts, however, are not sufficient to confirm the importance of sulfhydryl enzymes in the mechanism of drug action.

The beneficial effects of penicillamine on joint flexibility may be associated with its action on the deposition and cross-linking of collagen. Penicillamine blocks the formation of new insoluble collagen, degrades a certain fraction of recently synthesized insoluble collagen, and inhibits cross-linking (29,30). These properties may alleviate tendon rigidity and prevent infiltration of diseased muscle by collagen. The greater improvement in flexibility associated with the early treatment of the chicks may also be explained by the fact that the amount of degradation of preformed collagen in penicillamine-treated rats is greater in younger animals (30).

The improvement in wing flexibility may be extrapolated to considerations of muscle contractures observed in human dystrophy. If penicillamine treatment were also able to increase limb flexibility in human disease, it may be of clinical interest. Prevention of contractures in Duchenne patients might facilitate arm or hand movements and might also prolong the period of walking.

Although the mechanism of drug action has not been fully clarified, these studies may be useful for future investigations on the pathogenesis and treatment of human dystrophy.

IV. REFERENCES

(1). Chou, T.H., Hill, E.J., Bartle, E., Woolley, K., LeQuire, V., Olson, W., Roelofs, R., and Park, J.H., J. Clin. Invest. 56, 842-849 (1975).

(2). Cori, C.F., Slein, M.W., and Cori, G.T., J. Biol. Chem. 159, 565-566 (1945).

(3). Heyck, H., and Laudahn, G., in Exploratory Concepts in Muscular Dystrophy and Related Disorders (Milhorat, A.T., ed), pp. 232-244. Excerpta Medica Foundation, New York, 1967.

(4). Patnode, R., Bartle, E., Hill, E.J., LeQuire, V., and Park, J.H., J. Biol. Chem. 251, 4468-4475 (1976).

(5). Julian, L.M., and Asmundson, V.S., in Muscular Dystrophy in Man and Animals (G.H. Bourne and Golarz, M.N., eds), pp. 457-498. Hafner, New York, 1963.

(6). Shafig, S.A., Askanas, V., and Milhorat, A.T., Arch. Neurol. 25, 560-571 (1971).

(7). Park, J.H., and Koshland, D.E., J. Biol. Chem. 233, 986-990 (1958).

(8). Ehring, R. and Colowick, S.P., J. Biol. Chem.244, 4589-4599 (1969).
(9). Benitez, L.V. and Allison, W.S., J. Biol. Chem. 249, 6234-6243 (1974).
(10). Brown, J.R. and Hartley, B.S., Biochem. J.101, 214-228, (1966).
(11). Bradshaw, R.A., Ericsson, L.H., Walsh, K.A. and Neurath, H., Proc. Natl. Acad. Sci. 63, 1389-1394, (1969).
(12). Plummer, T.H. Jr. and Hirs, C.H.W., J. Biol. Chem. 239, 2530-2538 (1964).
(13). Liao, T.H., Salnikow, J., Moore, S. and Stein, W.H., J. Biol. Chem. 248, 1489-1493 (1973).
(14). Canfield, R. and Liu, A.K. J. Biol. Chem, 240, 1997-2002 (1965).
(15). DeHaas, G.H., Slotboom, A.J., Bonsen, P.P.M., Nieuwenhuizen, W., Van Deenen, L.L.M., Maroux, S., Dlouha, V., Desnuelle, P., Biochim. Biophys. Acta 221, 54-61 (1970).
(16). Harary, I. in Methods in Enzymology (Colowick, S.P. and Kaplan, N.O., Eds.) 6, 324-327 (1963).
(17). Ochoa, M. Jr. and Weinstein, I.B., J. Biol. Chem. 239, 3834-3842 (1964).
(18). Lehmann, I.R., Bessman, M.J., Simms, E.S. and Kornberg,A. J. Biol. Chem. 233, 163-170 (1958).
(19). Jovin, T. and Kornberg, A. J. Biol. Chem. 243, 250-259, (1968).
(20). Lynen, F. and Tada, M., Angew.Chem. 73, 513-519 (1961).
(21). Fletcher, M.J., Fluharty, A.L. and Sanadi, D.R., Biochim. Biophys. Acta 60, 425-427 (1962).
(22). Gutman, M., Mersmann, H., Luthy, J. and Singer, T.P., Biochemistry 9, 2678-2687, (1970).
(23). Christophersen, B.O. Biochemical J.106, 515-522, (1968).
(24). Scheinberg, I.H. and Sternlieb I., Amer. J. Med.29, 316-333, (1960).
(25). Bartter, F.C., Lotz, M., Thier, S., Roseberg, L.E. and Potts, J.T., Ann.Intern. Med.62, 796-822 (1965).
(26). Lorber, A., Chang, C.C., Masuoka, D. and Meacham, I., Biochem. Pharmacol. 19, 1551-1560 (1970).
(27). Hill, E.J., Chou, T., Roelofs, R., LeQuire, V. and Park, J.H., Proceedings of International Congress of Biochemistry, Hamburg, p. 591, (1976).
(28). Mahowald, T.A., Noltmann, E.A. and Kuby, S.A., J. Biol. Chem.237, 1535-1548 (1962).
(29). Nimni, M.E., J. Biol. Chem.243, 1457-1466, (1968).
(30). Nimni, M.E., Deshmukh, K., Gerth, N. and Bavetta, L.A., Biochem.Pharmacol.18, 707-714 (1969).

V. ACKNOWLEDGMENTS

This work was supported by grants from the Muscular Dystrophy Association, U.S. Public Health Service, & National Science Foundation.

KINETICS AND MECHANISM OF ACTIVATED LIVER ALCOHOL DEHYDROGENASE

R.T. Dworschack and B.V. Plapp

The University of Iowa

The mechanism of hydrogen transfer catalyzed by horse liver alcohol dehydrogenase with amidinated amino groups was studied with steady state kinetics. Hydroxybutyrimidylation of the enzyme increases the maximum velocities of the enzymatic reactions and the rates of dissociation of the enzyme-coenzyme complexes. Primary deuterium isotope effects obtained for oxidation of [1,1-D_2]-benzyl alcohol and for reduction of a series of para-substituted benzaldehydes indicate that the turnover numbers reflect the rates of interconversion of central complexes. The magnitudes and signs of the ρ values obtained for oxidation of p-substituted benzyl alcohols and for reduction of p-substituted benzaldehydes suggest that transfer of hydrogen occurs via concerted hydride and proton transfer in which only a small amount of charge develops in the transition state. The rate of oxidation of benzyl alcohol depends upon a group with a pK of 8.4, which must be unprotonated for maximum activity, but which allows partial activity in its protonated form. The maximum velocity for benzaldehyde reduction catalyzed by native enzyme required protonation of a group with a pK of 8.0. The pH effects for both enzymes can be explained by a coherent model in which a water (or hydroxide) molecule coordinated to the zinc ion acts as a proton donor (or acceptor) and the state of protonation of His-51 modulates the rate of transfer of hydrogen.

I. INTRODUCTION

The pioneering work of Klinman (1-3) on the mechanism of the interconversion of para-substituted benzyl alcohols and benzaldehydes catalyzed by yeast alcohol dehydrogenase (ED 1.1.1.1) has led to the conclusion that little or no charge develops on carbon one of the substrate during hydrogen transfer. Mechanisms involving either concerted general acid-base catalysis of a hydride transfer or a protonated radical intermediate and hydrogen atom transfer are consistent with the results. Similar studies with horse liver alcohol dehydrogenase have been more difficult because the chemical conversion of substrate to product catalyzed by the native liver emzyme cannot be observed with steady state techniques. Transfer of hydrogen is not rate-limiting for the turnover of the catalyst. Rather, the rate-limiting step for the oxidation of benzyl alcohol is dissociation of the enzyme-NADH complex (4), and for the reduction of benzaldehyde, release of benzyl alcohol is limiting (5).

Transient kinetic techniques have thus been used to study substituent effects with native enzyme (6,7). The interconversion of ternary complexes can be observed with these techniques, as indicated by large primary deuterium isotope effects, but the conclusions from these studies are not compatible. We have found that modification of the amino groups of liver alcohol dehydrogenase increases the rate of dissociation of enzyme-coenzyme complexes so that transfer of hydrogen becomes rate-limiting (8). Using a modified enzyme, we are now able to use steady-state kinetics to investigate the mechanism of catalysis.

II. DISCUSSION

A. Isotope and Substitutent Effects

Horse liver alcohol dehydrogenase (EE isozyme, Ref. 9) was partially acetimidylated in the presence of NAD^+ and pyrazole and the excess reagents were removed as described previously (8). The amino groups at the active sites were modified by reaction with 0.1 M 4-bromobutyramide (10) at pH 8 at 25° for 2 hr and the enzyme was activated 23-fold for the oxidation of ethanol.

This particular modification produces an enzyme that exhibits large primary isotope effects on steady-state turnover with various substrates. Over the range of pH from 8.0 to 9.9 the maximal rate of oxidation of benzyl alcohol by NAD^+ is 3.6-fold slower when benzyl alcohol dideuterated at carbon one is used. Likewise, the maximal rate of reduction of benzaldehyde by NADH is 2.4-fold slower when NADH deuterated in the 4-position of the nicotinamide ring is used. The isotope effects are about the same magnitude as those obtained from transient kinetic studies with the liver enzyme (5-7) and almost

C. Mechanism of Hydrogen Transfer and Explanation of pH Effects

Based mostly on X-ray crystallographic evidence but supported by some significant chemical and kinetic results, we propose that the structure shown in Scheme I can represent the active ternary complex just before hydrogen is transferred from an alcohol to the coenzyme. The structure incorporates all of the features proposed by Brändén *et al.* (15), but differs in that the alcohol is protonated (i.e., not an alcoholate) and that hydroxide ion is interposed between the alcohol and the serine hydroxyl group.

There are several reasons for the modification of the mechanism. Mechanistically, it is simpler and more reasonable to leave water on the zinc than to displace the water with the alcoholate. Displacing the water adds an extra step to the mechanism, which could be relatively slow if it required conformational alterations of the protein. Thermodynamically, formation of an alcoholate might require several kilocalories of binding energy and this certainly would not facilitate formation of the active complex. Experimentally, Sloan *et al.* (16) detected the presence of protons of water bound to Co^{II},

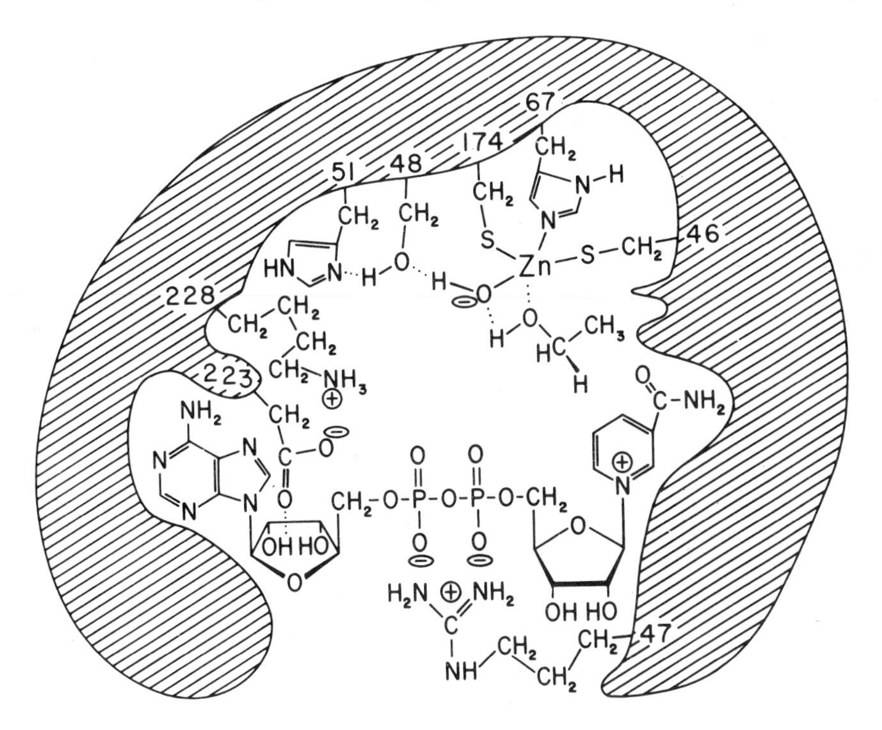

Scheme I

which was substituted for Zn^{II} in the enzyme, and showed that binding of coenzyme and a substrate analog immobilized some of the protons. Finally, as discussed below, placement of a water molecule as shown in Scheme I facilitates the explanation of the observed pH-dependencies.

As shown in Scheme I, the zinc ion has five ligands. This is consistent with the evidence from X-ray crystallography that orthophenanthroline binds to the zinc ion, displaces the water, and forms a penta-coordinate complex (15). Direct coordination of the alcohol (or aldehyde) to the zinc is supported by spectroscopic and kinetic evidence on the reactions of a chromophoric aldehyde with the enzyme (17,18).

Studies of the pH-dependencies of reactions of native alcohol dehydrogenase are relevant to the present discussion. The free enzyme has a group with a pK of about 9.0 (19,20), which shifts to about 7.5 in the enzyme-NAD^+ complex (19,20). From the pH dependence for the binding of trifluoroethanol (an inert substrate analog) or capric acid to the enzyme-NAD^+ complex, Shore *et al.* (21) determined that the enzyme-NAD^+ complex had a pK of 7.6.

They also found that when both NAD^+ and trifluoroethanol were bound to the enzyme in the pH range 5.5 to 8.5, exactly one proton was liberated per equivalent of enzyme. They proposed that the alcohol was bound directly to the unprotonated, or basic, form of the group that has the pK of 7.6 in the enzyme-NAD^+ complex.

(It appears that the pK of the group in the enzyme-NAD^+-alcohol complex must be less than 4.5) Taniguchi *et al.* (19) and Shore *et al.* (21) suggest that the group could be water bound to the zinc, as illustrated in Scheme I. It is then apparent that the hydroxide ion can act as the base to remove the proton from ethanol during the transfer of hydride ion from the ethanol to NAD^+. Moreover, it follows that during the reduction of an aldehyde, the protonated group, *i.e.* H_2O, could act as an acid to donate a proton to the substrate.

Evidence that the group can be protonated comes from the observation that capric acid binds preferentially to the protonated enzyme-NAD^+ complex and shifts the pK of the group in the ternary complex to a value that is probably above 9.0 (21). For the binding of an aldehyde, Dunn (22) has made the important observation that after the enzyme-NADH-aldehyde complex is formed, protons are not taken up during the pre-steady state reduction reaction, but one proton is taken up for each molecule of product NAD^+ that is displaced by NADH (at pH 8.8). This result implies that a group on the enzyme-NADH-aldehyde complex is protonated at pH 8.8, whereas the group in the resulting enzyme-NAD^+-alcohol complex is unprotonated at pH 8.8. Therefore, it appears that there is a group on the enzyme that acts as a proton donor or acceptor, in a manner exactly analogous to the imidazole group His-195 in lactate dehydrogenase (23). It is important to emphasize that the state of protonation of this group is coupled to the binding of substrates or substrate analogs (so that dissociation constants show pH dependencies), but that once the complex is formed it is not likely that the group changes its state of protonation over the pH range of 4.5 to 9 (so that maximum velocities could be pH-independent). From the X-ray crystallographic evidence, the water bound to the zinc ion is most likely to be the group.

If Scheme I represents the structure of a ternary complex poised for reaction, how then can we explain the effects of pH on the maximum velocities? If we assume that the hydroxide bound to zinc cannot be readily protonated at pH 8.0, the pH effects must be indirect. Based on the crystallographic evidence, we propose that the following forms may represent HES and ES. Protonation of the imidazole could disrupt the hydrogen bonding system and thereby decrease the rate of hydrogen transfer. But it should be noted that both the protonated and unprotonated forms are presumed to be catalytically active.

For the reduction of benzaldehyde catalyzed by native enzyme, turnover is largely controlled by the rate of dissociation of the enzyme-NAD^+-alcohol complex (5). Thus it is reasonable to assume that the pK of 8.0 corresponds to the ionization of the enzyme-NAD^+-alcohol complex and that the alcohol dissociates more rapidly from the protonated form. This is consistent with the pH dependencies for the apparent K_m for benzyl alcohol determined from transient kinetics and for steady state turnover of β-naphthaldehyde with native enzyme (14).

Rather than identify the pK of 8.4 with His-51 we should point out that it would be equally valid to interpret the results in terms of pH-dependent conformational changes. Indeed, Shore *et al.* (21) suggested that the pK of 6.4 seen in the transient oxidation of ethanol could result from an isomerization step. A further qualification of our interpretations is that the data do not demand that a water molecule be present during hydrogen transfer. Other mechanisms can explain the data. Results from more definitive investigations, such as X-ray crystallography are required to substantiate the present

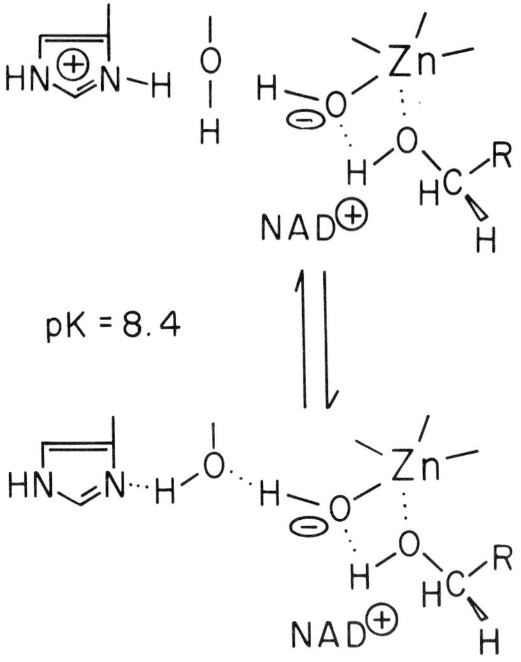

pK = 8.4

Scheme II

proposal.

III. ACKNOWLEDGEMENTS

This work was supported by USPHS Research Grant AA00279, Research Development Award AA10 (B.V.P.) and Training Grant GM550 (R.T.D.).

IV. REFERENCES

(1). Klinman, J.P., J. Biol. Chem. 247, 7977-7987 (1972).
(2). Klinman, J.P., J. Biol. Chem. 250, 2569-2573 (1975).
(3). Klinman, J.P., Biochemistry 15, 2018-2026 (1976).
(4). Wratten, C.C., and Cleland, W.W., Biochemistry 4, 2442-2451 (1965).
(5). McFarland, J.T., and Bernhard, S., Biochemistry 11, 1486-1493 (1972).
(6). Hardman, M.J., Blackwell, L.F., Boswell, C.R., and Buckley, P.D., Eur. J. Biochem. 50, 113-118 (1974).

(7). Jacobs, J.W., McFarland, J.T., Wainer, I., Jeanmaier, D., Ham, C., Hamm, K., Wnuk, M., and Lam, M., Biochemistry 13, 60-64 (1974).
(8). Plapp, B.V., Brooks, R.L., and Shore, J.D., J. Biol. Chem. 248, 3470-3475 (1973).
(9). Dworschack, R.T., and Plapp, B.V., Biochemistry, in press (1976).
(10). Fries, R.W., Bohlken, D.P., Blakley, R.T., and Plapp, B. V., Biochemistry 14, 5233-5238 (1975).
(11). Swain, C., Wiles, R., and Bader, R., J. Am. Chem. Soc. 83, 1945-1955 (1961).
(12). Blackwell, L.F., and Hardman, M.J., Eur. J. Biochem. 55, 611-615 (1975).
(13). Blankenhorn, G., Eur. J. Biochem. 67, 67-80 (1976).
(14). McFarland, J.T., and Chu, Y.-H., Biochemistry 14, 1140-1146 (1975).
(15). Brändén, C.-I., Jörnvall, H., Eklund, H., and Furugren, B., The Enzymes, 3rd Ed. 11, 104-190 (1975).
(16). Sloan, D.L., Young, J.M., and Mildvan, A.S., Biochemistry 14, 1998-2008 (1975).
(17). Dunn, M.F., and Hutchinson, J.S., Biochemistry 12, 4882-4892 (1973).
(18). Dunn, M.F., Biellman, J.-F., and Branlant, G., Biochemistry 14, 3176-3182 (1975).
(19). Taniguchi, S., Theorell, H., and Åkeson, Å., Acta Chem. Scand. 21, 1903-1920 (1967).
(20). Dalziel, K., J. Biol. Chem. 238, 2850-2858 (1963).
(21). Shore, J.D., Gutfreund, H., Brooks, R.L., Santiago, D., and Santiago, P., Biochemistry 13, 4185-4190 (1974).
(22). Dunn, M.F., Biochemistry 13, 1146-1151 (1974).
(23). Holbrook, J.J., Liljas, A., Steindel, S.J., and Rossmann, M.G., The Enzymes, 3rd Ed. 11, 191-292 (1975).

SOLVENT ISOTOPE EFFECTS IN THE YEAST ALCOHOL DEHYDROGENASE REACTION

Judith P. Klinman, Katherine M. Welsh and Donald J. Creighton

The Institute of Cancer Research
Philadelphia

Solvent isotope effects wereinvestigated in the yeast alcohol dehydrogenase reaction in an effort to distinguish among possible modes of acid-base catalysis. A determination of pK_a *and* k_{cat} *in* H_2O *and 96%* D_2O *indicates* $\Delta pK_a = pK_D - pK_H = 0.11$, $k_{H_2O}/k_{D_2O} = 1.20 \pm 0.09$ *in the direction of* p-CH_3O *benzyl alcohol oxidation, and* $k_{H_2O}/k_{D_2O} = 0.50 \pm .05$ *and* 0.48 ± 0.5 *for* p-CH_3O *benzaldehyde reduction by NADH and NADD. The surprisingly small effect of* D_2O *on pK, which contrasts with the common observa - tion that* $\Delta pK_a \simeq 0.4 - 0.6$, *may reflect the ionization of an active site Zn-*$\overline{O}H_2$. *The small isotope effect for* k_{cat} *in the direction of alcohol oxidation appears to rule out a mechanism involving concerted catalysis by an active site base of hydride transfer. The large inverse isotope effect on* k_{cat} *for aldehyde reduction implicates an intermediate formed prior to the rate determining transfer of hydrogen from coenzyme to substrate. The properties of this intermediate, and its relationship to the mode of hydrogen transfer between coenzyme and substrate, are discussed.*

I. INTRODUCTION.

Previous kinetic studies of the yeast alcohol dehydrogenase catalyzed interconversion of aromatic substrates indicate a rate limiting C-H bond cleavage step under steady state kinetic conditions, $V_H/V_D = 3-5$ (1,2), a distribution of charge at C-1 of substrate in the transition state which is similar to that of alcohol, $\rho^+ = 0$ for alcohol oxidation (1,2) and a role for an active site residue, pK = 8.25, in acid-base catalysis (3). The simplest chemical mechanism, consistent with the observed data, has been proposed to involve concerted catalysis by a protonic base of hydride transfer (Scheme I). Solvent isotope effects were investigated in an effort to distinguish this

Scheme I

Concerted Acid-Base Catalysis by a Protonic Base of Hydride Transfer

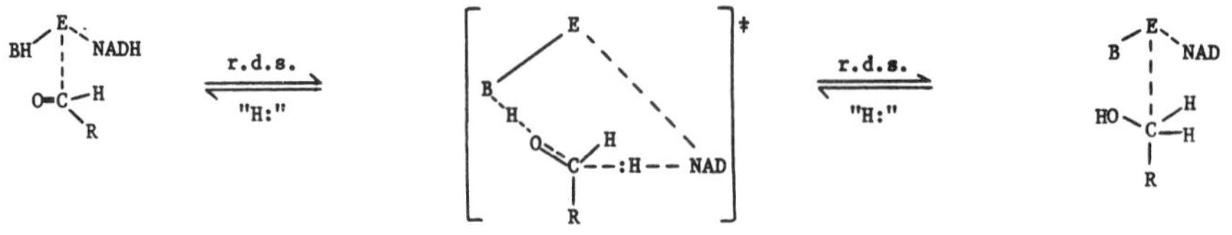

mechanism from ones in which proton transfer is uncoupled from and fast relative to C-H bond cleavage.

II. EXPERIMENTAL

Yeast alcohol dehydrogenase was obtained as an ammonium sulfate suspension from Boehringer, and dialyzed prior to use (1). The synthesis of isotopically labelled NADH has been described (1). NAD was purchased from P-L Biochemicals, chromato-pure and alcohol-free. p-CH_3O-Benzyl alcohol and p-CH_3O-benzaldehyde were vacuum distilled, and used immediately for kinetic studies. Solution of coenzymes and substrates were assayed enzymatically (1,2). Kinetic measurements were carried out at 25°, $\mu = 0.22$, by following the appearance of NADH or disappearance of NADH(D) at 340 nm. Intercepts from primary Lineweaver-Burk plots were replotted as a function of second substrate to obtain V_{max}; rate constants were calculated from V_{max} assuming four active sites per mole of enzyme and normalized to a specific activity of 100 U/mg (1,2). D_2O was purchased from Wilmad, and purified by bulb lyophilization. KP_i and KPP_i-glycine buffers were employed in the pH range 7.2 - 9.2 Buffer solutions were prepared in H_2O, and dried down and redissolved 2x in H_2O or D_2O. Coenzyme solutions were dried down and redissolved 2x in H_2O or D_2O. pD = pH (meter) + 0.4. The final concentration of H_2O in lyophilized D20 reaction mixtures was determined by nuclear magnetic resonance (Varian HA-1oo-15) using tetramethyl silane as an external standard.

III. RESULTS AND DISCUSSION

The separation of D_2O effects on k_{cat} and pK_a was achieved in the direction of p-CH_3O benzyl alcohol oxidation by plotting the reciprocal of the observed rate constants at a given pL as a function of the lyonium ion activity:[1]

$$v = k_{cat} \cdot E_B = \frac{k_{cat} \cdot E_T}{1 + L_3O^+/K_a} \qquad (1)$$

$$\frac{E_T}{v} = \frac{1}{k_{obs}} = \frac{1}{k_{cat}} + \frac{L_3O^+}{k_{cat} \cdot K_a} \qquad (2)$$

Equation (1) assumes that a single ionizing residue is required to be in a free base form for catalysis (3). Expression of E_B in terms of E_T, followed by rearrangement into a reciprocal form, leads to equation (2). The data in Figure 1-A and -B represent alcohol oxidation in D_2O and H_2O, respectively. Each data point is the result of duplicate or triplicate determinations of rate constants obtained from steady state kinetic data by extrapolation to infinite concentration of substrate and coenzyme. The intercepts of the plots in Figure 1 provide the limiting rate constant when $E_B = E_T$; from the log of the ratio of the slopes to intercepts one obtains pK_a for the ionizing residue.

In the direction of benzaldehyde reduction, the data have been analyzed on the assumption that the protonated form of a base is required for catalysis (3):

$$V = k_{cat} \cdot EBL = \frac{k_{cat} \cdot E_T}{1 + K_a/L_3O^+} \qquad (3)$$

$$\frac{E_T}{v} = \frac{1}{k_{obs}} = \frac{1}{k_{cat}} + \frac{K_a}{k_{cat} \cdot L_3O^+} \qquad (4)$$

Plots of $1/k_{obs}$ vs. $1/L_3O^+$ are illustrated in Figure 2-A and -B for p-CH_3O benzaldehyde reduction by NADH(D) in D_2O and H_2O respectively. The intercepts of these plots provide k_{cat} when $EBL = E_T$; and the negative log of the ratio of slopes to intercepts gives pK_a.

The effect of D_2O on pK_a is summarized in Table 1. The pK in water is within experimental error of a previously determined value of 8.25 for p-CH_3 benzyl alcohol oxidation and acetaldehyde reduction (3). Surprisingly, D_2O is found to have an

---

[1] Terms appearing in equations (1)-(7) are defined in the appendix.

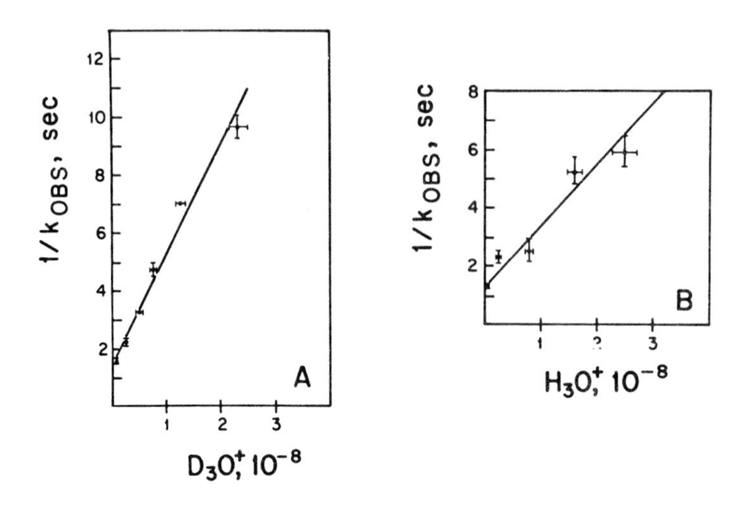

Fig. 1. Observed rate constants, obtained from steady state kinetic data by extrapolation to infinite concentration of substrate and coenzyme, for the oxidation of p-CH_3O benzyl alcohol in D_2O (A) and H_2O (B) as a function of pL. The data are plotted in reciprocal form according to equation (2) in

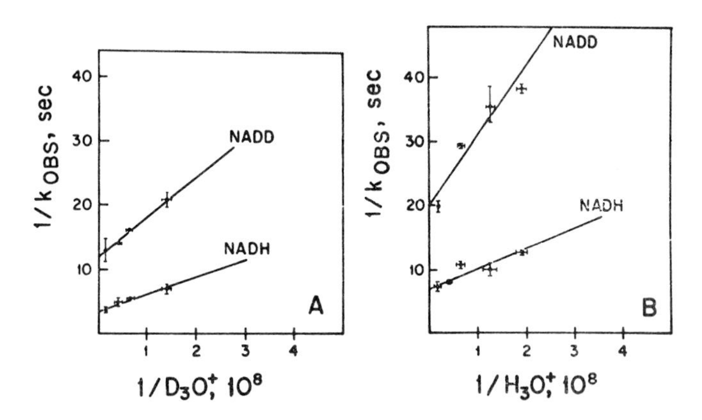

Fig. 2. Observed rate constants, obtained from steady satate kinetic data by extrapolation to infinite concentration of substrate and coenzyme, for the reduction of p-CH_3O benzaldehyde by NADH(D) in D_2O (A) and H_2O (B) as a function of pL. The data are plotted in reciprocal form according to equation (4) in text.

extremely small effect on the pK_a, $pK_D = 8.32 \pm 0.03$. This result contrasts with a commonly observed $\Delta pK_a = pK_D - pK_H \simeq 0.4 - 0.6$. The effect of D_2O on an ionization constant can be expressed in terms of fractionation factors, ϕ's, for the protonic species involved in the equilibria equations:

$$BH + H_2O \rightleftharpoons B + H_3O^+ \qquad (5)$$

$$BD + D_2O \rightleftharpoons B + D_3O^+ \qquad (6)$$

$$pK_D - pK_H = \log \frac{H_3O^+}{D_3O^+} \times \frac{B}{B} \times \frac{BD}{BH} \times \frac{D_2O}{H_2O} = \log \frac{\phi\ BH}{\phi\ H_3O^+} \qquad (7)$$

where $D_2O/H_2O = 1$ by definition (c.f. the Appendix). For many protonated oxygen and nitrogen bases, $\phi BH \simeq 1$ (4); and the effect of D_2O on the pK of these ionizing residues results primarily from the fractionation factor for H_3O^+, $\phi H_3O^+ = 0.33$ (5). The small $\Delta pK = 0.11$ observed in the YADH system suggests that ϕBH is similar to ϕH_3O^+; on the assumption that BH is monoprotic or that $\phi B = 1$, the fractionation factor for BH is calculated to be 0.43. Thiols are characterized by fractionation factors of about 0.5 (6,7), and two cysteines are known to be present at the active site of this enzyme (8-10). However, the 2.4 Å x-ray structure of horse liver alcohol dehydrogenase indicates a role for two cysteines as ligands for an active site

TABLE 1
Solvent Isotope Effect on pK_a

Substrate	pK H_2O	pK D_2O
p-CH_3O Benzyl alcohol oxidation	8.14	8.35
p-CH_3O Benzaldehyde reduction	8.27	8.29
Average =	8.21 ± 0.06	8.32 ± 0.03

zinc-water (11); the metal content (12,13) and homologous cysteine (10) in YADH are consistent with a similar active site configuration. The fractionation factor for $Zn\text{-}OH_2$ may not be very different from $H\text{-}\overset{+}{O}H_2$, and provides a possible explanation for the small observed effect of D_2O on the pK of the active site residue in YADH:

$$\overset{\delta+}{ZnOH_2} + H_2O \rightleftharpoons ZnOH + H_3O^+ \qquad (8)$$

The effects of D_2O on k_{cat} are summarized in Table 2. The kinetic isotope effect can be formulated as the ratio of fractionation factors for the transition state (ϕTS) and ground state (ϕGS), $k_{H_2O}/k_{D_2O} = \phi GS / \phi TS$. For alcohol oxidation ϕGS refers to the unprotonated base and alcoholic hydroxyl group, $\phi GS \simeq 1$ and $\phi TS \simeq 1/1.2 = 0.83$. In the direction of aldehyde reduction ϕGS refers to the protonated base and aldehyde carbonyl $\phi GS \simeq 0.43$ and $\phi TS \simeq 0.43/0.50 = 0.86$. Previous studies of primary hydrogen isotope effects and structure reactivity correlations have implicated a single rate determining C-H bond cleavage step in the YADH catalyzed interconversion of aromatic substrates (2). Consistent with a single rate determining step, similar transition state fractionation factors are observed for aldehyde reduction and alcohol oxidation, $\phi TS = 0.83 - 0.86$.

The solvent isotope effect on k_{cat} for alcohol oxidation is close to one, $k_{H_2O}/k_{D_2O} = 1.20 \pm 0.09$; this result appears to rule out the mechanism illustrated in Scheme I, since concerted acid-base catalysis is expected to be characterized by a normal solventisotope effect, $k_{H_2O}/k_{D_2O} \geq 2$ (4). In contrast, aldehyde reductions indicate a large inverse isotope effect, $k_{H_2O}/k_{D_2O} = 0.50 \pm 0.05$ for reduction by NADH. A three-fold decrease in k_{cat}, which results from substitution of hydrogen by deuterium at C-14 of the dihydronicotinamide ring of NADH, leads to a similar solvent isotope effect, $k_{H_2O}/k_{D_2O} = 0.58 \pm 0.06$. The dissimilarity between solvent isotope effects for alcohol oxidation and aldehyde reduction argues that an intermediate is formed in a fast step, prior to C-H cleavage, in the direction of aldehyde reduction; the inverse isotope effect observed upon formation of this intermediate is attributed to the loss of the unusual ground state fractionation factor for EBH, $\phi = 0.43$.

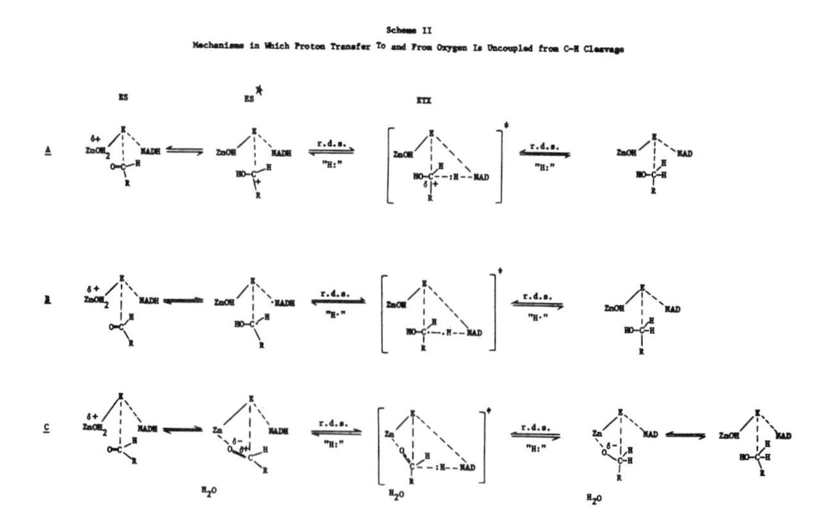

Scheme II
Mechanisms in Which Proton Transfer To and From Oxygen Is Uncoupled from C-H Cleavage

As illustrated in Scheme II, the properties of this intermediate relate to the mode of hydrogen transfer between coenzyme and substrate. The observed inverse isotope effect for aldehyde reduction is similar to values observed in model systems involving specific acid catalysis, e.g. $k_{H_2O}/k_{D_2O} = 0.54$ for the acid catalyzed enolization of acetone (14). A pre-equilibrium transfer of a protein to the carbonyl oxygen to generate a carbonium ion (Scheme II-A) appears unlikely in the yeast alcohol dehydrogenase reaction: however, in light of an approximately 13 pK unit difference between the active site base (pK = 8.25) and protonated aldehydes (pK =-3 to -7 (15)) and the finding that the distribution of charge at C-1 of substrate in the transition state is similar to that of product alcohol. Mechanisms B and C in Scheme II are proposed to be consistent with the available kinetic data. According to B, aldehyde reduction occurs by way of a pre-equilibrium transfer of a proton plus one electron to form a neutral, protonated radical intermediate, followed by a rate-determining hydrogen atom transfer (16). A comparison of pK's for the ionization of α-hydroxy radicals [(pK = 11.51 for $CH_3\text{-}\dot{C}H\text{-}OH$ $CH_3\text{-}\dot{C}H\text{-}O^-$ + H_3O^+ (17)] to the putative base indicates that a proton transfer from EBH to the radical anion would be thermodynamically favored, in contrast to a pre-equilibrium formation of a carbonium ion. In an alternate mechanism, C, the inverse isotope effect in the direction of aldehyde reduction results from a displacement of water from zinc by the carbonyl functional group of the aldehyde substrate, prior to a rate-determining hydride transfer. The amount of charge generated at the aldehyde carbonyl upon coordination to zinc will be considerably less that that resulting from a direct proton transfer. For example:

$$pK\ (ZN\text{-}\overset{\delta+}{O}H_2) = 8.7 \quad (18) \quad \textit{versus}$$
$$pK\ (H\text{-}\overset{+}{O}H_2) = -1.7$$

and mechanism C appears to be compatible with the observed charge properties at the reacting bonds of substrate in the transition state of this reaction.

Clearly, further experimental work is necessary to distinguish B from C. Appropriate models for the effect of D_2O on the ionization of zinc-bound water and the displacement of water from zinc by aprotic ligands would be extremely useful in interpreting solvent[2] isotope effects for YADH and other zinc containing enzymes. An as yet unanswered question concerning the mechanism

---

[2] Splinter *et.al.* (19) report a normal isotope effect for the ionization of aquopentamine cobalt (III), ΔpK = 0.48, in contrast to an earlier report of ΔpK = 0.18 (20).

TABLE 2
Solvent Isotope Effects on k_{cat}

Substrate	k_{cat}, sec^{-1} H_2O	k_{cat}, sec^{-1} D_2O	k_{H_2O}/k_{D_2O}
p-CH_3O benzyl alcohol oxidation	0.78 ± 0.02	0.65 ± 0.03	1.20 ± 0.09
p-CH_3O benzaldehyde reduction			
+ NADH	0.14 ± 0.007	0.28 ± 0.014	0.50 ± 0.05
+ NADD	0.049 ± 0.002	0.085 ± 0.005	0.58 ± 0.06

of alcohol dehydrogenase is the role of metals *vs.* metal-bound water in the chemical reaction. Whereas the available x-ray crystallographic data on horse liver alcohol dehydrogenase support direct coordination to zinc (11), nuclear magnetic resonance studies indicate second sphere complexes between substrate and metal (21). An important feature of mechanism C is the formation of an inner sphere coordination complex coupled to the loss of metal water, which occurs subsequent to the ES complex. Although the kinetic data implicate ES, a low concentration of this species relative to ES may preclude its detection via static probes.

IV. ACKNOWLEDGEMENTS

This work was supported by USPHS Grants GM-20627, RR-05539, CA-06927; by NSF Grant BMS73-00732; and also by an appropriation from the Commonwealth of Pennsylvania.

V. REFERENCES

(1). Klinman, J.P., J. Biol. Chem. 247, 7077 (1972).
(2). Klinman, J.P., Biochemistry 15, 2018 (1976).
(3). Klinman, J.P., J. Biol. Chem. 250, 2569 (1975).
(4). Schowen, R.L. Prog. in Phys. Org. Chem. 9, 275 (1972).
(5). Williams, jr., J.M., and Kreevoy, M.M., Advan. Phys. Org. Chem. 6, 63 (1968).
(6). Pohl, H.A., H. Chem. Eng. Data 6, 515 (1961).
(7). Hobden, F.W., Johnston, E.F., Weldon, H.D., and Wilson, C.L., J. Chem. Soc. 61 (1939).
(8). Belke, C.J., Chin, D.C.Q., and Wold, F., Biochemistry 13, 3418 (1974).

(9). Klinman, J.P., Biochemistry 14, 2568 (1975).
(10). Jornvall, A., Woenckhaus, C., and Johnscher, G., Eur. J. Biochem. 53, 71 (1975).
(11). Eklund, H., Nordstrom, B. Zeppezauer, E., Soderland, G., Ohlsson, I., Boiwe, T., Soderberg, B.-O., Tapia, O., Branden, C.J., and Akeson,A., J. Mol. Biol. 102, 27 (1976).
(12). Veillon, C., and Sytkowski, A.J., Biochem. Biophys. Res. Commun. 67, 1494 (1976).
(13). Klinman, J.P., and Welsh, K.,Biochem. Biophys. Res. Commun. 70, 878 (1976).
(14). Toullec, J., and Dubois, J.E., J. Am. Chem. Soc. 96, 3524 (1974).
(15). Stewart, R., Gatzke, A.L., Macke, M., and Yates, K., Chem. and Ind. 331,(1959).
(16). Williams, R.F., Shinkai, S., and Bruice, T.C., Proc. Natl. Acad. Sci., USA 72, 1763 (1975).
(17). Laroff, G.P., and Fessenden, R.W., J. Phys. Chem. 77, 1283 (1973).
(18). Chaberek, S., jr., Courtney, R.C., and Martell, A.E., J. Am. Chem. Soc. 74, 5057 (1952).
(19). Splinter, R.C., Harris, S.J., and Tobias, R.S., Inorg. Chem. 7, 897 (1968).
(20). Taube, H., J. Am. Chem. Soc. 82, 524 (1960).
(21). Sloan, D.L., Young, J.M., and Mildvan, A.S., Biochemistry 14, 1998 (1975).

VI. APPENDIX: Definition of Terms Appearing in Equations (1)-(7)

E_T: Total enzyme concentration

E_B: Concentration of enzyme in which an ionizing residue is unprotonated

E_{BH}: Concentration of enzyme in which an ionizing residue is protonated

k_{obs}: First order rate constant for the YADH-catalyzed reaction at a given pH

k_{cat}: First order rate constant for the YADH-cataylzed reaction under conditions of optimal pH. In the case of alcohol oxidation this occurs at high pH when enzyme has fully ionized, $E_T=E_B$. For aldehyde reduction, optimal activity results at low pH such that E_T = EBH.

K_a: The kinetically determined dissociation constant for the ionizing residue of YADH.

L_3O^+ : A general term referring both to H_3O^+ and D_3O^+.

ϕ : The fractionation factor, ϕ, describes the isotopic exchange reaction:

$$\text{X-H} + \text{H-O-D} \rightleftharpoons \text{X-D} + \text{H-O-H}$$

$$\phi = \frac{\text{H-O-H}}{\text{H-O-D}} \times \frac{\text{X-D}}{\text{X-H}}$$

By definition HOH/HOD = 1; and ϕ simplifies to give X-D/X-H. For $\phi > 1$, deuterium will be enriched in X-H relative to HOH, indicating that X-H is a tighter bond. Alternatively, when $\phi < 1$, the deuterium concentration in X-H will be depleted relative to HOH.

KINETICS OF NATIVE AND CHEMICALLY ACTIVATED HUMAN LIVER ALCOHOL DEHYDROGENASES[1]

A. Dubied and J.-P. von Wartburg
University of Bern

and D.P. Bohlken and B.V. Plapp[2]
University of Iowa

Acetimidylation and methylation of alcohol dehydrogenase from human livers of the normal phenotype (B_1B_1) increases activity and Michaelis and inhibition constants, suggesting that an amino group at the active site is modified, as was shown previously for the horse enzyme. The enzyme from atypical phenotype (B_1B_2) was only activated 2-fold by acetimidylation, which may indicate that substitution of Pro for Ala-230 or modification of Lys-228 is sufficient to fully activate the enzyme. Product inhibition patterns for native and modified human enzymes are consistent with an Ordered Bi Bi mechanism. However, the major isoenzyme of native human liver alcohol dehydrogenase B_1B_1 exhibits nonlinear kinetics over a wide range of ethanol concentrations, indicating the presence of subunits with different kinetic characteristics or negative cooperativity between subunits. Chemical modification makes the kinetics linear and alters the mechanism.

I. INTRODUCTION

When the ε-amino group of lysine residue 228 at the active site of horse liver alcohol dehydrogenase (E.C. 1.1.1.1) is alkylated or amidinated with substituents that retain the positive charge, the maximum velocities of the enzymatic reactions

[1]This work was supported by Research Grants AA00279 (Iowa City) and AA00233 (bern) from the United States Public Health Service, National Institute on Alcohol Abuse and Alcoholism, and Grant 3.441-0.74 from the Swiss National Foundation for Scientific Research. [2]Recipient of Research Scientist Development Award AA00010 from the National Institute on Alcohol Abuse & Alcoholism.

are increased up to 10-fold (1-5). The kinetic mechanisms of the modified enzymes appear to be the same as that of native enzyme, that is, Ordered Bi Bi (1,2,6). The increased activity of the modified enzymes is due to increased rates of dissociation of the enzyme-coenzyme complexes, the rate-limiting steps in the reactions catalyzed by the native enzyme (2,7-9). With the horse liver enzyme, small, positively-charged substituents, such as methyl or acetimidyl, increase turnover numbers and Michaelis constants for ethanol, but do not greatly change the Michaelis constants for NAD^+ (3); such modified enzymes could be more active *in vivo* (10).

The main isoenzymes of horse and human liver alcohol dehydrogenases have been shown to be highly homologous (11,12). On the other hand, they differ somewhat in catalytic properties, such as turnover number, pH optimum and substrate specificity. In order to compare their reaction mechanisms and the effects of modifications of amino groups, we have studied the kinetic mechanisms of native and chemically modified human enzymes. We were also interested in determining whether the genetically determined "atypical" variant of the enzyme, which has higher specific activity and a lower pH optimum than the normal form (13,14), would be activated by modification of amino groups. The atypical form has been shown to have a proline substituted for alanine at residue number 230, two residues away from lysine residue 228 at the active site (12).

II. DISCUSSION

A. Steady-State Kinetic Studies

Product inhibition studies on the predominant form of native and chemically activated enzyme from normal livers were carried out in order to determine the mechanism of the enzyme and the magnitudes of the kinetic constants. As found previously for the horse enzyme (1,6), an Ordered Bi Bi mechanism is the simplest mechanism consistent with the results. The product inhibition studies for the acetimidylated and methylated enzymes were carried out as with native enzyme except that somewhat higher (e.g., 2 to 10-fold) concentrations of substrates and products were used in order to accomodate the higher kinetic constants of the acetimidylated enzyme. The inhibition patterns also appear to fit the Ordered Bi Bi mechanism.

Some of the kinetic constants derived from the product inhibition studies are presented in Table 1. The Michaelis and inhibition constants for the native human enzyme are larger, by up to 5-fold, than the constants for the horse enzyme

TABLE 1

Kinetic Constants for Native and Modified Human Liver Alcohol Dehydrogenase[a]

Constant	*Native*	*Acetimidyl*	*Methyl*
K_m^{NAD}, μM	17	40	40
$K_m^{Ethanol}$, mM	1.8[b]	33	5.8
K_i^{NADH}, μM	49	150	580
$V_1/E_t^{NAD + Ethanol}$, s-1	1.4	4.7	7.4
Units/mg, pH 9[c]	3.0	22	28

[a]*The enzyme was purified by fractionation with ammonium sulfate and chromatography on DEAE-cellulose and CM-cellulose by procedures adapted from the literature (15,16), and acetimidylated twice at pH 8 and 25° with 0.1 M ethyl acetimidate (1) or methylated (17) and freed of excess reagents by gel filtration. The values were computed by fitting all of the data for a product inhibition experiment to the equation describing the type of inhibition (1,18) and corrected for subsaturating concentrations of nonvaried substrates on the assumption of an Ordered Bi Bi mechanism. The buffer used was 0.1 μ sodium phosphate, pH 8, at 25°.*

[b]*This is an apparent value, obtained with a high, but narrow range of substrate concentrations.*

[c]*The turnover numbers were calculated on the assumption that the pure human enzyme had a specific activity of 2.5 units/mg in the pH 8.8 assay described by Lutstorf et al. (19) or 3.0 units/mg in the pH 9 (1) assay and an equivalent weight of 40,000.*

at the same pH (3,7). These kinetic constants are increased by acetimidylation and methylation, and importantly, the relative increases are about the same using either the horse or human enzymes as references for their modified enzymes. This observation strongly suggests that a lysine residue homologous to number 228 in the horse enzyme is also present in human enzyme, which has been previously suggested on the basis of sequence work (12). Furthermore, it suggests that the mechanisms of the two enzymes and the role of the amino group in the active site are similar.

It also appears that human liver enzyme modified with small, positively-charged substituents might be useful *in vivo* for accelerating ethanol metabolism, as was suggested previously on the basis of studies with the horse enzyme (10). The basis for this suggestion was that the modified enzymes are more active as indicated by the 3 to 10-fold larger turnover numbers, the Michaelis constants for NAD^+ are nearly the same as native enzyme so that the modified enzyme could compete for the NAD^+ available in the cell, and the Michaelis constants for ethanol are increased so that the modified enzyme would become saturated at more highly intoxicating concentrations of ethanol than is native enzyme (10,21). This suggestions is quite hypothetical and speculative, of course, but does indicate the potential value of these results for developing methods to accelerate ethanol metabolism. It may also be noted that the 4-hydroxybutyrimidylated enzyme (22) has such high kinetic constants that it may not be more active under *in vivo* conditions.

B. Activation of Atypical Enzyme

The native isoenzyme BB of atypical phenotype is a mixture of the normal subunit B_1 and the more active subunit B_2 (12). In contrast to the normal B_1B_1 enzyme, the acetimidylated B_1B_2 enzyme was prone to precipitation, and it lost about 50% of its activity per day. Nevertheless, acetimidylation increased the maximum activity by only about 2-fold. This might indicate that the substitution of proline in subunit B_2 for Ala-230 in subunit B_1 alters the local structure of the enzyme in a manner similar to that obtained by the acetimidylation of Lys-228. The net effect is that alteration of residues 230 or 228 increases activity, but alteration of both residues does not activate additively. On the other hand, the acetimidylated B_2 subunit is apparently less stable than the acetimidylated B_1 subunit, so it is possible that the 2-fold activation observed is lower than it should be simply because the acetimidylated B_2 subunit was denatured and only the activity of the acetimidylated B_1 subunit remained to be expressed.

C. Nonlinear Kinetics of Native Human Enzyme

Although the B_1B_1 native enzyme has a mechanism consistent with Ordered Bi Bi when results obtained over a narrow range of concentrations are considered, use of much wider ranges of ethanol as substrate reveals nonlinear kinetics. The Lineweaver-Burk plots show downward curvature with concentrations of ethanol up to 0.1 M; at higher concentrations substrate inhibition becomes apparent. A product inhibition pattern plotted according to Eadie and Hofstee for the best representation of nonlinear kinetics, is given in Fig. 1. It is apparent that acetaldehyde gives competitive inhibition against varied concentrations of ethanol. It is also noteworthy that when limited ranges of ethanol are considered (0.2 to 1.0 mM, 4 to 20 mM, or 20 to 100 mM), each set of data fitted a competitive pattern. The data at each concentration of acetaldehyde were fitted successfully to a 2/1 function.

In order to interpret the apparent kinetic constants, a mechanism explaining the nonlinear kinetics must be assumed. One possibility is a Random Bi Bi mechanism in which rapid binding steps are not assumed, but where free enzyme can first bind ethanol or NAD^+. Evidence for the occurrence of such a mechanism, especially at high concentrations of ethanol, has been obtained for the horse liver enzyme (23). However, nonlinear kinetics with the human enzyme were obtained with

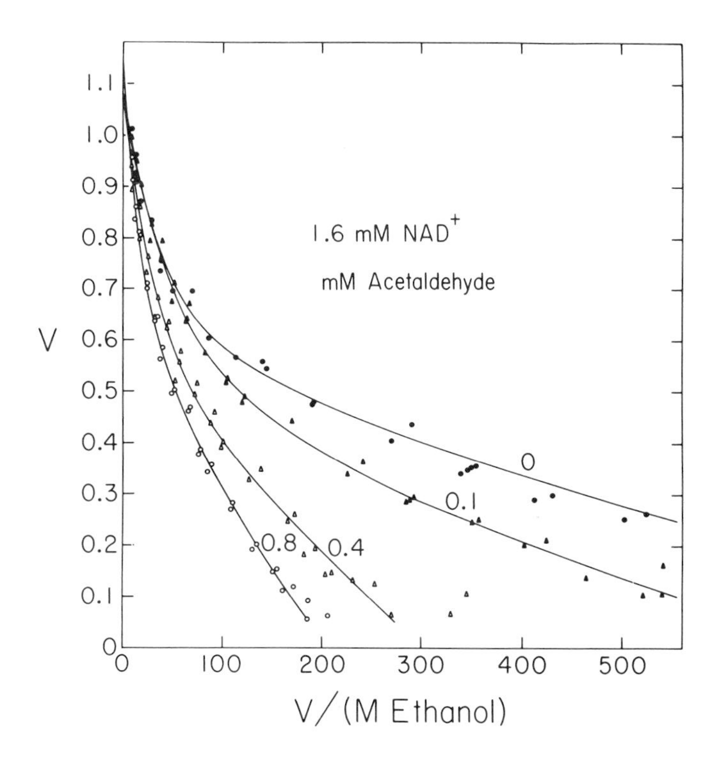

Fig. 1. Nonlinear kinetic behavior of native human enzyme acting on ethanol. The studies were carried out with a Gilford spectrophotometer. The buffer was 33 mM sodium phosphate, pH 8.0, and the temperature was 25°C. The concentrations of substrates are indicated on the figure. $\underline{V}$ *has arbitrary units. The points are experimental and the lines are computed fits to 2/1 function (25):* $v = V(S^2 + DS)/(S^2 + BS + C)$.

saturating concentrations of NAD^+ (1.6 mM, or 20 mM-- not shown), which would effectively order a random mechanism by forcing the binding of NAD^+ to free enzyme. Thus the results are not reasonably explained by a simple random mechanism. A random mechanism with an enzyme-NADH-alcohol complex that dissociates more rapidly than the enzyme-NADH-alcohol complex (24) is another possibility, but then inhibition by higher concentrations of ethanol is not simply explained.

One explanation that is consistent with the available data is that the preparation contains two forms of enzyme that differ in kinetic characteristics. Although the preparation is chromatographically and electrophoretically homogeneous (12,14), microheterogeneity, analogous to the valine/alanine exchange at position 43 in the horse enzyme (11), cannot be excluded. Substitution of an amino acid residue at a critical location at the active site could greatly affect activity, as is found with the substitution of proline for alanine at position 230. If it is assumed that there are only two different kinds of sites and that these act independently, the activities observed should simply be the sum of the individual activities (25): $v = V_1S/(K_1 + S) + V_2S/(K_2 + S)$. The equation for the sum is a 2/1 function and can be solved for the kinetic constants pertaining to each form of enzyme. The constants given in Table 2 suggest that the postulated forms differ greatly in magnitudes of Michaelis constants for ethanol, but that they have approximately equal maximum velocities. This might suggest that there are equal amounts of the two forms.

TABLE 2
Kinetic Constants for Oxidation of Ethanol Calculated on the Assumption that Two Forms of Human Enzymes are Present in the Normal Phenotype

	Form 1	*Form 2*
K_m, mM	0.52	8.3
V_1, U/ml	0.52	0.56

Among other explanations for the nonlinear kinetics is that the dimeric enzyme exhibits negative cooperativity, in that the binding of the first molecule of ethanol causes decreased affinity for the second molecule of ethanol. A more extreme form of negative cooparativity, called "half-of-the-sites reactivity" has been proposed previously for the action of horse liver alcohol dehydrogenase on NADH and aromatic aldehydes (26). If the native human enzyme does exhibit negative cooperativity, it is interesting that the acetimidylated enzyme does not, at least over the range of ethanol concentrations from 0.1 mM to 0.1 M. This result may indicate that modification of amino groups disrupts cooperative interactions between subunits and also that acetimidylation activates by altering the rates of conformational isomerizations, which may occur during the binding of the coenzymes (1,6). The relationship between the nonlinear kinetics of the native human enzyme and the activity *in vivo* remains to be elucidated.

III. REFERENCES

(1). Plapp, B.V., J. Biol. Chem. 245, 1927-1735 (1970).
(2). Plapp, B.V., Brooks, R.K., and Shore, J.D., J. Biol. Chem. 248, 3470-3475 (1973).
(3). Zoltobrocki, M., Kim, J.C., and Plapp, B.V., Biochemistry 13, 899-903 (1974).
(4). Sogin, D.C., and Plapp, B.V., J. Biol. Chem. 250, 205-210 (1975).
(5). Dworschack, R.T., Tarr, G., and Plapp, B.V., Biochemistry 14, 200-203 (1975).
(6). Wratten, C.C., and Cleland, W.W., Biochemistry 2, 935-941 (1963).
(7). Dalziel, K., J. Biol. Chem. 238, 2850-2858 (1963).
(8). Silverstein, E., and Boyer, P.D., J. Biol. Chem. 239, 3908-3914 (1964).
(9). Dalziel, K., and Dickinson, F.M., Biochem. J. 100, 34-46 (1966).
(10). Plapp, B.V., in Alcohol and Aldehyde Metabolizing Systems (R.G. Thurman, T. Yonetani, J.R. Williamson, and B. Chance, Eds.), pp. 91-100. Academic Press, New York, 1974.
(11). Jörnvall, H., and Pietruszko, R., Eur. J. Biochem. 25, 283-290 (1972).
(12). Berger, D., Berger, M., and von Wartburg, J.-P., Eur. J. Biochem. 50, 215-225 (1974).
(13). von Wartburg, J.-P., Papenberg, J. and Aebi, H., Can. J. Biochem. 43, 889-898 (1965).
(14). Schenker, T.M., Teeple, L.J., and von Wartburg, J.-P., Eur. J. Biochem. 24, 271-279 (1971).
(15). von Wartburg, J.-P., Bethune, J.L., and Vallee, B.L., Biochemistry 3, 1775-1782 (1964).
(16). Blair, A.H., and Vallee, B.K., Biochemistry 5, 2026-2034 (1966).
(17). Means, G.E., and Feeney, R.E., Biochemistry 7, 2192-2201 (1968).
(18). Cleland, W.W., Advan. Enzyml. 29. 1-32 (1967).

(19). Lutstorf, U.M., Schürch, P.M., and von Wartburg, J.-P., Eur. J. Biochem. 17, 497-508 (1970).
(20). Westerfeld, W.W., Texas Rep. Biol. Med. 13, 559-577 (1955).
(21). Lundquist, F., in Biological Basis of Alcoholism (Y. Israel, and J. Mardones, Eds.), pp. 1-52. Wiley-Interscience, New York, 1971.
(22). Fries, R.W., Bohlken, D.P., Blakley, R.T., and Plapp, B.V., Biochemistry 14, 5233-5238 (1975).
(23). Hanes, C.S., Bronskill, P.M., Gurr, P.A., and Wong, J. T.F., Can. J. Biochem. 50, 1385-1413 (1972).
(24). Dalziel, K., and Dickinson, F.M., Biochem. J. 100, 491-500 (1966).
(25). Cleland, W.W., The Enzymes, 3rd Ed., 2, 1-65 (1970).
(26). Bernhard, S.A., Dunn, M.F., Luisi, P.L., and Shack, P., Biochemistry 9, 185-192 (1970).

SUBSTRATE SPECIFICITY AND THE HYDROPHOBIC SITE OF LIVER ALCOHOL DEHYDROGENASE (L-ADH)

A.D. Winer

University of Kentucky

Horse liver alcohol dehydrogenase (L-ADH) has a very broad specificity for primary, secondary and cyclic alcohols, as does ADH from a variety of sources such as rat and human. The hydrophobic nature of the substrate binding site has been confirmed in the present study by the use of n-alkyl substituted aminoethanols, compounds which are known to be incorporated into liver and brain phospholipids. These substrates show higher reactivity as the groups around the n-alkyl substituted base become more hydrophobic. As observed with ethanol as substrate, formation of ternary abortive complexes of enzyme-NADH-aminoethanol are seen at high concentrations of aminoethanols. The lipophilic nature of the association of aminoethanols with the binary enzyme-NADH and enzyme-NAD$^+$ complexes has been studied and compared with the substrate analogue inhibitors, amides (isobutyramide) and acids (caproic). As with acids and amides as inhibitors, the association of aminoethanol substrates with the binary complexes increases as the number and size of the substituted n-alkyl group increases.

It is well established that horse liver alcohol dehydrogenase (L-ADH)[a] from a variety of mammalian sources (i.e., horse, rat and human) have broad substrate specificites for the alkyl group of primary, secondary, and cyclic alcohols (1-8). The lipophilic nature of the substrate binding site has been confirmed by kinetic experiments (9-12) as well as by

---

[a] The following abbreviations are used:

ADH; Alcohol dehydrogenase (E)

NAD$^+$ (O) and NADH (R); oxidized and reduced nicotinamide adenine dinucleotide, respectively.

3-Acetyl-Pyridine AD, deamino AD, pyridine-3-aldehyde AD; the corresponding 3-acetylpyridine, deamino and pyridine-3-aldehyde derivatives of NAD$^+$.

X-ray crystallographic studies (13,14). The question as to the identity of the physiological substrate(s) of ADH is still unanswered although it was posed very early by Theorell and his co-workers (11). In an attempt to help answer this question, the substrate properties of a number of n-alkyl substituted bases such as monomethyl and dimethylethanolamine, which are known to be incorporated into tissue phospholipids (15-17) have been studied. It has been found that not only are some of these n-alkyl substituted bases good substrates for the "EE" isoenzyme of L-ADH but more importantly, the substrate reactivity increases as the groups around the n-alkyl substituted base becomes more hydrophilic. The lipophilic nature of the association of alcohol and aldehyde substrates with the binary enzyme-NAD^+ and enzyme-NADH complexes has been confirmed since aliphatic acids and amides are more inhibitory on the n-alkyl substituted aminoalcohols as the chain length is increased.

I. MATERIALS AND METHODS

Horse liver alcohol dehydrogenase was prepared according to the method of Dalziel (15) and the "EE" isoenzyme obtained by chromatography on carboxymethylcellulose (16). The enzyme is crystallized from $\mu = 0.1$ phosphate, pH 7.4 buffer containing 8% ethanol and stored in 30% ethanol at -15°. The enzyme solution is prepared by dissolving the crystals in $\mu = 0.1$ phosphate pH 7.4 and dialyzing against the same buffer for 3 days at 0°, with frequent changes of dialyzing buffer. The crystalline enzyme was treated with NAD^+ to remove traces of endogenous ethanol as previously described (17). The enzyme concentration was determined by fluorimetric titration with NADH in the presence of 0.15 M isobutyramide (12). The pyridine nucleotides, NAD^+ and NADH were purchased from Boehringer and Son, Mannheim and purified by DEAE-cellulose chromatography and crystallized as the free acid according to Winer (18). The pyridine nucleotide analogues were purchased from Pabst Biochemicals and used without further purification. Isobutyramide was recrystallized 3X with water and the n-alkylaminoalcohols redistilled twice prior to use. The fatty acids and the alkylaminoalcohols were obtained from Eastman Chemical Co. as was the isobutyramide. The fatty acids were solubilized after lyophilization of the potassium salts.

Fluorimetric measurements were performed on a Farrand recording spectrofluorimeter equipped with a water-jacketed cell compartment. All fluorescence measurements were made at 27° in glycine-NaOH buffer, $\mu = 0.1$, pH 9. The excitation wavelength was 325 nm or 320 nm and no corrections were made for either excitation or emission wave-lengths. A final volume of 2.0 ml in 1 x 1-cm quartz cuvettes was used for all fluorimetric measurements with addition of reactants from Lang-Levy micropipettes. Spectrophotometric measurements were made in a Cary Model 118c spectrophotometer.

II. RESULTS AND DISCUSSION

That L-ADH from horse, rat and human sources have very broad substrate specificity for primary and secondary alcohols is shown in Table 1.

TABLE 1
Substrate Specificity of Alcohol Dehydrogenase

Alcohol	*Horse (8)*	*Rat (19)*	*Human (7)*
1-butanol	1.6	1.9	1.7
1-hexanol	1.3	2.0	1.5
1-pentanol	1.2	5.9	--[a]
1-propanol	1.1	1.6	1.4
Ethanol	1.0	1.0	1.0
1-octanol	1.0	0.7	--[a]
2-propanol	0	0.5	0.4
methanol	0.1	0.2	1.2

[a] *not reported.*

Higher chain length alcohols such as butanol, pentanol and hexanol are as good or better substrates for ADH from all three sources when compared to ethanol. The rat liver enzyme utilizes pentanol about six times better than ethanol whereas hexanol is oxidized at about twice the rate as is ethanol. The horse liver enzyme utilizes butanol better than all substrates tested. That other alcohol substrates such as isoamyl, allyl and cyclohexanol are all oxidized as well or better than ethanol by L-ADH ahs been reported (3,8). In all the specificity studies reported here, the "EE" isoenzyme has been used. A number of steroid substrates are known to be oxidized by the "SS" isoenzyme (4,19). It has been reported that a number of enzymatically active human ADH isoenzymes can be separated on the basis of reactivity with the substrates, butanol, benzyl alcohol and cyclohexanol.

The specificity of the "EE" isoenzyme of L-ADH for n-alkyl substituted aminoalcohols is shown in Table 2.

TABLE 2
N-Alkyl Substituted Ethanolamines as Substrates for L-ADH

	K_m (mM)	V_{max} (% of Ethanol)
Ethanol	1.2	100
Monomethylethanolamine (MMEA)	13.0	26
Dimethylethanolamine (DMEA)	1.2	22
Diethylethanolamine (DEEA)	11.0	66
Dibutylethanolamine	150.0	32
Ethanolamine		0
Choline		0

Of the aminoalcohol substrates tested, dimethylethanolamine has a Michaelis constant similar to that found with ethanol, but with a maximal velocity one fifth that found with ethanol as substrate under the conditions used in the experiment. All substrates were tested in glycine-NaOH buffer, μ = 0.1, pH 9.5 at 27° with an NAD^+ concentration of 250 μM and an enzyme concentration of 5×10^{-2} μN[a]. The substrate concentration reanged from 0.5 to 25 mM except for DBEA which has relatively poor solubility and was used in the concentration range of 0.1 to 10 mM. Ethanolamine and choline are not oxidized by L-ADH nor will yeast alcohol dehydrogenase show reactivity with any of the listed substrates. All n-alkyl substituted aminoalcohols studied are reversibly oxidized to the corresponding aldehyde as determined by gas liquid chromatography (20). The low K_m of DMEA relative to MMEA and DEEA may reflect the similarity of structure of DMEA and isobutyramide, an inhibitor of the ER binary complex. Isobutyramide as an inhibitor has a smaller inhibitor constant for the binding of inhibitor to the ER complex than does n-butyramide or n-valeramide. The lower V_{max} for DMEA may reflect the alteration in dissociation rate of R from the ER binary complex.

The reactivity of DMEA as a function of pH is shown in Fig. 1.

---

[a]The enzyme concentration is expressed as micronormality (μN) of NADH binding sites per liter.

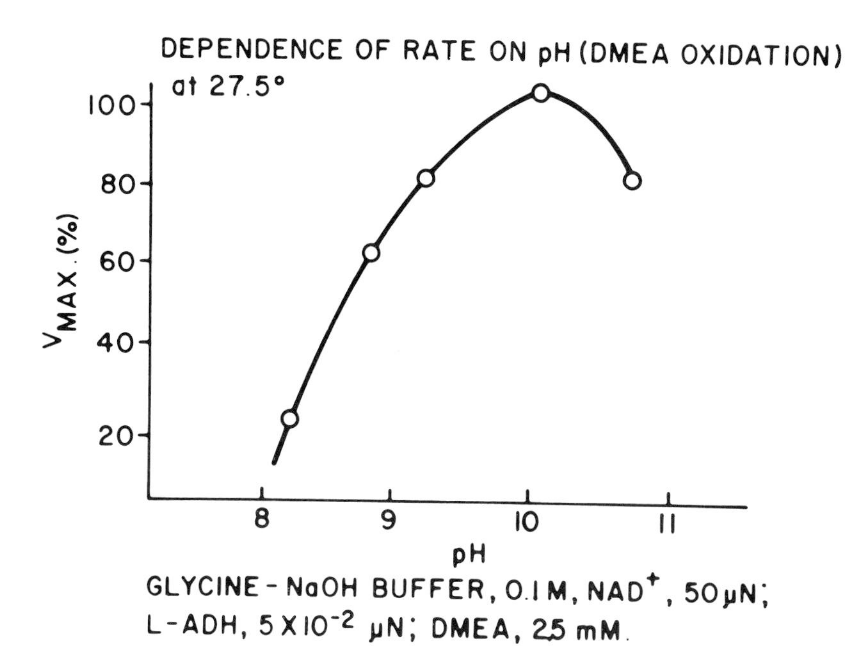

Fig. 1. Dependence of rate of DMEA oxidation on pH.

The coenzyme specificity is shown in Table 3.

TABLE 3
Comparison of Reaction Rates of NAD+ and NAD+ Analogues with L-ADH

Coenzyme	Relative Rate (%)[a] Ethanol	DEEA
NAD^+	100	100
3-Acetyl-Pyridine AD	330	87
Deamino AD	138	75
Pyridine-3-Aldehyde AD l	31	15

[a]Glycine-NaOH buffer, μ = 0.1, pH 9.5 at 27°; coenzyme 10^{-3} M, Alcohol 3.3 mM and L-ADH 10^{-2} μN.

As indicated in TABLE 3, the 3-Acetyl-Pyridine AD is utilized about three times better than NAD^+ with ethanol as substrate. However, it is not as good a coenzyme with DEEA as substrate. With Deamino AD as coenzyme, the rate is somewhat greater than NAD^+ with ethanol as substrate and somewhat less than NAD^+ with DEEA as substrate.

Substrate inhibition at high concentrations of ethanol have been reported by several investigators (10,11). In the present studies, the aminoalcohol substrate DMEA shows substrate inhibition at concentrations larger than 2 mM as shown in Fig. 2.

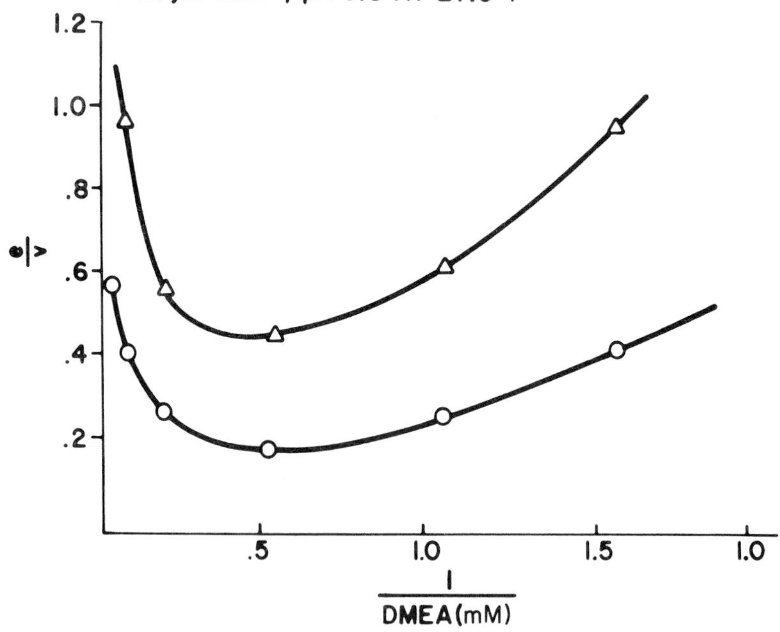

Fig. 2. Substrate inhibition by DMEA.

It is concluded, as suggested earlier (10,20), that this inhibition is due to the formation of the abortive E-NADH-alcohol ternary complex. Substrate activation with increasing concentrations of some alcohols such as 2-propanol, 2-butanol and cyclohexanol has also been observed (22).

Stable ternary complexes of fatty acid amides, NADH and L-ADH was first reported by Winer and Theorell (12). From equilibrium fluorescence titration studies, as well as kinetic experiments, it was shown that fatty acid amides form ternary complexes with E_2R_2 but not with E_2O_2 and that they compete kinetically with aldehyde but not with alcohol for the same binding site. It was also concluded that these ternary complexes possess a similar structure to the reactive complexes, containing aldehyde. Likewise, fatty acids were shown to form ternary complexes with EO but not ER and that they compete kinetically with the alcohol but not the aldehyde binding site. Studies with DMEA as substrate and increasing chain length of fatty acids and fatty acid amides is shown in Table 4.

TABLE 4
The Effect of Fatty Acids (I_a) and Fatty Acid Amides (I_m)[a]

Chain Length	K_{EO}, I_a (μM)	K_{ER}, I_m (μM)
C_2	9500	5000
C_3	2700	505
C_4	1600	64
C_5	1400	25
C_6	300	11

[a]Glycine-NaOH buffer, μ = 0.1, pH 9 at 27°, coenzyme 10^{-3} M, DMEA 3.3 mM and L-ADH 10^{-2} μN.

It can be seen that the dissociation constants describing the effect of acid or amide inhibitor on the E-coenzyme binary complex (K_{EO}, I_a) and (K_{ER}, I_m) decrease as the chain length increases from C_2 to C_6. Thus, with DMEA as substrate, as observed with ethanol as substrate, strong inhibitory effects are found with both acids and amides. The lipophilic nature of the association of the aminoalcohol substrate DMEA with the E-coenzyme binary complex indicates that the concept of a "hydrophobic pocket" at the active site of L-ADH has validity and will be of help in further studies on the substrate specificity of ADH.

III. REFERENCES

(1). Dickerson, F.M., and Dalziel, K., Biochem. J. 104, 65 (1967).

(2). Graves, J.M.H., Clark, A., and Ringold, H.J., Biochemistry 4, 2655 (1965).

(3). Merritt, A.D., and Tomkins, G.M., J. Biol. Chem. 234, 2778 (1958).

(4). Pietruszko, R., Ringold, H.J., Kaplan, N.O., and Everse, J., Biochem. Biophys. Res. Commun. 33, 503 (1965).

(5). Prelog, V., Pure Appl. Chem. 9, 119 (1964).

(6). Theorell, H., and Bonnichsen, R., Acta Chem. Scand. 5, 1105 (1951).

(7). Wartburg, J.P. von, Behune, J.L., and Vallee, B.L., Biochemistry 3, 1175 (1964).
(8). Winer, A.D., Acta Chem Scand. 12, 1695 (1958).
(9). Dalziel, K., and Dickerson, F.M., Biochem. J. 100, 34 (1966).
(10). Shore, J.D., and Theorell, H., Arch. Biochem. Biophys. 117, 375 (1966).
(11). Sund, H., and Theorell, H., in "The Enzymes" (P.D. Boyer, H. Lardy, and K. Myrback, Eds.), Vol. 7, p. 25. Academic Press, New York, 1963.
(12). Winer, A.D., and Theorell, H., Acta Chem. Scand. 14, 1729 (1960).
(13). Brändén, C.I., Zeppezauer, E., Boiwe, T., Söderlund, G., Süderberg, B.O., and Nordström, B., in "Pyridine Nucleotide Dependent Dehydrogenases" (H. Sund, Ed.), p. 129. Springer, Berlin, 1970.
(14). Ansell, G.B., and Chojnacki, T., Biochem. J. 98, 303 (1966).
(15). Dalziel, K., Acta Chem. Scand. 12, 459 (1958).
(16). Artom, C., and Lofland, Jr., H.B., Biochem. Biophys. Res. Commun. 3, 244 (1960).
(17). Hall, M.O., and Nyc, J.F., J. Lipid Res. 2, 321 (1961).
(18). Winer, A.D., J. Biol. Chem. 239, 3600 (1964).
(19). Arslanean, M.J., Pascoe, E., and Reinhold, J.G., Biochem. J. 125, 1039 (1971).
(20). Pietruszko, R., Ringold, H.J., Li, T.K., Vallee, B.L., Åkeson, Å., and Theorell, H., Nature 221, 440 (1969).
(21). Unpublished observations, A.D. Winer.
(22). Dickerson, F.M., and Dalziel, K., Nature 214, 31 (1967).

PROPERTIES OF HORSE LIVER ALCOHOL DEHYDROGENASE SS

C.N. Ryzewski and R. Pietruszko

Center of Alcohol Studies, Rutgers University

A purification scheme to obtain homogeneous SS isozyme of alcohol dehydrogenase is summarized and some kinetic properties of SS isozyme are presented. SS isozyme has values of Michaelis constants for two and three carbon substrates almost one order of magnitude larger than those for EE isozyme. The maximal turnover numbers for aliphatic substrates with SS isozyme are constant and independent of the substrate structure, although the values are 30% of the values previously determined for EE isozyme. The Km for ethanol is lower at pH 10 than at pH 7, an unusual finding for an alcohol dehydrogenase. In other respects SS isozyme is similar to EE isozyme, i.e., the Km values for substrates decrease with increasing chain length and NAD(H) is the preferred coenzyme. These properties of SS isozyme are also compared to AA isozyme and to rat liver and human liver alcohol dehydrogenase. Some evidence is presented which suggests that SS isozyme may function by two kinetic mechanisms: one for nonsteroidal substrates and another for steroidal substrates.

I. INTRODUCTION

Horse liver alcohol dehydrogenase (ADH) has been separated into nine bands by starch or agar gel electrophoresis (1). The band with the fastest cathodal migration can be separated from the other horse liver ADH isozymes by DEAE chromagography (2). The composition of this isolated band depends on whether an A-type or an S-type horse liver is used. Preparations from A-type livers are heterogeneous consisting of AA, AS and SS isozymes despite their homogeneous appearance and identity to S-type preparations on starch gel electrophoresis (2). Preparations from S-type livers consist only of SS isozyme (2).

Activity with 3β-hydroxysteroids and 3-ketosteroids is found only with those isozymes which possess an S subunit (3). The other two horse liver ADH subunits, E and A, are devoid of steroid activity (2). A difference of six amino acids exists between the E and S subunits (4) and makes the EE and SS isozymes useful in studies of structure-function relationships of dehydrogenases. For example, the steroid activity which distinguishes SS isozyme from EE isozyme has been attributed to only one of the six amino acid differences between the E and S subunits (5). Further comparisons between the EE and SS isozymes are limited by the paucity of information concerning SS isozyme and by the lack of a published method for complete purification. The development of a method for purifying SS isozyme appeared to be the prerequisite for characterization of this isozyme and for structure-function comparisons with EE isozyme and with other horse liver ADH isozymes.

II. METHODS

A. Purification

A purification scheme, which is summarized in Table 1, utilizing techniques previously used to obtain partially purified SS isozyme was devised to obtain this isozyme free from protein and isozyme contaminants. Details of the purification procedure are presented elsewhere (6). SS isozyme is located during fractionation procedures by starch gel electrophoresis and by the cyclohexanone/5β-dihydrotestosterone (cyclo/5βDHT) activity ratio utilizing previously described enzyme assays (2) with cyclohexanone at a concentration of 12.3 mM and 5βDHT at 114 μM. The cyclo/5βDHT ratio is 1.0 for SS isozyme in contrast to 9.7 for ES isozyme or at least 20 for A-type preparations (2). Only one horse liver is used per preparation to avoid the possibility of combining the more complex isozymes from A-type livers with SS isozyme. Horse liver ADH is precipitated from the homogenate with ammonium sulfate and, after extensive dialysis, SS isozyme is separated from other horse liver ADH isozymes by DEAE chromatography. Elution of SS isozyme from the 5'AMP affinity gel requires the simultaneous presence of cholate and NAD as reported previously for a 5'AMP gel with a denser affinity group content (7). The procedure for elution of the CM cellulose column is also a modification of a published method (8). The purified isozyme is free from protein and isozyme contaminants (6).

III. RESULTS AND DISCUSSION

A. Kinetic Properties: Substrate and Coenzyme Specificity

Km values for SS isozyme with aliphatic substrates decrease with increasing chain length (Table 2) and in this respect SS

TABLE 1

Purification of SS Isozyme[a]

Step	Description[b]
1. Homogenate	1 g liver per ml 5 mM sodium phosphate pH 7.5
2. Ammonium sulfate fractionation and dialysis	ADH is precipitated between 35 and 80% ammonium sulfate followed by exhaustive dialysis
3. DEAE cellulose chromatography	SS isozyme elutes ahead of other horse liver ADH isozymes and is collected free of them.
4. 5'AMP Sepharose chromatography	SS isozyme is eluted by addition of sodium cholate and NAD to the elution buffer after washing to remove contaminants.
5. CM cellulose chromatography	SS isozyme is adsorbed to the column and subsequently eluted with a salt gradient after washing to remove contaminants.

a. For details see reference 6.
b. All purification procedures are carried out at 4°.

isozyme resembles EE isozyme. However, the large Michaelis constants for 2 and 3 carbon substrates with SS isozyme (Table 2) are in sharp contrast to those for EE isozyme (9) which are almost one order of magnitude lower.

SS isozyme displays lower Km values for NADH than for NADPH when either a steroidal or a nonsteroidal substrate is kept constant (Table 3) indicating that SS isozyme like EE isozyme is NAD(H)-dependent.

The maximal turnover numbers for nonsteroidal substrates are constant and independent of substrate structure for SS isozyme (Table 2) as for EE isozyme (9), although the values for SS isozyme are 30% of those for EE isozyme. Since the maximal turnover in the oxidative direction (1.09 sec^{-1}) for SS isozyme approaches the value of the off-velocity constant for NADH (1.47 sec^{-1}) (10), the overall rate of the oxidation of aliphatic alcohols is probably limited by the rate of dissociation of reduced coenzyme.

TABLE 2
Substrate Michaelis Constants and Maximal Turnover Numbers for SS Isozyme[a]

Substrate	*Km (mM)*	*Turnover Number*[b]
Ethyl alcohol	40.0	1.27
Propyl alcohol	2.9	1.16
Butyl alcohol	0.09	0.94
Hexyl alcohol	0.03	0.99
Acetaldehyde	6.0	31.8
Propionaldehyde	1.24	41.9
Butyraldehyde	0.06	35.4
Hexaldehyde	0.012	56.6
5βDHT	0.03	2.05
5β-pregnan-21-ol-3,20-dione hemisuccinate	0.047	0.53

a. Assays were performed at 25° with either 170 μM NADH or 500 μM NAD in 0.1 M sodium phosphate pH 7.0.

b. Expressed as the number of molecules of substrate per second per active site. Enzyme concentration was determined according to (12).

TABLE 3
Coenzyme Michaelis Constants for SS Isozyme[a]

Substrate	*Constant Substrate Conc. (mM)*	*Coenzyme*	*Coenzyme Km (μM)*
Nonsteroidal			
Acetaldehyde	48	NADH	3.6
Acetaldehyde	24	NADPH	970
Steroidal			
5βDHT[b]	0.114	NADH	0.75
5βDHT[c]		NADPH	140

a. Determined in 0.1M sodium phosphate pH 7.0 at 25°.
b. Determined in cuvettes with a 10 cm light path.
c. From (13).

A constant value for maximal turnover of aliphatic aldehydes (Table 2) implies that the rate of dissociation of oxidized coenzyme may also be rate-limiting at least for aliphatic substrates. Lower values for maximal turnover of steroidal ketones suggest that dissociation of coenzyme is not rate-limiting for these substrates. The lower Km for coenzymes with steroidal substrates (Table 3), in addition to the lower maximal turnover of steroids, indicates that the kinetic mechanism with steroidal substrates differs from that with nonsteroidal substrates for SS isozyme.

B. Comparison of SS Isozyme with Other ADH Enzymes

It is unusual to find that for SS isozyme the Km for ethanol at pH 10 is lower than at pH 7 (Table 4) since the opposite is true for EE isozyme (9). Rat liver ADH is thought to be functionally similar to SS isozyme in steroid activity as a result of similar amino acid substitutions at positions related to this activity (5). However, for rat liver ADH, values of Km for ethanol at pH 10 and at pH 7 resemble those for EE isozyme (Table 4). Hence it is not possible to assume that large Km values for two and three carbon substrates are the result of amino acid substitutions associated with steroid activity.

All human liver ADH isozymes show steroid activity (11), in contrast to horse liver ADH for which steroid activity is associated only with those isozymes which have an S subunit. In this regard, the SS isozyme may be more useful than the EE isozyme for structure-function comparisons with human liver ADH isozymes. However, like rat ADH, the human liver ADH isozymes have Km values for ethanol at pH 10 and at pH 7 which are similar to those for EE isozyme (Table 4).

Since AA isozyme is similar to SS isozyme in mobility on starch gel electrophoresis (2), it is possible that AA isozyme may be a polymorphic form of SS isozyme and may show a higher degree of sequence homology to SS isozyme than shown by EE isozyme. However, AA isozyme is not active with the steroids which are substrates for SS isozyme. One possible explanation for this apparent lack of steroid activity is an amino acid substitution in the A subunit in residues related to steroid activity. Preliminary investigations indicate that AA isozyme resembles EE isozyme more than SS isozyme in several other kinetic properties (Table 4), in stability (2), and in the degree to which NADH fluorescence is enhanced in the ternary complex (10) with isobutyramide (unpublished data).

TABLE 4
Kinetic Properties: Comparison of SS Isozyme with Other Alcohol Dehydrogenase Enzymes

Enzyme	*Steroid Activity*	*Km (mM)* Ethanol pH 10	*Km (mM)* Ethanol pH 7	*Km (mM)* Acetaldehyde pH 7
Horse liver ADH				
SS isozyme	+	6.7[a]	40.0	6.0
EE isozyme	none	2.0[a]	0.76[b]	0.23[b]
AA isozyme	none	0.71[c,d]	0.18	0.08
Rat liver ADH	+	2.13[e]	0.50[f]	0.12
Human liver ADH	+	1.3-2.5[g]	0.64-0.96[g]	0.53

a. *Performed in 0.1M glycine pH 10.0 with 500 μM NAD.*
b. *From (9).*
c. *From (14).*
d. *Value at pH 9.5.*
e. *From (15).*
f. *From (16).*
g. *Range of values for human liver ADH isozymes in (11).*

Although the physiological role of the steroid activity associated with alcohol dehydrogenase isozymes is unknown, steroid activity is important in structure-function studies of the isozymes. Of the three homodimers of horse liver ADH, (EE, SS and AA), only the SS isozyme is steroid-active. Sequence comparisons between the E subunit, the S subunit and rat alcohol dehydrogenase have led to the suggestion that steroid activity may be correlated with the presence of a single amino acid substitution (5). Further studies of the A subunit which has no associated steroid activity should be useful in determining the accuracy of this correlation.

IV. ACKNOWLEDGEMENTS

The authors wish to thank Charles and Johanna Busch Memorial Fund and NIAAA (00AA186) for financial support.

V. REFERENCES

(1). Pietruszko, R., and Theorell, H., Arch. Biochem. Biophys. 131, 288 (1969).

(2). Pietruszko, R., and Ryzewski, C., Biochem. J. 153, 249 (1976).
(3). Pietruszko, R., Clark, A., Graves, J., and Ringold, H.J., Biochem. Biophys. Res. Commun. 23, 526 (1966).
(4). Jörnvall, H., Eur. J. Biochem. 16, 41 (1970).
(5). Eklund, H., Brändén, C-I., and Jörnvall, H., J. Mol. Biol. 102, 61 (1976).
(6). Ryzewski, C.N., and Pietruszko, R., Eur. J. Biochem. (submitted for publication).
(7). Andersson, L., Jörnvall, H., Åkeson, Å., and Mosbach, K., Biochem. Biophys. Acta 364, 1 (1974).
(8). Lutstorf, U.M., Schurch, P.M., and von Wartburg, J.P., Eur. J. Biochem. 17, 497 (1970).
(9). Pietruszko, R., Crawford, K., and Lester, D., Arch. Biochem. Biophys. 159, 50 (1973).
(10). Theorell, H., Åkeson, Å., Liszka-Kopeć, B. and de Zalenski, C., Arch. Biochem. Biophys. 139, 241 (1970).
(11). Pietruszko, R., Theorell, H., and de Zalenski, C., Arch. Biochem. Biophys. 153, 279 (1972).
(12). Winer, A.D. and Theorell, H., Acta Chem. Scand. 14, 1729 (1960).
(13). Pietruszko, R., Biochem. Biophys. Res. Commun. 54, 491 (1973).
(14). Pietruszko, R., Biochem. Biophys. Res. Commun. 60, 687 (1974).
(15). Marković, O., Theorell, H., and Rao, S., Acta Chem. Scand. 25, 195 (1971).
(16). Reynier, M., Acta Chem. Scand. 23, 1119 (1969).

ROOM AND LOW TEMPERATURE SPECTRAL IDENTIFICATION OF THE COBALT BINDINGS SITES IN LIVER ALCOHOL DEHYDROGENASE

Henry R. Drott

University of Pennsylvania

Visible spectra of cobalt substituted alcohol dehydrogenase (ADH) samples have been recorded at 298°K and 77°K. Low temperature spectroscopy enhances the spectral resolution of the electronic transitions of the metal for the enzyme and its ternary complexes. The increased spectral resolution permits assignment of spectral bands to the coordination geometry of each distinct metal and analysis of the kinetics of incorporation of the metal into the enzyme. Reconsideration of the exchange kinetics theoretically has been performed and gave values of 0.48 hr^{-1} and 0.053 hr^{-1} for k_1 and k_2, respectively. Using these values curves can be generated for calculated absorption of cobalt in ADH in order to compare it to measured absorption as a function of total cobalt concentration. The results allow the conclusion that the fast exchanging metal substitutes for the zinc is at the structural site and the slow exchanging cobalt substitutes for the zinc at the catalytic site. These results are supported by the fluorescence quenching data, epr dipolar experiments and enzymatic activity measurements. Optical experiments with model tetrahedral cobalt complexes with four thio ligands as well as two thio ligands and two coordinated anions substantiate the spectral assignments.

I. INTRODUCTION

Horse liver alcohol dehydrogenase (EC 1.1.1.1) is a metalloenzyme and contains 4 g. atoms of zinc per molecule (1,2). Initially two zinc ions per subunit were established by use of chelating agents (3-5) and subsequently, one of the zinc ions per subunit was demonstrated as essential for catalytic activity while the other zinc ion was projected to be involved in the structural stabilization of the enzyme (6). The X-ray crystal structure of ADH at 2.4 A resolution confirmed these conclusions, in addition to defining the amino acids groups ligated to the metals

(7,8). Moreover, the kinetics of isotope-exchange of the zinc ions distinguished the metal binding sites (6). The kinetically distinct rates of exchange of the zinc ions coupled with the formation of catalytically active cobalt and cadmium substituted has provided the stimulus for studies with "metal hybrids" of the enzyme, leading to conflicting assignments between the "fast" and "slow" exchanging metal and the metal binding sites (9-18). Recently, the spectral and kinetic properties of the totally substituted cobalt alcohol dehydrogenase and of complexes of the enzyme with various ligands have been described (14,16,18). This finding provides an alternate pathway for delineation and assessment of the kinetics of the exchange reaction, and hopefully resolution of the conflicting assignments.

In this study, the optical properties of the cobalt substituted alcohol dehydrogenases have been monitored in an effort to definitely establish the correlation between the fast and slow exchanging metal with the "catalytic" and "structural" sites. The visible spectra of the substituted enzyme and the binary and ternary complexes of the enzyme with coenzymes and substrate inhibitors were measured as a function of the cobalt ions incorporated in the enzyme. Optical spectra were recorded at 77° in addition to 298° to increase the spectral resolutions, and were compared to spectra of model compounds involving cobalt (II) in order to make specific spectral assignments and to predict the geometry and composition of the ligand field of the metal ions.

II. MATERIAL AND METHODS

Horse liver alcohol dehydrogenase was purchased from Boehringer -Mannheim (7313238). The assay method of Dalziel was used at pH 10 (19) and NADH titrations of the enzyme in the presence of isobutyramide were carried out spectrofluorometrically (20). As a periodic check of the NADH titrations, spectrophotometric titrations of the enzyme with NAD^{+} in the presence of pyrazole were performed (21). NAD+ and NADH, Grade III, were obtained from Sigma Chemical Company, isobutyramide was supplied by Eastman Chemical Company, the remaining materials were analyzed-grade reagents and certified standards.

The procedure for exchanging the zinc ions with cobalt ions is described in the literature (14), except that about 7 ml of a 180-200 uN enzyme solution was employed in the exchange reaction. Samples (0.5-0.7 ml) of the enzyme undergoing exchange at room temperature (27°) were withdrawn, dialyzed against 4 changes of 0.2 M Tris-Acetate buffer, pH 7.0 at $4^{o}C$, filtered (Millipore type HA) into sterile polypropylene capped tubes and stored at 0° under a nitrogen atmosphere. The polyethylene exchange vessel, dialysis flasks and transfer pipettes used were previously leached with dilute nitric acid and thoroughly rinsed with distilled deionized water. The dialysis tubing was boiled in 3 changes of distilled, deionized water to remove any contaminants and preservatives.

Optical spectra were recorded at 298° in a Cary 118C spectrophotometer using 10 mm semimicro cells, and at 77° in a split-beam spectrophotometer constructed at the Johnson Foundation using 3 mm path length cells together with a Dewar flask system for liquid N_2 or He (22). Spectrofluorometric titrations were carried out with a Perkin Elmer MPF-2A spectrofluorometer; cobalt concentrations were measured with a Varian AA5 atomic absorption spectrophotometer.

III. RESULTS

A. Optical Spectra of Totally Substituted Cobalt Enzyme at 298°

The visible and near ultraviolet spectra of totally cobalt substituted alcohol dehydrogenase and complexes of the enzyme have been re-examined. Several subtle but new features were discovered. As previously reported, absorption bands associated with the cobalt were measured at 740, 655 and 343 nm for the totally cobalt substituted enzyme and similar shifts in the spectrum for the enzyme were observed on formation of binary and ternary complexes with coenzymes and inhibitors (9,14,16 18), The most striking feature discovered was the fact that the 740 nm absorption peak was not split nor shifted on binary or ternary complex formation (Figure 1) (16,17). Secondly, isobestic points were noted when spectra were compared for various combinations of coenzyme and inhibitor with the enzyme at various cobalt levels. Thirdly, hypo- and hyper chromic changes were also observed (Figure 1) on addition of isobutyramide to the enzyme-NADH binary complex.

B. Optical Spectra of Totally Substituted Cobalt Enzyme at 77°

Spectra of the totally cobalt substituted alcohol dehydrogenase and the complexes of the enzyme with coenzymes and inhibitors in the visible region have also been recorded at low temperature for increased spectral resolution of the electronic transition of the metal. The broad hand centered at 655 nm (298°) is clearly resolved into two bands centered at 635 and 655 nm at 77° as shown in Figure 2a. The absorption maximum located at 740 nm is unaffected by the decrease in temperature. Greater resolution, however, arises for the spectra of the ternary complexses when measured at 77° as seen in Figure 2a. Optical spectra were also recorded at 4.2°K, but failed to show

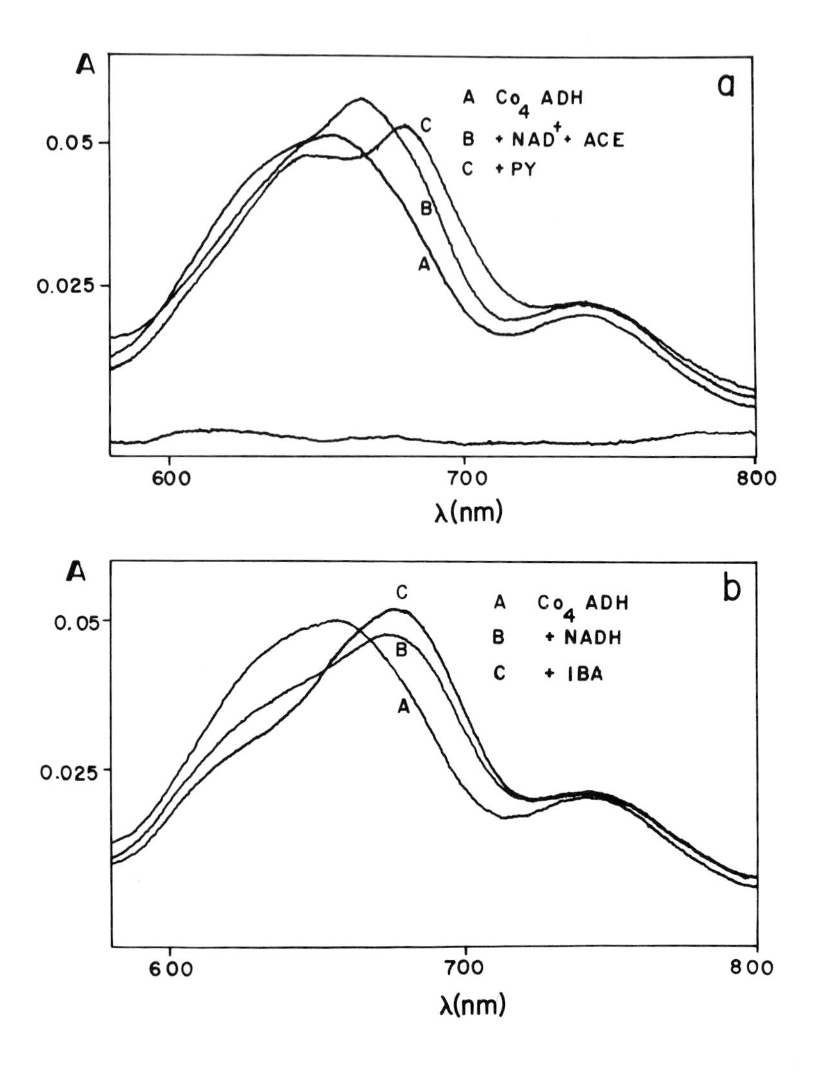

Figure 1. Optical Spectra of CO_4 ADH & Complexes at 298°.

further enhancement in the resolution so that all spectra were subsequently recorded at 77°K.

C. <u>Kinetics of the Exchange of Cobalt Ions for Zinc Ions in Alcohol Dehydrogenase</u>

Previous studies have documented the multiphasic nature of the exchange kinetics of the substitution of the zinc ions by either zinc-65 or cobalt ions or the reverse substitution of cobalt ADH with labelled zinc (9-11,15-18). In this study, the

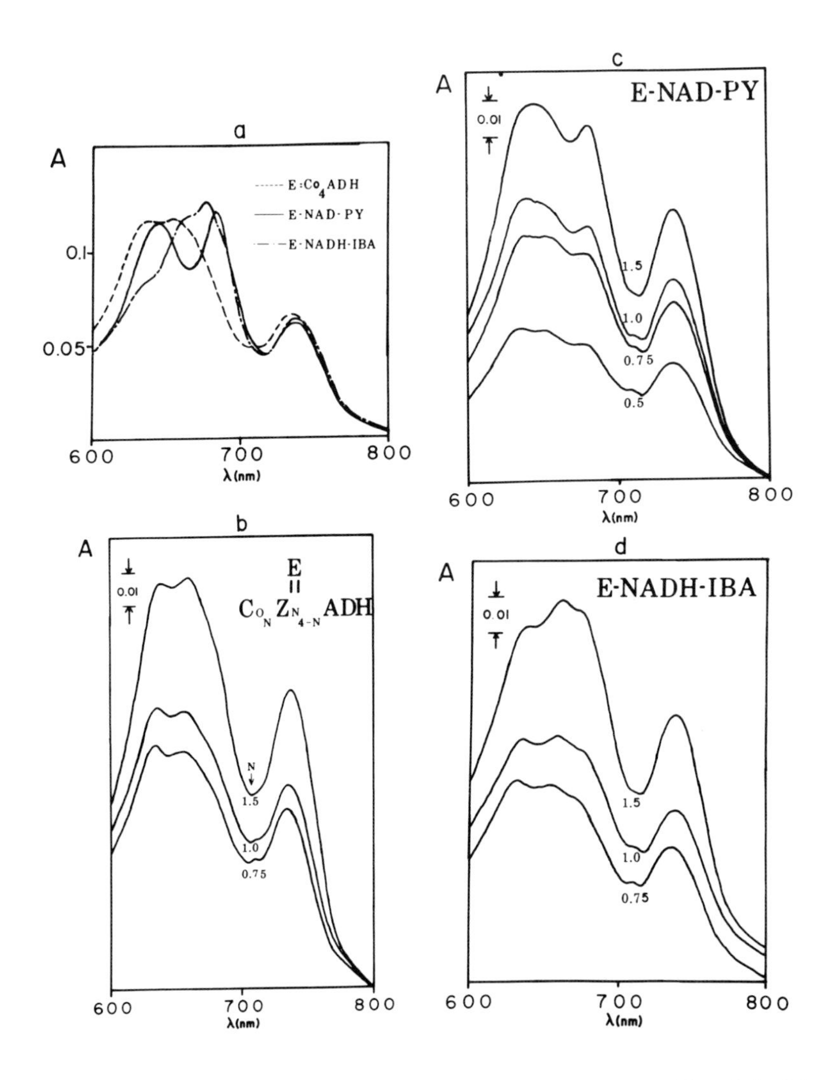

Figure 2. Optical Spectra of Partially & Totally Cobalt Substituted ADH & Complexes at 77°

rate of incorporation of the cobalt into the enzyme has been followed by atomic absorption and visible light absorption spectroscopy. Semilogarithmic plots of our data verify the biphasic nature of the exchange reaction.

Because the exchange reaction involves two distinct sites per subunit and is carried out under pseudo first order condition, a general reaction scheme for the exchange can be illustrated as:

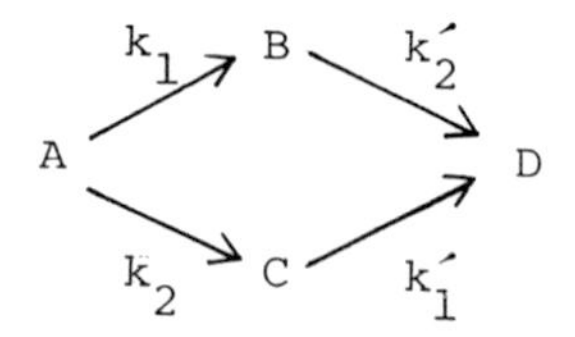

where A represents one subunit of the enzyme with two distinct metal binding sites. B and C correspond to the subunit having one of the two sites substituted, D is the totally substituted subunit of the enzyme, k_1, k_2, k_1' and k_2' are the corresponding rate constants, and a_o the initial concentration of A. The concentrations of A, B, C and D can be calculated from suitable rate expressions and the law of conservation $[A_o] = [A] + [B] + [C] + [D]$. Moreover, a function F is defined as equal to $[B] + [C] + 2[D]$, and is the experimental quantity measured as discussed by Swain (23):

$$F = A_o[2-e^{-kt}(2 + \frac{k_1}{k_2'-k} + \frac{k_2}{k_1'-k_1}) + \frac{k_1}{k_2'-k} e^{-k_2't} + \frac{k_2}{k_1'-k} e^{-k_1't}] \quad (1)$$

where $k = k_1 + k_2$. Rearranging this expression and converting to logarithmic form yields Equation 2.

$$\ln(1- \frac{F}{2A_o}) = \ln \frac{1}{2} + \ln k\ (t) \quad (2)$$

$$\text{where } k(t) = e^{-kt}(2 + \frac{k_1}{k_2'-k} + \frac{k_2}{k_1'-k}) - \frac{k_1}{k_2'-k} e^{-k_2't} - \frac{k_2}{k_1'-k} e^{-k_1't}$$

From Equation 2, it is readily seen that the initial slope is given by:

$$\left.\frac{d[\ln(1- \frac{F}{2A_o})]}{dt}\right|_{t=o} = \frac{k_1 + k_2}{2}$$

If $k_1 >> k_2$, then $\frac{k_1 + k_2}{2} \sim \frac{k_1}{2}$ and not simply k_1 (6). In addition, the final slope is $\frac{k_2}{2}$, if $k_1' >> k_2'$. In all probability k_1 equals k_1' and k_2 equals k_2' for the exchange of zinc by the cobalt in the enzyme. Thus, the intercept for the extrapolation of final slope to zero time is $\frac{1}{2}$, but the initial and final slopes are unchanged by this simplification. The fact that a biphasic exchange plot and an intercept of 0.5 are observed cannot rule out the proposed reaction scheme (15,18). Furthermore, this treatment of the kinetics would imply that the values of k_1 calculated previously (10) and used subsequently in solving for the different distances (13) is incorrect.

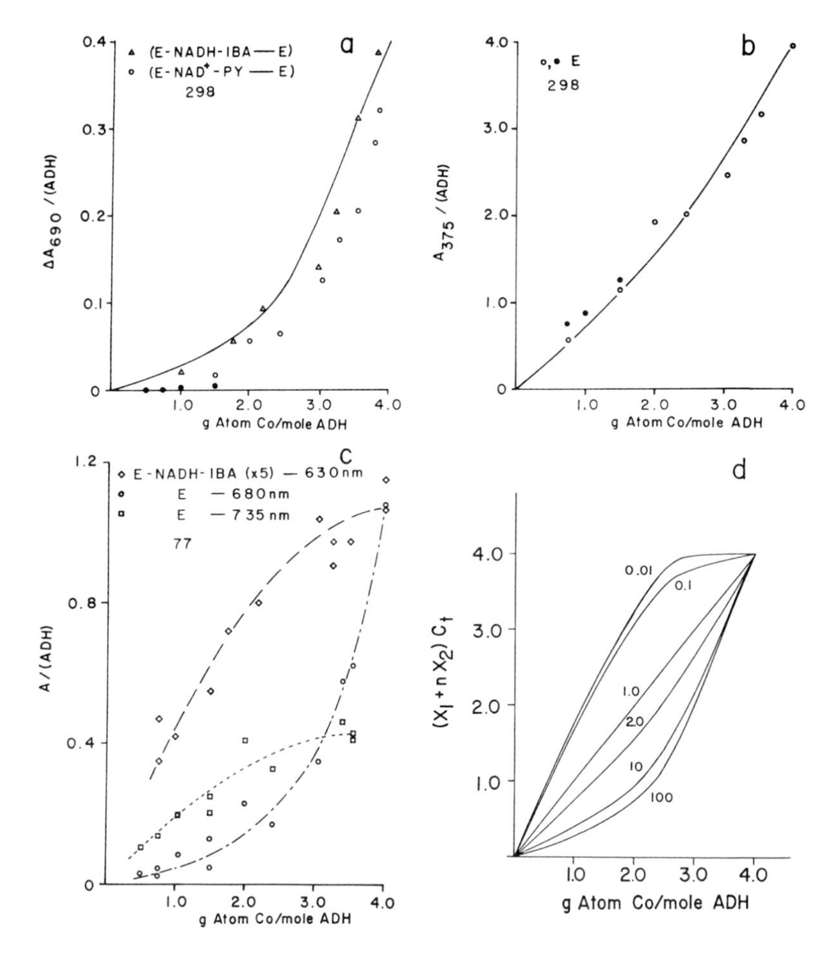

Figure 3. Measured and Calculated A & ΔA for ADH Complexes at Various Levels of Cobalt Incorporated.

D. Effect of Differing Cobalt Levels on Optical Spectra at 77°

Because of increased spectral resolution, the optical spectra of the partially substituted enzyme and ternary complexes of the enzyme with conenzymes and inhibitors were recorded at 77° during the entire reaction. The initial substitution of the cobalt into the enzyme exhibits different absorption properties than the totally cobalt substituted alcohol dehydrogenase and is shown in Figure 2b-2d. The specific features are the change in intensity

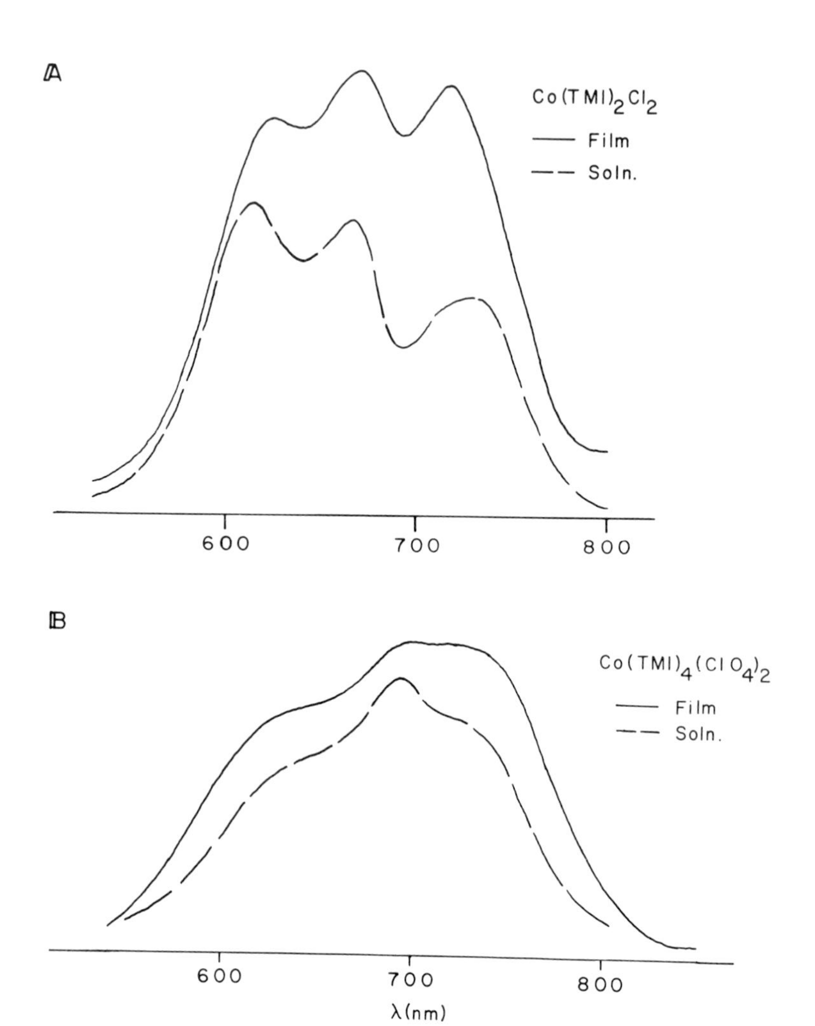

Figure 4. Spectra of Tetrahedral Co (II) with Four Thio Ligands. TMI is 2-thiono-4,4,5,5,-tetramethylimidazolidine.

of the 635 nm band relative to the 650 nm absorption peak for the enzyme (Figure 2b), the loss of fine structure in the enzyme-pyrazole-NAD^+ complex coupled with the growing intensity at 680 nm (Figure 2c) and the almost static absorption at 635 nm with an increasing absorption at 674 nm in the enzyme-isobutyramide-NADH complex (Figure 2d). It was observed that the 740 nm band was also unperturbed in these spectra as noted earlier.

The increase in optical density as well as the differences in optical density between enzyme and enzyme with coenzyme and

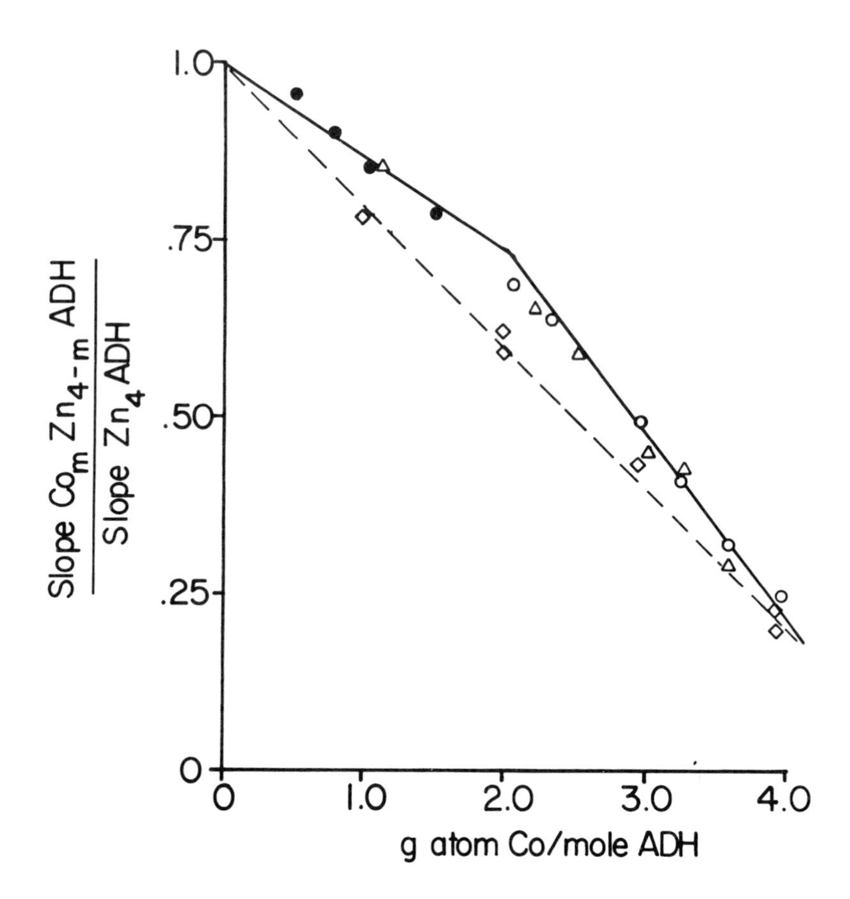

Figure 5. Quenching of NADH Fluorescence as a Function of Co Content Solid line (● △ o): Samples of ADH containing different levels of Co (II) as a consequence of equilibrium exchange dashed line (◇): Samples prepared by mixing proper molar proportions of Co_4 ADH and Zn_4 ADH.

inhibitor are recorded in Figure 3 as a function of total cobalt incorporated for selected wavelengths at 298°. These data clearly show that optical changes are not only sensitive to the amount of cobalt incorporation but are also diagnostic of the exchange pattern. The significance of the curvature of the data will be discussed later in detail.

IV. DISCUSSION

The visible absorption spectrum of cobalt substituted alcohol dehydrogenase gives insight not only into the geometry of the ligand field but also into the composition of the ligand

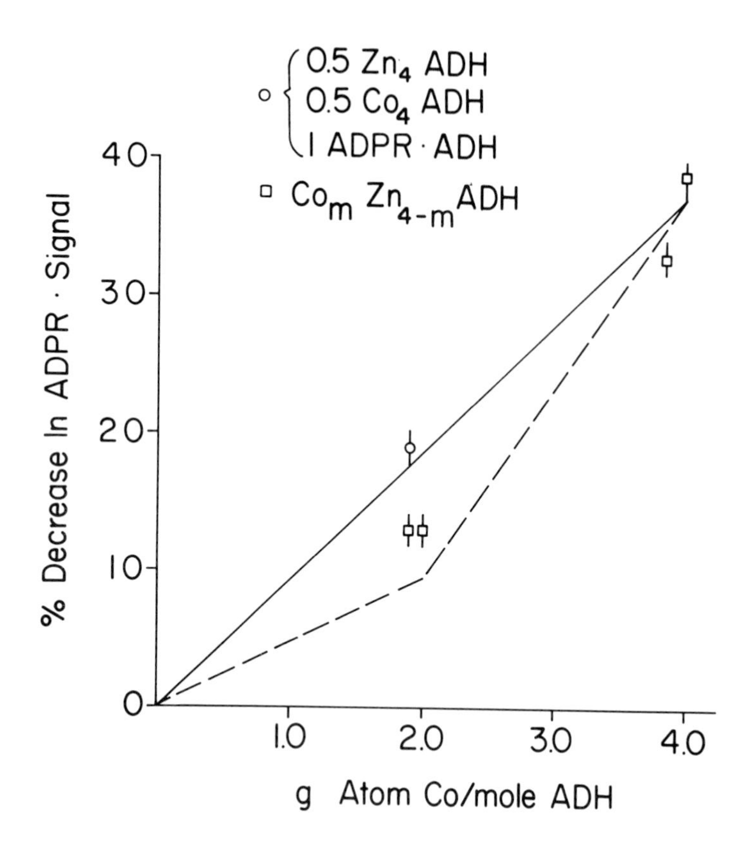

Figure 6. Dipolar Interaction Dependence on Cobalt Level. Dashed line (□): samples of ADH containing different levels of Co (II) as a consequence of equilibrium exchange; solid line (o): sample prepared by mixing equal molar quantities of Co_4 ADH and Zn_4 ADH.

field that the cobalt metal ion experiences. Past interpretations of the optical data for the electronic transitions of cobalt as to a tetrahedral environment based on the wavelength maxima, the peak intensities, and the bandwidth of the absorption peak as well as the charge transfer bands are confirmed by this study (9,16, 24). However, other spectral features have been employed to separate the otherwise almost identical spectra of the two different cobalt ions and to resolve the conflicting assignment of the kinetically fast and slow cobalt to the structural and catalytic sites, respectively.

The measurement of a new spectrum for the total cobalt substituted enzyme in Tris-Acetate buffer on addition of NAD^+, can be easily accounted for by the formation of a ternary complex,

Table 1

Summary of Assignment of Structural and Catalytic Sites

λ(nm) 298°	77°	FORM	KINETICS	ASSIGNMENT
740	740	E	FAST	STRUCTURAL
670	682	EPYO	SLOW	CATALYTIC
650	655	E	SLOW	CATALYTIC
	630	E	FAST	STRUCTURAL
375*		EPYO-E	SLOW	STRUCTURAL & CATALYTIC

* ISOBESTIC POINT

E-NAD^+-Acetate (Figure 1B). It was estimated that the fraction of the active sites occupied by acetate in the absence of NAD^+ was 0.18 (25). At that level of saturation, no difference between spectra recorded in phosphate buffer and those recorded in Tris-Acetate buffer was detected, suggesting that acetate does not bind or interact with either cobalt ions. The addition of NAD^+ to the cobalt substituted enzyme in Tris-Acetate buffer causes spectral perturbations and it was estimated that 95% of the enzyme was in the ternary complex (26). It was noted by Shore and Santiago (14) that NAD^+ fails to produce spectral shifts in phosphate buffer, even at saturating concentrations of the oxidized coenzyme. From these facts it is concluded that a ternary complex of the enzyme oxidized coenzyme and acetate is formed and that the presence of NAD^+ induces an environment which causes the acetate to perturb the spectral properties of the cobalt ions. These data also serve as an optical verification that aliphatic acids and NAD^+ form ternary complexes with ADH (25).

When pyrazole, a competitive inhibitor for ethanol (21)

was added to the sample in Tris-Acetate buffer (Figure 1C) a further red shift of the 650 nm peak was recorded and was identical to the spectrum of the ternary complex of the enzyme, coenzyme and pyrazole in phosphate buffer. This further red shift is a spectral demonstration of the greater affinity of enzyme-NAD^+ for pyrazole as compared to acetate. In a multiple inhibitor study with caprate, a fatty acid, and pyrazole, Yonetani has argued that these two inhibitors were acting at the same site (27). Even when trifluoroethanol, another competitive inhibitor for ethanol (25) was added prior to pyrazole, a spectrum corresponding to the enzyme-coenzyme-trifluoroethanol complex was measured as well. Thus, the magnitude of the red shift is believed to represent the greater affinity of the inhibitor for the enzyme-coenzyme complex.

The absorption measured in the 740 nm region could not be perturbed by either ligation (Figure 1) or cooling (Figure 2a) (16,17). This finding suggests that the origin of this particular transition is related to the cobalt bound to the structural site. Evidence supporting this conclusion is the well documented fact that cobalt complexes containing four tetrahedral-thio ligands have absorption maxima in this region (Figure 4), but when only two thio ligands are available, the maxima lie to higher energy (28,29), and the X-ray analysis of the native enzyme has clearly shown four sulfur ligands coordinated with the structural zinc (8).

Mathematical expressions for calculating the absorption profile as a function of total cobalt incorporated have been used to obtain a family of curves in which two parameters relate the theoretical values to the measured values. A scaling factor between the calculated and measured absorption is equal to the extinction coefficient for one metal binding site (i.e., E_1 or E_2) and a "best-fit" parameter (i.e., α or β) is equal to the ratio of the extinction coefficients for the two sites. In Figure 3d, the theoretical curves exhibit three characteristic features. First, for α or $\beta < 1$, that is $E_1 > E_2$, the curves lie <u>above</u> the line for α or $\beta = 1$, and increase rapidly with the initial metal concentration followed by a tapering off at high metal concentrations. Thirdly, for α or $\beta > 1$, the calculated curves <u>lie</u> below the line for α or $\beta = 1$ and increase slowly with metal concentration and rise rapidly at high levels of cobalt. Only when <u>both</u> the rate constants and α or β differ by a factor of 1000 or more do the theoretical curves have a distinct inflection point. In Fig. 3B, a plot of absorbance at an isobestic point as function of cobalt concentration at room temperature is found to correspond approximately to the $\alpha = 2$ curve, predicting a $E_1 = 0.67\ mN^{-1}cm^{-1}$ and $E_2 =$

1.34 $mN^{-1}cm^{-1}$. Similarly the optical data was analyzed for several wavelengths to note the dependence of the absorption peak on the cobalt concentration. As illustrated in Figure 3a, the value of β = 1000 closely approximates the experimental data for the pairs (ERI-E) and (EOI-E). The significance of a large β and the size of the scaling factor makes ΔE_1 (= $0.2 \times 10^{-3} mN^{-1}cm^{-1}$) very small suggesting the extinction for the fast exchanging metal is unperturbed upon ligation and only the slow exchanging metal is responding to the effect of complexation.

With the aid of this analysis of the dependence on the metal ion concentration, and spectral properties of cobalt complexes containing thio ligands, specific assignment of the fast and slow exchanging cobalt to the structural and catalytic sites is now possible. Beginning with 740 nm band, it correlates with the calculated curves for $\alpha < 1$, (Figure 3c) it is unpreturbed on ligation or cooling, and it has almost identical spectral properties of tetrahedrally coordinated complexes of cobalt with four thio ligands. Thus it is concluded that the origin of 740 nm peak is from a cobalt bound to structural site of alcohol dehydrogenase, (16,17) that has been shown by X-ray crystallography to have four cystine residues coordinating to the zinc metal (8). Furthermore, the absorption band at 630 nm (Figure 3c) also exhibits this characteristic dependence on the level of cobalt incorporated in the enzyme. However, the absorption in the 682 nm (Figure 3c) region displays an opposite trend, implying that it is coupled to the slow exchanging cobalt. Furthermore, when the progress course for this region is compared to the difference traces in Figure 3a, both sets of data are characterized by a lag phase. Hence this region and peaks arising from complexation must be associated with the slow exchanging cobalt. Although the identification of the slow exchanging site as the catalytic site in alcohol dehydrogenase is not so well-defined as for the structural site, it can be argued that the assignment is correct not only from the difference in the kinetics and from spectral studies of tetrahedrally coordinated complexes of cobalt containing two thio ligands, but also from the direct assignment of the fast exchanging site as the structural site. These conclusions are summarized in Table 1.

Recently, nuclear magnetic resonance studies (13) argue that the readily replaced cobalt ions replaced zinc at the "catalytic site" and the slowly exchaning cobalt ions replaced zinc at the "structural site." This assignment is, hence, completely opposite of the conclusions reached from the present study. Thus, at present, these differences are unresolved.

However, both electron paramagmetic resonance and fluorescence data more directly related to distance measurements support the conclusion reached by this optical study.

The paramagnetic effect of cobalt substituted in ADH on the enhanced fluorescence of NADH in presence of isobutyramide and the electron paramagnetic resonance of the spin label in ADPR (12) was examined as a function of total cobalt incorporated and is seen in Figure 5 and Figure 6, respectively. One can clearly see from these data that greater paramagnetic effect is manifested in the substitution of the final two cobalt ions. Although it can be argued that one does not know the κ factor in the fluoresence quenching results, the electron paramagnetic resonance experiments do not have anglular limitations. Moreover, the crystallographic structure unequivocally demonstrates that any paramagnetic effect observed in the electron paramagnetic resonance experiments is a direct result of substitution of cobalt at the catalytic site.

V. ACKNOWLEDGEMENTS

Research supported by Grant AA 00292 of the National Institute of Alcohol Abuse and Addiction, National Institutes of Health. The author is a Research Career Development Awardee, National Institute of General Medical Sciences (1 KO4 GM 70470).

VI. ABBREVIATIONS

ADH, horse liver alcohol dehydrogenase; IB, isobestic point; ERI, enzyme-reduced coenzyme-isobutyramide; EOI, enzyme-oxidized coenzyme-pyrazole; ADPR', adenosine-5-diphsophate-4-(2,2,6,6-tetramethylpiperidine-1-oxyl).

VII. REFERENCES

(1). A. Akeson, Biochem. Biophysics Res. Comm., 17, 211 (1964).
(2). D.E. Drum, T.K.L. and Vallee, B.K., Biochemistry, 8, 3783 (1960).
(3). Theorell, H., Nyguard, A.P., and Bonnichson, R., Acta. Chem. Scand. 9, 1148 (1955).
(4). Vallee, B.L., Hock, F.L., J. Biol. Chem. 225, 185 (1957).
(5). Vallee, B.L., Coombs, T.L., Williams, R.J.P., J. Am. Chem. Soc., 20, 397 (1958).
(6). Drum, D.E., Li, T.K., Vallee, B.L., Biochemistry, 9, 3792 (1969).

(7). Branden. C.I.. Eklund. H.. Nordstrom. B.. Boiwe. T.. Soderlund, G., Zeppezauer, E., Ohlsson, I., and Akeson, A., Proc. Natl. Acad. Sci. U.S., 70, 2439 (1973).
(8). Eklund, H., Nordstrom, B., Zeppezauer, E., Soderlund, G., Ohlsson, I., Boiwe, T., and Branden, C.I., FEBS Letters, 44, 200 (1974).
(9). Drum, D.E., and Vallee, B.L., Biochem. Biophys. Res. Comm. 41, 33 (1970).
(10). Young, J.M., and Wang, J.A., J. Biol. Chem. 246, 2815 (1971).
(11). Takahashi, M., and Harvey, R.A., Biochemistry 12, 4743 (1973).
(12). Drott, H.R., Santiago, D., and Shore, J.D., FEBS Letters 39, 21 (1974).
(13). Sloan, D.L., Young, J.M., and Mildvan, A.S., Biochemistry, 14. 1998 (1975).
(14). Shore, J.D., and Santiago, D., J. Biol. Chem., 250,2008 (1975).
(15). Harvey, R.A., and Barry, A., Biochem. Biophys. Res. Comm. 66, 935 (1975).
(16). Sytkowski, A.J., and Vallee, B.L., Biochem. Biophys. Res. Comm. 67, 1488 (1975).
(17). Sytkowski, A.J. and Vallee, B.L., Proc. Nat. Acad. Sci., U.S.A. 73, 344 (1976).
(18). Harvey, R.A., and Barry, A., Biochem. Biophys. Res. Comm. 72, 886 (1976).
(19). Dalziel, K., Acta Chem. Scand. 11, 397 (1957).
(20). Winer, A.D., and Theorell, H ., Acta. Chem. Scand. 14, 1729 (1960).
(21). Theorell, H., and Yonetani, T., Biochem. Z. 338, 537 (1963).
(22). Hagihara, B., and Iizuka, T., J. Biochem. (Tokyo) 69, 355 (1971).
(23). Swain, C.G., J. Am. Chem. Soc. 66, 1696 (1944).
(24). Vallee, B.L., Drum, D.E., and Kennedy, F.S. in "Alcohol and Aldehyde Metabolizing Systems" (R.G. Thurman, T. Yonetani, J.R. Williamson, and B. Chance, Eds.), p. 55. Academic Press, New York, 1974.
(25). Sigman, D.S., J. Biol. Chem. 242, 3815 (1967).
(26). Winer, A.D. and Theorell, H., Acta.Chem. Scand. 14, 1729 (1960).
(27). Yonetani, T., in "Symposium on pyridim-nucleotide-dependent dehydrogenases?" (G. Schwert and A.D. Winer, eds.), p. 27. University of Kentucky Press, 1970.
(28). Devore, E.C., and Holt, S.L., J. Inorg. Nucl. Chem. 34, 2303 (1972).
(29). King, T.M., Ph. Dissertation. University of Tennessee, Knoxville, Tenn., 1967.

HETEROGENEITY IN THE RAPIDLY EXCHANGING METALS OF HORSE LIVER ALCOHOL DEHYDROGENASE

R.A. Harvey and A. Barry

CMDNJ - Rutgers Medical School

The zinc atoms of horse liver alcohol dehydrogenase have been partially replaced by cobalt to give the fully active hybrids, $CoZn_3$-LADH and Co_2Zn_2-LADH. The rate of metal substitution was triphasic with one, two and four gram atoms of cobalt incorporated after one, eight and one hundred forty hours of exposure to cobalt salts. The hybrid enzyme containing one gram atom of cobalt has a characteristic visible absorption spectrum which was not perturbed by NADH or 1,10-phenanthroline. In contrast, titration of Co_2Zn_2-LADH with 1,10-phenanthroline resulted in a stoichiometric decrease in the absorbance of the hybrid at 650 nm which reached an end point when the concentration of chelator was half the total cobalt content of the hybrid. This spectral change was accompanied by an irreversible loss of approximately half the enzymic activity. These data indicate that cobalt substitution occurred initially at a non-catalytic site followed by incorporation of a second gram atom of cobalt at a catalytic site.

I. INTRODUCTION

Liver alcohol dehydrogenase is a metalloenzyme containing four gram atoms of zinc. Two of the zinc atoms are at each of the two catalytic sites while two of the metal atoms are located 20 Å from the catalytic site (1,2). The zinc atoms of the enzyme can be partially or completely replaced by cobalt with retention of enzymatic activity (3-8). The characteristics of the metals of Co_4-LADH represent an average of metal at the catalytic sites and at the non-catalytic sites and, hence, are not useful in differentiating the properties of the two classes of metal in the enzyme. In contrast, cobalt/zinc LADH hybrids in which the site of metal substitution is known are potentially valuable in defining the functional roles of the two classes of metal. The present work examines the spectral and kinetic properties of two hybrid enzymes, $CoZn_3$-LADH and Co_2Zn_2-LADH in an effort to establish the site(s) of metal substitution.

II. MATERIALS AND METHODS

Crystalline horse liver alcohol dehydrogenase (Boehringer Mannheim Corp.), 10 mg/ml, was dialyzed against 0.2 M sodium phosphate, pH 8.0 and then against two changes of 0.1 M Na_2SO_4, pH 7.0. Cobalt-substituted enzymes were prepared by a modification of the method of Shore and Santiago (9); Zn_4-LADH was dialyzed against 0.2 M $CoCl_2$-0.1 M Na_2SO_4-0.1 M NaAc, pH 5.5. Unbound metal was removed by dialysis against 0.025 M Hepes-0.1 M KCl, pH 7.0. All dialysis steps were carried out in 600 ml air-tight cylinders on rapid dialysis racks (10) at room temperature. Cobalt and Hepes-KCl buffers were degassed under vacuum before use and saturated with nitrogen at the start of dialysis. All other chemicals and methods have been described previously (8).

III. RESULTS AND DISCUSSION

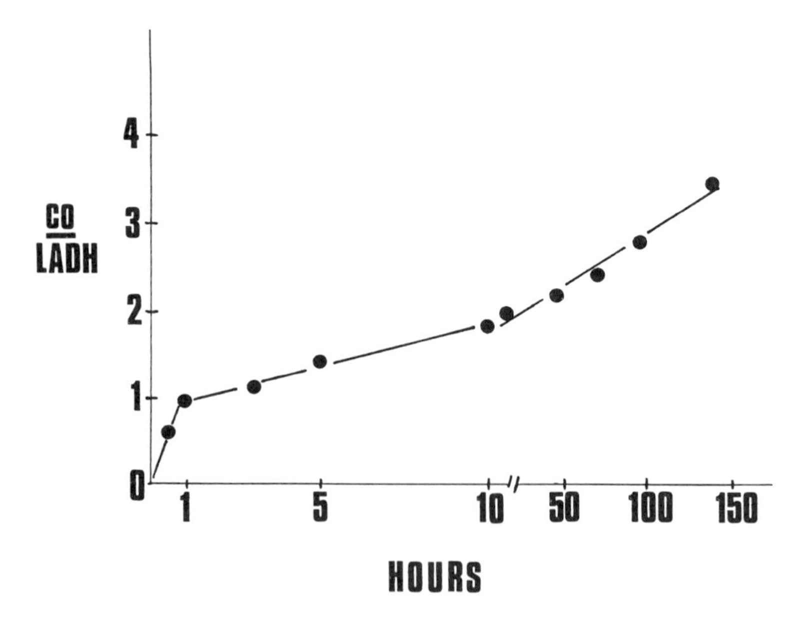

Fig. 1. Time course of cobalt incorporation.

A. Kinetics of Cobalt Substitution

Under the conditions of metal substitution described above, the rate of cobalt substitution is clearly triphasic (Fig. 1). Incorporation of the first cobalt occurs in approximately one hour while the rate of substitution of the second cobalt is substantially slower, requiring about 8 hours of dialysis. Exchange of the last two zinc atoms is essentially complete after 4-5 days dialysis against several changes of cobalt buffer. These differences in the rates of substitution of the first and second cobalt atoms suggest incorporation occurs at different sites or, possibly, that the first cobalt incorporated alters the structure of the enzyme so that subsequent metal substitition is retarded.

B. Spectral Perturbation by 1,10-Phenanthroline

Crystallographic studies (2) have shown that 1,10 phenanthroline (OP) binds to the active site metals of LADH. The visible absorption spectrum of the hybrid, which corresponds to the d-d electronic transition of the cobalt, should be altered by OP if the chelator forms an inner sphere complex with cobalt at the active site. Fig. 2 shows that the spectrum of $CoZn_3$-LADH is essentially unaltered by OP while the chelator significantly perturbs the 650 nm absorption maximum of Co_2Zn_2-LADH.

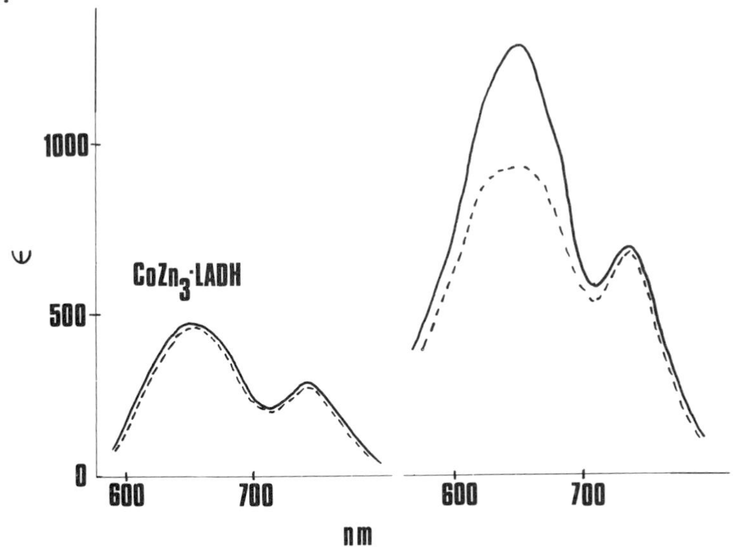

Fig. 2. Absorption spectrum of Co_2Zn_2-LADH in absence (——) and presence (----) of 1 mM 1,10-phenanthroline. Extinction coefficients based on molarity of dimeric enzyme.

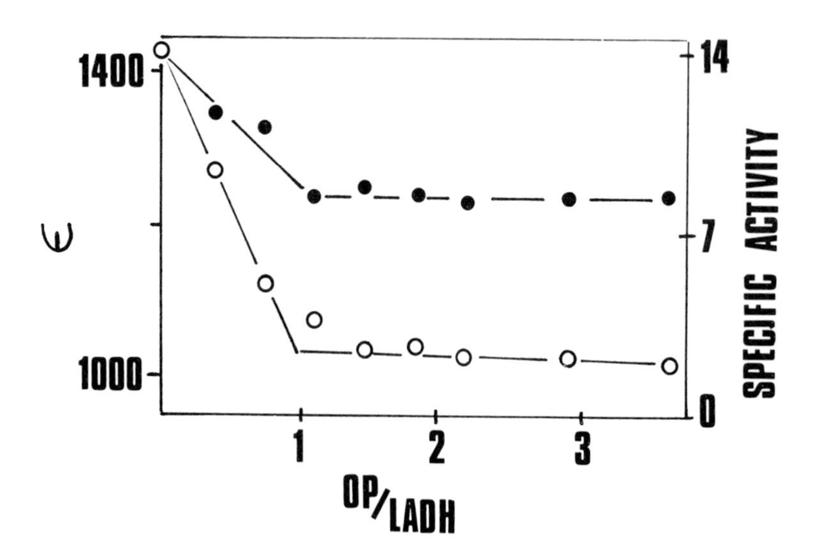

Fig. 3. Titration of Co_2Zn_2-LADH with 1,10-phenanthroline. (o), ε650 nm; (•), specific activity of LADH.

Fig. 3. shows that the titration of Co_2Zn_2-LADH with successive additions of OP result in a stoichiometric decrease in the absorbance of the enzyme at 650 nm and a concomitant, irreversible loss of almost half the enzymic activity. These changes, which probably reflect the high affinity of OP for cobalt at the active site exhibit an end point when OP/LADH = 1.

The magnitude and the stoichiometry of the spectral and kinetic changes which accompany OP binding to $CoZn_3$-LADH and Co_2Zn_2-LADH suggest that cobalt substitution occurs initially at a non-catalytic site, followed by incorporation of the second gram of cobalt at a catalytic site. This pattern of substitution is different from that proposed by Sloan *et al.*, (11) (two cobalts initially substituting at catalytic sites) and from that suggested by Vallee and Sytkowski (12,13) (two cobalts initially substituting at non-catalytic sites). It is possible that the cobalt/zinc hybrids described here and in the literature may each show a unique pattern of cobalt substitution, as a result of subtle but important differences in the conditions of metal substitution used by the various authors. However, recent data seem to indicate that the cobalt hybrid enzymes prepared by three different methods in three laboratories all exhibit a number of similar, if not identical, spectral and magnetic properties (see Barry *et al.*, these Proceedings). Thus the divergent conclusions concerning the site(s) of cobalt substitution in LADH seem to arise from differences in experimental approach and in interpretation of the data, rather than from variability in the LADH hybrid preparations.

IV. REFERENCES

(1). Brändén, C.I., Eklund, H., Nordström, B., Boiwe, T., Söderlund, G., Zeppezaurer, E., Ohlsson, I., and Akeson, Å., Proc. Natl. Acad. Sci. 70, 2439-2442 (1973).

(2). Branden, C.I., Jornvall, H., Eklund, H., and Furugren, B., in "The Enzymes" (P.D. Boyer, Ed.), Vol. 11, Part A, pp. 203-190. Academic Press, New York , 1975.

(3). Drum, D.E., and Vallee, B.L., Biochem. Biophys. Res. Commun. 41, 33-39 (1970).

(4). Young, J.M., and Wang, J.H., J. Biol. Chem. 246, 2815-2821 (1971).

(5). Takahashi, M., and Harvey, R.A., Biochemistry 12, 4743-4750 (1973).

(6). Drott, H.R., Santiago, D., and Shore, J.D., FEBS Letters 39, 21-23 (1974).

(7). Harvey, R.A., and Barry, A., Biochem. Biophys. Res. Commun. 66, 935-941 (1975).

(8). Harvey, R.A., and Barry, A., Biochem. Biophys. Res. Commun. 72, 886 (1976).

(9). Shore, J.D., and Santiago, D., J. Biol. Chem. 250, 2008-2012 (1975).

(10). Englander, S.W., and Crowe, D., Anal. Biochem. 12, 579-584 (1965).

(11). Sloan, D.L., Young, J.M. and Mildvan, A.S., Biochemistry 14, 1998-2008 (1975).

(12). Sytkowski, A.J., and Vallee, B.L., Biochem. Biophys. Res. Commun. 67, 1488-1493 (1975).

(13). Sytkowski, A.J., and Vallee, B.L., Proc. Natl. Acad. Sci. 73, 344-348 (1976).

MAGNETIC RESONANCE STUDIES OF THE INTERACTIONS OF IMIDAZOLE AND OTHER LIGANDS WITH COBALT SUBSTITUTED ALCOHOL DEHYDROGENASE FROM LIVER

J. Maitland Young and Albert S. Mildvan

The Institute for Cancer Research
Philadelphia

The paramagnetic effects of fully Co(II) substituted alcohol dehydrogenase, where both structural and catalytic metal ion sites have been replaced on the longitudinal relaxation rates of the protons of imidazole and isobutyramide were determined. These values were used to calculate lower limit distances from the catalytic Co(II) to the protons of imidazole and isobutyramide on the enzyme. These distances are too great by 2-3Å for the direct coordination of these ligands by the catalytic Co(II) but are appropriate for second sphere complexes with an intervening water ligand. The paramagnetic effects of a hybrid $(Co)_{1.3}(Zn)_{2.7}$ alcohol dehydrogenase on the relaxation rates of the protons of water, imidazole and isobutyramide correlate with the extent of occupancy of the fast exchanging metal site, suggesting this to be the catalytic metal site. This correlation is maintained in water relaxation studies with hybrid enzymes containing 1 to 2 Co(II) ions prepared in other laboratories . No paramagnetic effects of Co(II) at either the structural or catalytic sites were detected on the EPR spectrum of spin-labeled o-phenanthroline, indicating a Co(II) to nitroxide distance ≥ 9 Å . We conclude that the kinetically labile, hence the phyiologically active complexes of alcohol dehydrogenase are second sphere enzyme-metal-(H_2O_2)-substrate complexes rather than inner sphere enzyme-metal-substrate complexes.

I. INTRODUCTION

Alcohol dehydrogenase from liver is a dimeric metalloprotein which has two Zn(II) ions per subunit. As determined by X-ray diffraction, the Zn(II) ions are of two types (1,2). A "catalytic Zn(II)" located near the binding site for ADP-ribose, retaines water ligand and receives two cysteine and one imidazole ligand from the protein (2). The other, a "structural Zn(II)" is 20 Å away, receives four cysteine ligands from the protein and retains no water ligands (1,2). Two types of metal sites had previously been detected in solution by the biphasic kinetics of replacement of Zn(II) by ^{65}Zn(II) (3) or by Co(II) (4). Disagreement exists in the assignment of the fast-exchanging Zn(II) and slowly exchangig Zn(II) detected in solution, to the catalytic and structural Zn(II) sites detected by X-ray (5-8). To examine this issue, and to study the interaction of substrate analogs with the catalytic metal, magnetic resonance studies of water, imidazole, isobutyramide, and spin-labeled o-phenanthroline (OP-SL) in the presence of the Co(II)-substituted enzyme have been carried out.

II. MATERIALS AND METHODS

Partially Co(II) substituted alcohol dehydrogenase ($Co_{1.3}Zn_{2.7}$ ADH) was prepared by an 8-hour exchange using procedure 3 of Young and Wang (4). The fully substituted enzyme was prepared as described by Shore and Santiago (9). Spin labeled o-phenanthroline (10) was a gift from Professor L. Piette.

The longitudinal ($1/T_1$) and transverse ($1/T_2$) relaxation rates of the protons of water were measured at 24.3. MHz using the NMR specialities PS60W instrument. The relaxation rates of the protons of imidazole and isobutyramide were measured at 100 MHz using the Varian XL-100-Ft system and at 220 MHz using the Varian HR-220-FT system. All measurements were made by pulse methods as previously described (5). The EPR studies of OP-SL were made at 9.14 GHz. The paramagnetic contribution to the longitudinal relaxation rate of each ligand L, ($[L]/T_{1p}[Co(II)]$) in the presence of the Co(II) containing hybrids was calculated by subtracting the $1/T_1$ value determined in the presence of $(Zn)_4$ADH, and was used in equation (1) to determine the Co(II) to proton distance (r) (see reference (5)).

$$\frac{\{L\}}{T_{1p}\,\{Co(II)\}} = \frac{(895)^6}{r^6} \times f\,(\tau_c) \qquad (1)$$

where

$$F\,(\tau_c) = \frac{3\tau_c}{1+\omega_I^2\tau_c^2} + \frac{7\,\tau_c}{1+\omega_S^2\tau_c^2} \qquad (1a)$$

The correlation times (τ_c), which are dominated by the short electron spin relaxation time of Co(II), were determined by the frequency dependence of $[L]/T_{1p}[Co(II)]$ in each case.

III. RESULTS AND DISCUSSION

A. Paramagnetic Effects of Fully Substituted CO_4 ADh

To avoid ambiguities in the nature of the sites occupied, the effects of CO_4ADH on the relaxation rates of various ligands were determined (Table I). With this preparation both the catalytic and structural metal sites are occupied by the paramagnetic Co(II). The paramagnetic effects on $1/T_1$, in all cases, are not limited by chemical exchange, as established by the frequency dependence of $[L]/T_1[Co(II)]$ and by the >3 fold greater effects on $1/T_2$ than on $1/T_1$ (Table II). Hence the values of $[L]/T_{1p}[Co(II)]$ may be used in equation (1) for distance calculations. Assuming that only the catalytic Co(II) increases the relaxation rates of the protons of imidazole and isobutyramide (IBA), lower limit Co(II) to proton distances are calculated which are too great by 2-3 Å for direct coordination of these analogs by the catalytic Co(II) (Table II). The calculated distances are, however, appropriate for second sphere complexes in which a water ligand on Co(II) (∿2.8 Å thick) intervenes (Ref. 5). Any corrections made for the small paramagnetic contribution of Co(II) bound at the more distant structural sites would increase the calculated distances between these ligands and the catalytic Co(II).

Although X-ray studies of the ternary enzyme-NADH-isobutyramide complex or of any of the ternary substrate complexes have not been reported, Branden and colleagues have detected by X-ray methods the direct coordination of imidazole by the catalytic Zn(II), after prolonged (∿ 1 week) soaking of the ADH crystals in imidazole (11). These results differ from those found in solution by NMR studies (5) which require only 2-4 hours of data collection and in which the enzyme remains fully active. These differences in coordination may, therefore, result from the widely differing experimental conditions. Alternatively, the NMR studies may miss a slowly exchanging inner sphere imidazole ligand for which sufficient space exists. However, the presence of a slowly exchanging inner sphere isobutyramide ligand is unlikely from the limited space available. Similarly, in the abortive enzyme-NADH-ethanol complex, second sphere distances in solution were calculated from NMR studies (5). Hence the results, at present, suggest that complexes which more closely approach the active complex yield second sphere distances, although no structural studies of a functioning ternary complex have yet been made either by NMR or by X-ray.

B. Paramagnetic Effects of Partially Substituted Co(II) ADH

The paramagnetic effects of $Co_{1.3}Zn_{2.7}$ ADH on the protons of

TABLE I

Paramagnetic Effects of $(Co)_4ADH$ *on the Longitudinal Relexation Rates of Various Ligands*

Ligand	*Frequency* (MHz)	[NADH] (μM)	$\frac{[Ligand]}{T_{1p}[Co]total}$ (sec^{-1})	$\frac{[Ligand]}{T_{1p}[Co]catalytic}$ (sec^{-1})
Imidazole (20 mM)	100	500	29.6 $(H\text{-}C_2)$ 11.5 $(H\text{-}C_{4,5})$	59.2 $(H\text{-}C_2)$ 23.0 $(H\text{-}C_{4,5})$
		0	39.2 $(H\text{-}C_2)$ 35.6 $(H\text{-}C_{4,5})$	78.4 $(H\text{-}C_2)$ 71.2 $(H\text{-}C_{4,5})$
H_2O	24.3	0 500[a,b]	5065 ± 525 1730	10,125 ± 1045 3460
Isobutyramide	100	500[b]	26.6 $(CH_3\text{-})$ 35.1 $(H\text{-}C_2)$	53.2 $(CH_3\text{-})$ 70.2 $(H\text{-}C_2)$

The errors in the relaxtion rates are ± 10%. Solutions contained 0.5 mM enzyme bound Co(II), 0.1 m Na^+ *phosphate buffer, pH 7.2 T = 21 ± 2°.* $(Co)_4ADH$ *was prepared by the procedure of reference (9).*

[a] *Solution also contained 1 mM isobutyramide.* [b] *Data from reference (5).*

TABLE 2

Calculation of Distances from Enzyme Bound (Co)II to the Protons of Imidazole and Isobutyramide in $(Co)_4ADH$ Complexes.

Complex	Interaction	$\frac{[Ligand]}{T_{1p}[Co]}$ (100 MHz) $(sec)^{-1}$	τ_c (100 MHz) sec x 10^{12}	$f(\tau_c)$ (100 MHz) (sec x 10^{12})	$r_{calculated}$ Å	Expected r for Inner Sphere Å	Δr ($r_{calculated}$ − $r_{Inner\ Sphere}$) Å
$m(Co)_4ADH$[a]	$Co...H\text{-}C_2$	78.4	0.8 - 6.8[b]	7.1 - 25.8	6.7 ± 0.7	3.27	3.4 ± 0.7
	$Co...H\text{-}C_{4,5}$	71.2	0.2 - 5.0[b]	2.0 - 21.6	6.1 ± 1.2	3.88	2.2 ± 1.2
$Im(Co)_4ADH$-NADH[a]	$Co...H\text{-}C_2$	59.2	0.3 - 2.4[b]	3.1 - 15,6	6.4 ± 0.9	3.27	3.1 ± 0.7
	$Co...H\text{-}C_{4,5}$	23.0	0.2 - 5.0[b]	2.0 - 21.6	7.4 ± 1.4	3.88	3.5 ± 1.4
$IBA(Co)_4ADH$-NADH[c]	$Co...CH_3$-	53.2	1.0	9.0	7.6 ± 0.2	4.2	3.4 ± 0.2
	$Co...H\text{-}C_2$	70.2	1.0	9.0	7.2 ± 0.3	4.0	3.2 ± 0.3

[a] *From $1/T_{2p}$, the exchange rates of imidazole into these complexes exceed 400 sec^{-1} at 21 ± 2°.*

[b] *Extreme range calculated from the ratio of [Imidazole]/T_{1p} [Co] at 220 MHz to that at 100 MHz of 2.25 to 3.43.*

[c] *From reference (5).*

ligands studied (Table III) were normalized in two alternative ways: (a) by assuming that the relaxation resulted only from the fast exchanging Co(II) (Table III, Col. 3) and (b) by assuming that relaxation resulted only from the small amount of slowly exchanging Co(II) (Table II, Col. 4). The relaxation rates resulting from these alternative assumptions are compared (Table III) with the relaxation rates of the catalytic Co(II) based on the data obtained with the fully substituted enzyme from Table I. The paramagnetic effects of the fast exchanging Co(II) on the $1/T_1$ values of the protons of water, imidazole, and isobutyramide (Table III) correspond to those of the catalytic Co(II). A possible exception is the case of imidazole in the absence of NADH. Moreover, the addition of NADH and isobutyramide to the hybrid enzyme produces a large reduction in the relaxation rate of water protons. Since $1/T_1$ values are functions of distance, these results indicate that the fast exchanging Co(II) is closer to water, imidazole and isobutyramide than is the slowly exchanging Co(II), a property expected of the active site metal.

Essentially identical relaxation rates of water, which were eliminated by NADH and isobutyramide, were obtained with a preparation of Co_1Zn_3ADH prepared by Dr. H.R. Drott (12) using a more rapid exchange procedure (9). A study of two different preparations of Co_2Zn_2ADH prepared by H.R. Drott and by R.A. Harvey indicated that the incorporation of a second Co(II) did not further increase the relaxation rate of water but that like the Co_1Zn_3ADH hybrids, the paramagnetic effects on $1/T_1$ were eliminated by the presence of NADH and isobutyramide (12). These results indicate consistent relaxation effects of hybrid enzymes prepared under differing conditions and rates of exchange in three laboratories (12). They further indicate that when a total of 2 of the 4 Zn(II) ions have been replaced by Co(II), discrimination between catalytic and structural sites is no longer apparent (8), in contrast to the results reported by Sytkowski and Vallee (6,7).

C. Comparison of ZnADH and Co Substituted ADH on the EPR Spectrum of Spin-Labeled o-phenanthroline (OP-SL)

The EPR spectra of OP-SL (0.1 mM) in the presence of Zn_4ADH (6.0 mg/ml) the partially substituted hybrid $Co_{1.3}Zn_{2.7}ADH$ (5.9 mg/ml) and the fully substituted Co_4ADH (5.9 mg/ml) all showed the broadening typical of an immobilized spin label as originally reported for Zn_4ADH (10). No differences in the amplitudes or line shapes were detected with all three preparations indicating that the electron spin relaxation time of the bound nitroxide was unaffected by either the fast or slowly exchanging Co(II)ion. The addition of ADP-ribose (0.36 mM) did not significantly alter the spectra. Using our experimental uncertainty of ± 7% in the amplitude together with the appropriate dipolar equation (13) yields a lower limit Co(II) to nitroxide distance > 9 Å. These results indicate an extended conformation for OP-SL and ADH, but provide no further information on the identity of the fast and slowly exchanging Co(II) ions.

TABLE 3

Correlation of Paramagnetic Effects of $(Co)_{2.7}$ ADH on Ligands with Occupancy of fast Exchanging Metal Sites

Ligand	[NADH] μM	$\frac{[Ligand]}{T_{1p}[Co]fast}$ (sec^{-1})	$\frac{[Ligand]}{T_{1p}[Co]slow}$ (sec^{-1})	$\frac{[Ligand]^a}{T_{1p}[Co]catalytic}$ (sec^{-1})
Imidazole ($H-C_2$) (100 MHz)	500	31.6	325	59.2
($H-C_{4,5}$)		17.4	179	23.0
Imidazole ($H-C_2$) (100 MHz)	-	28.5	294	78.4
($H-C_{4,5}$)		19.7	204	71.2
H_2O[b] (24.3 MHz)	-	$1.8x10^4$	$16.0x10^4$	$1.0x10^4$
	500[c]	$0.98x10^4$	$9.4x10^4$	$0.35x10^4$
Isobutyramide[b] (CH_3-) (100 MHz)	500	51.1	460	53.2
($H-C_2$)		62.2	560	70.2

The hybrid $(Co)_{1.3}(Zn)_{2.7}$ ADH was prepared by procedure 3 of reference (4). The errors in the relaxation rates are ± 10%.

[a]From Table I. [b]From Reference (5). [c]Also obtained 1 mM isobutyramide.

IV. CONCLUSIONS

We conclude that the most rapidly entering Co(II) in liver alcohol dehydrogenase occurs at the active site, and that it retains a water or hydroxyl ligand as judged by its magnetic resonance effects. More importantly, the kinetically labile, hence the physiologically active complexes of the subtrate analogs, isobutyramide and imidazole, and of the substrates (5) are second sphere enzyme-metal-(H_2O_2)-substrate complexes rather than inner sphere enzyme-metal-substrate complexes.

V. ACKNOWLEDGEMENTS

We are grateful to Mr. Richard Freedman for his expert technical assistance. This work was supported by National Institutes of Health Grant AM-13351, by National Science Foundation Frant PCM74-03739, by Grants CA-06927 and RR-05539 to this Institute from the National Institutes of Health and by an appopriation from the Commonwealth of Pennsylvania.

Vl. REFERERENCES

(1). Branden, C.I., Eklund, H., Nordstrom, B., Boiwe, T., Soderlund, G., Zeppezauer, E., Ohlsson, I., and Akeson, A., Proc. Nat. Acad. Sci., USA 70, 2439 (1973).

(2). Nordstrom, B., Zeppezauer, E., Soderlund, G., Ohlsson, I., Boiwe, T., Soderberg, B.O., Tapia, O., Branden, C.I., and Okeson, A., J. Mol. Biol. 102,27 (1976).

(3). Drum, D.E., Li, T.K., and Vallee, B.L., Biochemistry 8, 3792 (1969).

(4). Young, J.M., and Wang, J.H., J. Biol. Chem. 246,2815 (1971).

(5). Sloan, D.L., Young, J.M., and Mildvan, A.S., Biochemistry 14, 1998 (1975).

(6). Sytkowski, A.J., and Vallee, B.L., Biochem. Biophys. Res. Commun. 67, 1488 (1975).

(7). Sytkowski, A.J., and Vallee, B.L., Proc. Nat. Acad. Sci., USA 73, 344 (1976).

(8). Harvey, R.A., and Barry, A., Biochem. Biophys. Res. Commun. 66, 935 (1975).

(9). Shore, J.D., and Santiago, D., J. Biol. Chem. 250, 2008 (1975).

(10). Spallholz, J.E., and Piette, L.H., Arch. Biochem. Biophys. 148, 596 (1972).

(11). Branden, C.I., Jornvall, H., Eklund, H., and Furugren, B., in: "The Enzymes", 3rd edition, XI, 103 (1975).
(12). Drott, H.R., Harvey, R.A., Mildvan, A.S., and Young, J.M. (This volume).
(13). Drott, H.R., Santiago, D., and Shore, J.D., FEBS Letter 39, 21 (1974).

WATER PROTON RELAXATION AND SPECTRAL STUDIES OF ALCOHOL DEHYDROGENASE PARTIALLY SUBSTITUTED WITH COBALT

H.R. Drott, R.A. Harvey, A.S. Mildvan, & J.M. Young

University of Pennsylvania
Rutgers Medical School
Institute for Cancer Research
Bryn Mawr College

Consistent water proton relaxation effects and optical spectra have been obtained with cobalt containing alcohol dehydrogenases. These hybrids, containing from 1 to 2 Co(II) ions per mole of enzyme, were prepared by different procedures in three laboratories.

I. INTRODUCTION

While there is a general agreement that the exchange of CO(II) for Zn(II) in Zn_4ADH occurs in several kinetically distinct steps (1-4) disagreement exists about the location of the fast and slowly entering Co(II) ions. Thus, from optical spectra (2,4-8) and from EPR data (7) it has been argued that the most rapidly entering Co(II) ion occupies the structural site, while from nuclear relaxation data (9,10) it has been argued that this Co(II) occupies the catalytic site. Since different workers have used widely differing conditions of Co(II) substitution, the present nuclear relaxation and optical studies were jointly undertaken by three laboratories to determine whether hybrids obtained in different laboratories showed structural differences.

II. MATERIALS AND METHODS

The hybrid enzymes were prepared in the laboratories of Young, Drott, and Harvey under the conditions summarized in Table 1 and discussed in detail elsewhere (2, 7-11) and were

TABLE 1

Comparison of Water Proton Relaxation Rates and Optical Absorbancies of Various Hybrids from Various Laboratories

Authors	*Method of Preparation of Hybrid*	*Metal Content*	$\frac{[H_2O]}{T_{1p}[Co]_{total}} \times 10^{-4}$ sec^{-1}	$\frac{Abs(655\ nm)}{[ADH]}$	$\frac{Abs(740\ nm)}{[ADH]}$
Sloan *et al.* (9)	Procedure III of reference (2) $[Co^{2+}]$=50 mM; (8 hrs.)	$Co_{1.3}Zn_{2.4}ADH$	1.4±0.2		
	Ionic Strength = 0.15 pH = 5.5	+NADH+IBA	0.3±0.1		
Young and Mildvan (10)	Procedure III of reference (2) $[Co^{2+}]$=50 mM; (8 hrs.) Ionic Strength = 0.15 pH = 5.5	$Co_{1.4}Zn_{2.3}ADH$	1.0±0.3	650	430
Harvey and Barry (4,8)	Modification of Procedure of reference (11). $[Co^{2+}]$=200 mM; (8 hrs). Ionic strength = 0.07. pH = 5.5	Co_2Zn_2ADH	0.54±0.07	1300	700
		+IBA	0.31±0.06		
		+NADH	0.25±0.06		
		+NADH + ETOH	0.20±0.06		
		+NADH + IBA	0.08±0.06		
Drott (3,7)	Procedure of reference (11) $[Co^{2+}]$= 100 mM (1 hr.)	Co_1Zn_3ADH	1.2±0.3	360	200
		+NADH + IBA	≤ 0.1		
	(5 hrs.)	Co_2Zn_2ADH	0.40±0.12	1160	700
		+NADH + IBA	≤ 0.1		
	Procedure III of reference (2) (30 hrs.)	Co_2Zn_2ADH	0.55±0.07		

The relaxation rates of water protons were determined at 24.3 MHz. All studies were done at 21±2°.

then traded. Water proton relaxation rates were measured (9, 10) and optical spectra at room temperature and low temperature were obtained (4,7,8) as previously described.

III. RESULTS AND DISCUSSION

The results are summarized in Table 1 which also includes previously reported parameters for comparison. An impressive general finding is that the water proton relaxation rate (normalized to the total Co(II) content ($[H_2O]/T_{1p}[Co]_{total}$) as well as the extinction coefficients depend primarily on the extent of incorporation of Co(II) rather than on either the laboratory of origin, or on the widely differing conditions or rates of Co(II) incorporation. It is also clear from the data on three preparations that when <1.4 Co(II) ions are exchanged for Zn(II), the predominant occupancy of a single type of site occurs (Table 1). However when two Co(II) ions are exchanged, the relaxation rates and the optical data indicate that two types of sites are highly populated, as first pointed out by Harvey (4). These results differ from those of Sytkowski and Vallee (5,6) who upon incorporating two Co(II) ions under conditions similar to those of Harvey (4,8,11) and Drott (7,11), interpret their optical spectra as reflecting solely the occupancy of the structural metal sites by Co(II).

From Table 1, it can be seen that the first site occupied by Co(II) is almost solely responsible for the relaxation of water protons, and this water relaxation is reduced by NADH and substrate analogs. These are properties expected of the catalytic metal which retains a water of hydration (12). However the rapidly entering Co(II) also shows a high extinction coefficient at 740 nm as does a model $Co(S = CR_2)_4$ complex (7). The Co(II) in this model, like the structural metal in the enzyme, has four sulfur ligands. However, the enzyme ligands are saturated cysteine sulfur atoms. Further, the optical spectrum of the most rapidly entering Co(II) does not respond to azide (2), pyrazole (2) or o-phenanthroline (2,4,8). This indicates that this Co(II) is either buried, i.e., structural (2,4-8), or alternatively that it is catalytic with the added ligands forming second sphere complexes (9,10). Hence disagreement persists as to the nature of the first site occupied by Co(II), but it is clear from these studies that the three laboratories involved are preparing Co(II)-containing hybrids of indistinguishable structure.

IV. ACKNOWLEDGEMENTS

This work was supported by National Institutes of Health Grants AA-00292, and AM-13351, by National Science Foundation Grant PCM74-03739, by Grants CA-06927 and RR-05539 to this Institute from the National Institutes of Health and by an appropriation from the Commonwealth of Pennsylvania.

V. REFERENCES

(1). Drum, D., Li, T.K., and Vallee, B.L., Biochemistry 8, 3792 (1969).
(2). Young, J.M., and Wang, J.H., J. Biol. Chem. 246, 2815 (1971).
(3). Drott, H.R., Santiago, D., and Shore, J.D., FEBS Letters 39, 21 (1974).
(4). Harvey, R.A., and Barry, A., Biochem. Biophys. Res. Commun. 66, 935 (1975).
(5). Sytkowski, A.J., and Vallee, B.L., Biochem. Biophys. Res. Commun. 67, 1488 (1975).
(6). Sytkowski, A.J., and Vallee, B.L., Proc. Nat. Acad. Sci., USA, 73, 344 (1976).
(7). Drott, H.R., This Volume.
(8). Harvey, R.A., and Barry, A., This Volume.
(9). Sloan, D.L., Young, J.M., and Mildvan, A.S., Biochemistry 14, 1998 (1975).
(10). Young, J.M., and Mildvan, A.S., This Volume.
(11). Shore, J.D., and Santiago, D., J. Biol. Chem 250, 2008 (1975).
(12). Brändén, C.I., Jornvall, H., Eklund, H., and Furugren, B., in "The Enzymes", 3rd Edition XI 103 (1975).

INTERACTION AND ACTIVITIES OF NAD^+ AND NAD^+ ANALOGS WITH FOUR DEHYDROGENASES

Robert J. Suhadolnik and Michael B. Lennon

Temple University, Philadelphia

The effects of six NAD^+ analogs on the K_m, V_{max}, K_D, fluorescence quenching, k_1, and decomposition of the E·NADH complex with four dehydrogenases are reported. The interaction of the adenine ribose and adenine of NAD^+ of reactions done in solution, where catalysis <u>*is complete*</u> *is compared with the x-ray maps where catalysis* <u>*is not complete*</u>. *It appears that the 2'- and 3'hydroxyl groups of NAD^+ are not essential for bonding, but are essential for proper orientation of the nicotinamide ring to produce a productive complex. NAD^+, 2'dNAD^+, and 3'dNAD^+ have the same affinity with the enzymes. It may be that the 2'hydroxyl group of NAD^+ does not result in a typical 3-5 kcal/mole hydroxyl-carboxylate hydrogen bond. Alternatively, the 2'- and 3'dNAD^+ may not be positioned correctly in the conenzyme domain for maximum catalysis. This requires that the intrinsic binding energy for NAD^+, 2'-and 3'-dNAD^+ be the same. Because yeast ADH and presumably horse liver ADH, GAPDH and LDH have compulsory ordered kinetics, we observed a decrease in both k_1 and the rate of decomposition of the E·NADH complex. Replacement of adenosine of NAD^+ with tubercidin, formycin, inosine or ε-adenosine produced changes in the K_m, K_D, and V_{max} that are discussed in terms of their accomodations within the coenzyme domain.*

I. INTRODUCTION

The use of nicotinamide adenine dinucleotide (NAD^+)[1] as a common substrate for a large number of enzymes has generated wide-spread interest in the use of this compound as a cofactor and substrate in a number of cellular reactions that regulate protein, RNA, and DNA synthesis (Scheme I). NAD^+ analogs have been extremely useful as biochemical probes to increase our understanding of the mechanism of dehydrogenase action, especially since the three dimensional alignment of the common binding structure of the coenzyme domain of the dehydrogenases has been determined by electron density maps (1). The x-ray data clearly show that (1) the aspartate residue which is conserved in all dehydrogenases binds with the 2'-hydroxyl group of NAD^+, (2) the 3'-hydroxyl of NAD^+ is hydrogen bonded to the polypeptide chain carbonyl with halo-GAPDH and in the ternary complex of LDH, and (3) the adenine ring binds in the hydrophobic crevice.

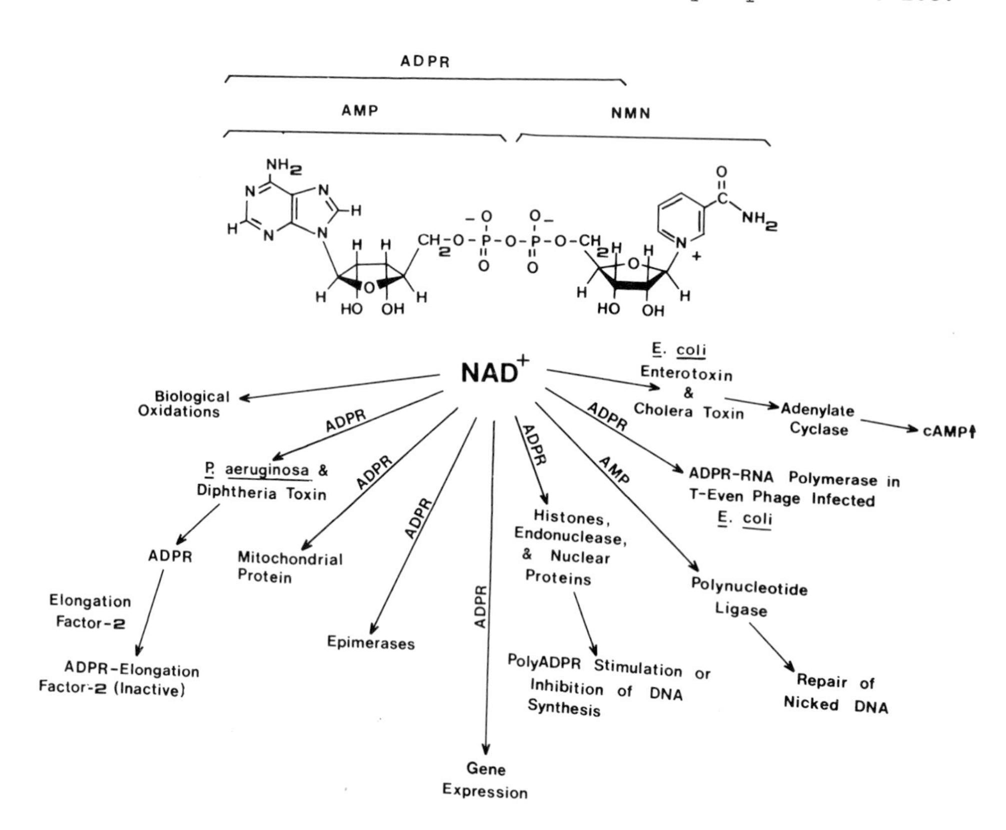

Scheme 1. The Role of NAD^+ as an Electron Carrier and a Substrate for Non-Oxidative Reactions.

The relative rates of reduction of NAD^+ analogs in which the adenine ring has been replaced with other aglycones have been described (2-4). The convenient enzymatic synthesis of NAD^+ with specific modifications in the adenine ring and the adenine ribose is shown in Fig.1. These syntheses provide the opportunity to further test and compare structural-functional-hydrogen bonding requirements of the cofactor in the coenzyme domain of the dehydrogenases. From these results, we are able to contrast the NAD^+ binding of the holo and ternary complexes of the dehydrogenases in crystalline structures, where catalysis *is not complete*, with reactions in solutions, where catalysis *is complete*.

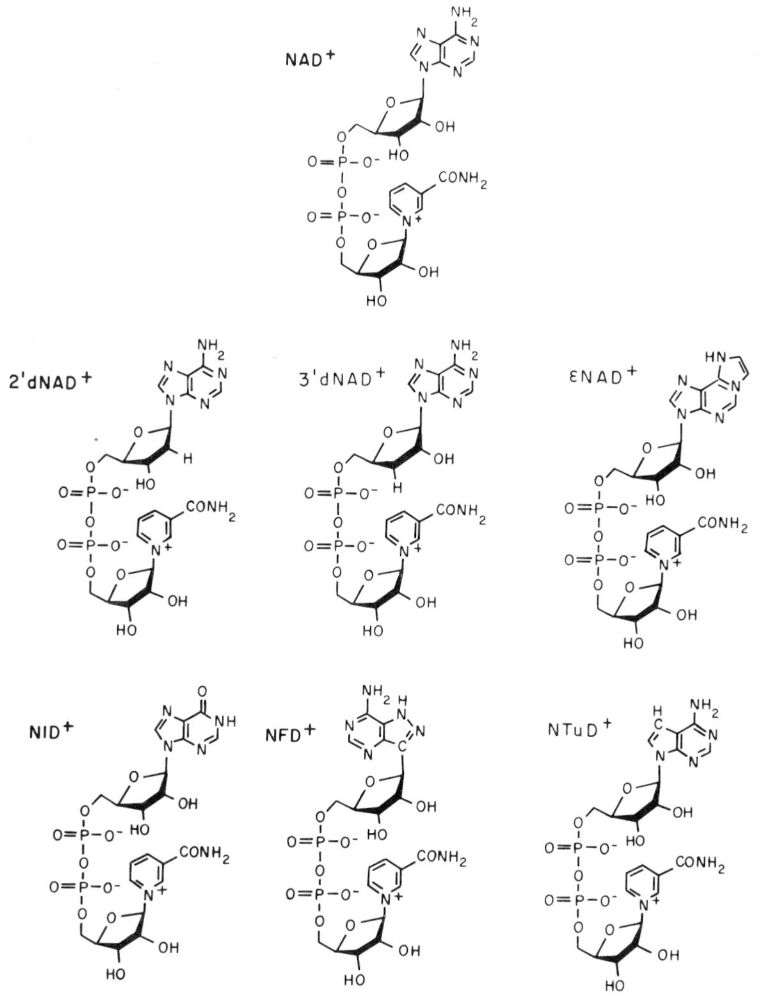

Figure 1. Structures of NAD^+ Analogs Modified in the Adenine Ring or the Adenine Ribose.

II. RESULTS

By using hog liver NAD^+ pyrophosphorylase with NMN, a number of compounds including 2'dATP, 3'dATP, FTP and TuTP become substrates for the formation of NAD^+ analogs (Fig. 2). Characterization of NAD+ and NAD+ analogs, the uv spectral data, the molar extinction coefficients, and proof of structure have been described (5)

The interaction of NAD+ and NAD+ analogs with horse liver ADH is described in Fig. 3 A Molecular representation of the coenzyme domain for horse liver ADH from the x-ray data is included with the enzyme results for ease of comparison. Changing the imidazole ring of NAD^+ ($NTuD^+$, NFD^+, or NID^+) affects the K_m, the K_D and the V_{max}. Greenfield *et al.* (3) reported similar findings with εNAD^+. When either the 2'-hydroxyl or 3'-hydroxyl group is removed (2'dNAD^+ or 3' NAD^+), catalysis is markedly affected. However, affinity of 2'dNAD^+ or 3' NAD^+ is the same as that for NAD^+ since the K_m's and K_D's are similar.

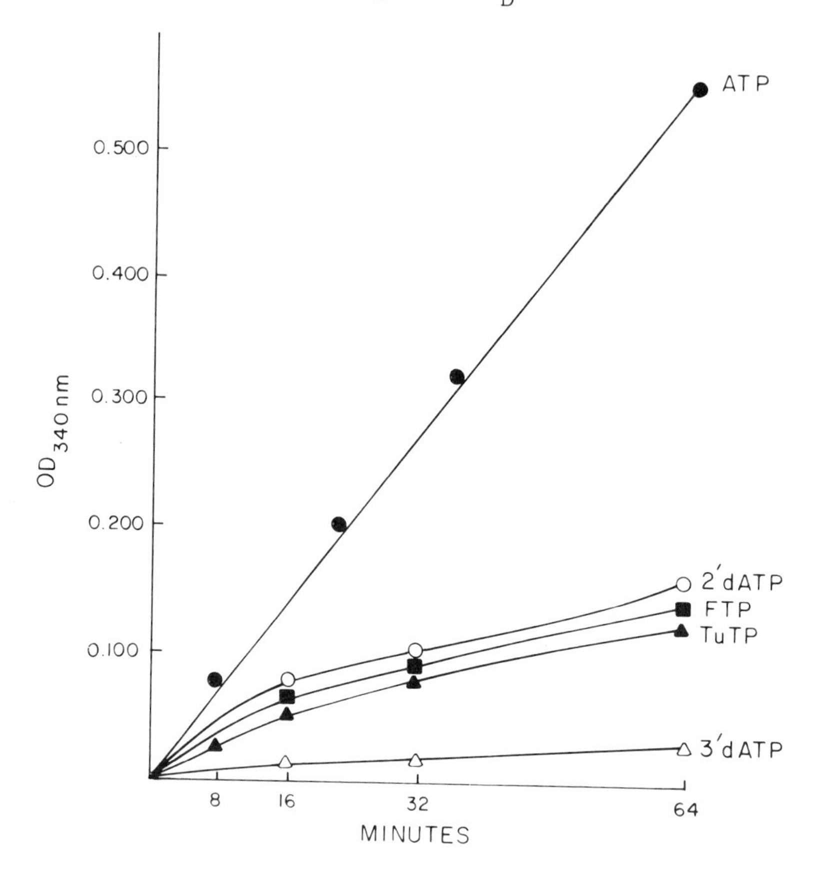

Figure 2. Formation of NAD^+ and NAD^+ Analogs by NAD Pyrophosphorylase. For Experimental Conditions, refer to Ref. 5.

Interaction of NAD^+ and NAD^+ analogs with yeast ADH has also been studied. Based on x-ray and primary structural data available for mammalian, yeast, and bacillar alcohol dehydrogenases, aspartate is an invariant residue that hydrogen bonds with the 2'-hydroxyl of NAD^+ (6 and personal communication, Dr. H. Jornvall). With $NTuD^+$ and NFD^+ a small increase in K_m is observed when the imidazole ring is changed to a pyrrolo ring or pyrazolo

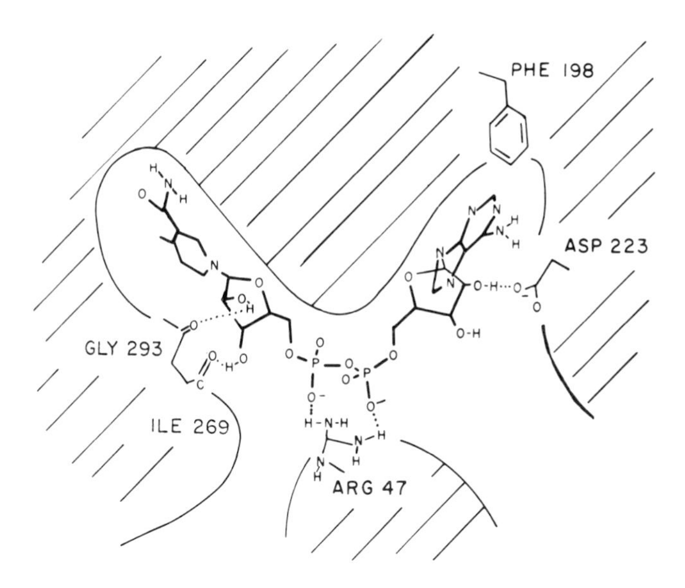

	K_m	V_{max} [a]	Percent Quenching	Fluorescent K_D	Equilibrium K_D
	μM			μM	μM
NAD^+	20	3.5	42	17.9	19.4
$NTuD^+$	23	2.8	48	18	--
NFD^+	13	1.4	--	--	--
εNAD^+	41	2.1	enhancement	58	--
NID^+	54	1.9	40	61	--
$2'dNAD^+$	24	0.7	14	40	38
$3'dNAD^+$	14	0.6	21	28	26
ADPR	--	--	20	40	--

Figure 3. Diagrammatic Representation of the NAD^+ Binding Site for Horse Liver ADH and Related Kinetic and Binding Data.

ring (Table I). The decrease in V_{max} with $NTuD^+$ and NFD^+ is similar to that observed for horse liver ADH. With εNAD^+ and NID^+, there is a 13 and 8.3 fold increase in K_m but an 87 and 86% decrease in V_{max}. The $2'dNAD^+$ and $3'dNAD^+$ analogs show little change in the K_m and K_D; however, the V_{max} decreases markedly.

GAPDH is unique among the dehydrogenases. The crystalline tetrameric enzyme contains two molecules of firmly bound coenzyme which cannot be removed by dialysis or recrystallization, but can be removed by treatment with charcoal (7). The *anti*

TABLE 1

Analog	Horse Liver ADH		Yeast ADH		GAPDH		LDH	
	K_m	V_{max}[a]	K_m	V_{max}	K_m	V_{max}	K_m	V_{max}
	μM		μM		μM		μM	
NAD^+	20	3.5	54	111	66	0.7	13	54.1
$NTuD^+$	23	2.8	58	60	77	11.8	13	18.0
NFD^+	13	1.4	74	32	83	6.3	3	2.1
εNAD^+	41	2.1	710	14	840	8.0	87	21.2
NID^+	54	1.9	450	16	1000	3.0	32	3.5
$2'dNAD^+$	24	0.7	65	4.8	60	5.3	15	11.5
$3'dNAD^+$	14	0.6	42	0.9	74	3.9	8	4.6

[a] *units =nmoles/min/mg protein*

TABLE II

Determination of Absolute V_{max}/K_m and V_{max}/K_m Relative to NAD^+ with the NAD^+ Analogs and the Dehydrogenase Studied

Analog	Horse Liver ADH		Yeast ADH		GAPDH		LDH	
	V_{max}/K_m		V_{max}/K_m		V_{max}/K_m		V_{max}/K_m	
	Absolute	Relative to NAD^+	Absolute	Relative to NAD^+	Absolute	Relative to NAD^+	Absolute	Relative to NAD^+
	$nm^{-1}min^{-1}$ x 10^{-5}	fold decrease	$nm^{-1}min^{-1}$ x10^{-6}	fold decrease	$nm^{-1}min^{-1}$ x10^{-5}	fold decrease	$nm^{-1}min^{-1}$ x10^{-6}	fold decrease
NAD^+	1.8	- - -	2.1	- - -	2.6	- - -	4.2	- - -
$NTuD^+$	1.2	1.5	1.0	2.1	1.5	1.73	1.4	3.0
NFD^+	1.1	1.6	0.43	4.9	0.76	3.4	0.7	6.0
εNAD^+	0.5	3.6	0.02	104	0.095	28	0.24	17.5
NID^+	0.35	5.1	0.04	52.5	0.03	80	0.11	38
$2'dNAD^+$	0.29	6.2	0.07	30	0.88	3.0	0.76	5.5
$3'dNAD^+$	0.43	4.2	0.02	105	0.53	5.0	0.58	7.2

form favors the 2' and 3'-hydroxyl groups of NAD^+ (8). The slight increases in K_m and decreases in V_{max} for $NTuD^+$ and NFD^+ are very similar to those shown with horse liver ADH and yeast ADH (Table I). The conenzyme domain of GAPDH is more sensitive to the replacement of the N^6-amino group with the 6-oxo group (i.e. NID^+) than is yeast ADH.

LDH is a tetrameric enzyme with independent binding sites for NAD^+ (9). The x-ray diffraction data for LDH predicts: (1) hydrogen bonding of the carboxyl group of aspartate 53 and the carbonyl group of amino acid 29 with the 2' and 3'-hydroxyl groups of NAD^+ (9). With NFD^+ the K_m decreases while the V_{max} decreases by 96%. Replacement of adenine with ε-adenine allows binding to the enzyme and function as a coenzyme analog. The data in Table I for LDH show that the binding of the coenzyme and analogs (2'dNAD^+ and 3'dNAD^+) does not change appreciably as evidenced by the K_m's that are similar to the K_m for NAD^+; however, catalysis is markedly affected as evidenced by the change in V_{max} for NAD^+ from 54.1 to 11.5 and 4.5 for 2'dNAD^+ and 3'dNAD^+, respectively.

Because K_m values do not always have simple physical relationships in compulsory ordered reactions (10), fluorescence and equilibrium binding studies were performed. The protein fluorescence quenching of liver ADH by NAD^+, 2'dNAD^+, 3'dNAD^+, and ADPR is shown in Figure 4. Whereas the percent quenching with 2'dNAD^+ and 3'dNAD^+ is similar to ADPR, i.e. 13,22 and 19% respectively, but different from the 42% quenching observed for NAD^+, the K_m and K_D values for the 2'dNAD^+ and 3'dNAD^+ are similar.

Determination of association constant (k_1), changes in association constants with NAD^+ analogs, and a comparison of changes in decomposition of E·NADH analog complexes is shown in Table II. For many dehydrogenases, it is known that the mechanism follows compulsory ordered kinetics (10-14). If one makes that assumption, the k_1 can be determined from the observed V_{max} and K_m values. The V_{max}/K_m values for the analogs and the absolute values of V_{max}/K_m for NAD^+ are presented in Table II. With all of the NAD^+ analogs studied, V_{max}/K_m decreased with the four dehydrogenases . Removal of the 2' and 3'-hydroxyl groups of NAD^+ resulted in a slow down in the rate of addition (k_1) of the conenzyme analog to the enzyme. Yeast ADH showed the greatest slow down, i.e., 30 and 105 fold, respectively, with 2'dNAD^+ and 3'dNAD^+.

III. DISCUSSION

<u>Studies of coenzyme analogs.</u> According to current electron

density maps from x-ray crystallographic studies, the nucleotide binding domain for NAD^+ consists of a central six-stranded parallel sheet. Each strand is joined in a defined sequence with α-helices (1). By synthesizing NAD^+ analogs modified in the adenine and the adenine ribose, we have been able to compare the interaction of NAD^+ with four dehydrogenases *in the crystalline form* with reactions *in solution*. In the case of the dehydrogenases, the steady state kinetic parameters can be utilized to obtain binding and dissociation rate constants for the reactions as shown in equation 1. V_{max}/K_m

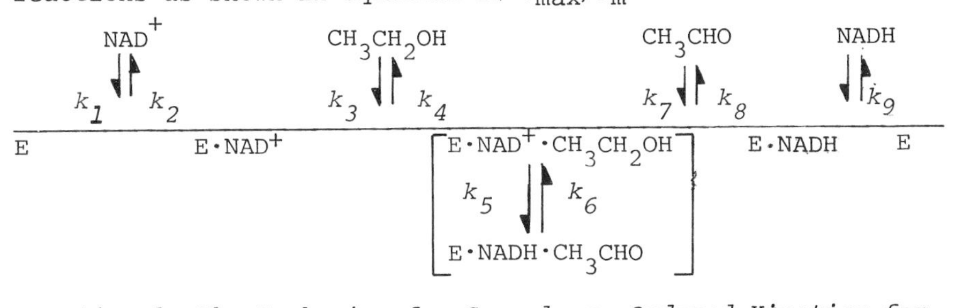

Equation 1. The Mechanism for Compulsory Ordered Kinetics for Yeast ADH (13).

can be calculated where K_m is the Michaelis constant for the conenzyme determined at infinite second substrate concentration. With horse liver ADH the rate limiting step has been clearly demonstrated to be the dissociation of the coenzyme from the enzyme. For this dehydrogenase, and possibly the other dehydrogenases, the V_{max} reflects the "off-rate" or the dissociation rate constant. In addition, the recent x-ray diffraction data add additional support for an ordered mechanism by showing that the binding of NAD^+ creates the binding site for the second substrate by inducing a conformational change in the enzyme (10).

For compulsory ordered kinetics, V_{max}/K_m is an estimate of k_1 (or "k_{on}"), the second order rate constant of the enzyme reaction with NAD^+. At infinite concentrations of second substrate, this ratio can be quite informative in terms of comparing the effect of rate of addition of NAD^+ with the rate of addition of NAD^+ analogs. Unexpectedly, the reaction of the four dehydrogenases with the six NAD^+ analogs shows a slow down in the rate of addition to the enzyme (k_1 or "k_{on}") (Table II). On the assumption that k_9 reflects V_{max}, there is also a slow down in the rate of decomposition of the E·NADH complex. In the case of horse liver ADH, the kinetic data of Klinman (13) implicate k_9 and support this assumption. With horse liver ADH and 2'dNAD^+, there is a six fold decrease in the rate of addition of conenzyme to enzyme (k_1). Similarly, V_{max} decreased five fold and, to the extent that this decrease reflects k_9, these data or replaced with ε-adenine (εNAD^+). Replacement of the adenine ring with pyrimidine bases produced NAD^+ analogs that were not as good coenzymes as εNCD^+ (2,19,20). The kinetic data described here with εNAD^+ are in good agreement with the report of Greenfield *et al.* (3). The greatest increase in K_m values with εNAD^+ was observed with yeast ADH and GAPDH (Figs. 4,5). Apparently, the more crowded condition at the binding site is manifest when ε-adenine replaced adenine.

In summary, this report compares the interactions of the adenine ribose and the adenine ring of NAD^+ with the conenzyme domain as determined by x-ray data with the K_m, K_D, V_{max}, and fluorescence quenching with NAD^+ analogs of enzyme reactions done in solution. When the adenine ring is modified, there is good agreement with the x-ray data. When the adenine ribose is modified there are differences with the x-ray data. Three possibilities can be offered to explain the differences observed with the 2'dNAD^+ and 3'dNAD^+: (1) the contribution of the hydrogen bonds in the NAD^+ molecule between the 2' and 3'-hydroxyl positions is so small that removal of these bonds does not produce changes in the affinity of the coenzyme for the dehydrogenases; (2) the loss of the hydrogen bond between the 2'-hydroxyl of NAD^+ and the β-carboxyl of aspartate, but retention of the same affinity, is compensated for by the binding of 2'dNAD^+ with the peptide backbone. The objection to this possibility is that it only explains the data obtained with 2'dNAD^+ and not the data for 3'dNAD^+ with horse liver ADH and yeast ADH where hydrogen bonding does not involve the 3'-hydroxyl group. However, as with 2'dNAD^+, the affinity of 3'dNAD^+ remains the same but the V_{max} decreases and there is only a small decrease in the fluorescence quenching. Therefore, some other group(s) in the polypeptide chain would be required to interact with 2'dNAD^+ such that the intrinsic binding energy lost by removal of the 2' or 3'-hydroxyl group would be equally compensated for by other interactions; (3) there is no significant interaction between the hydroxyls of ribose and the binding domain of these dehydrogenases.

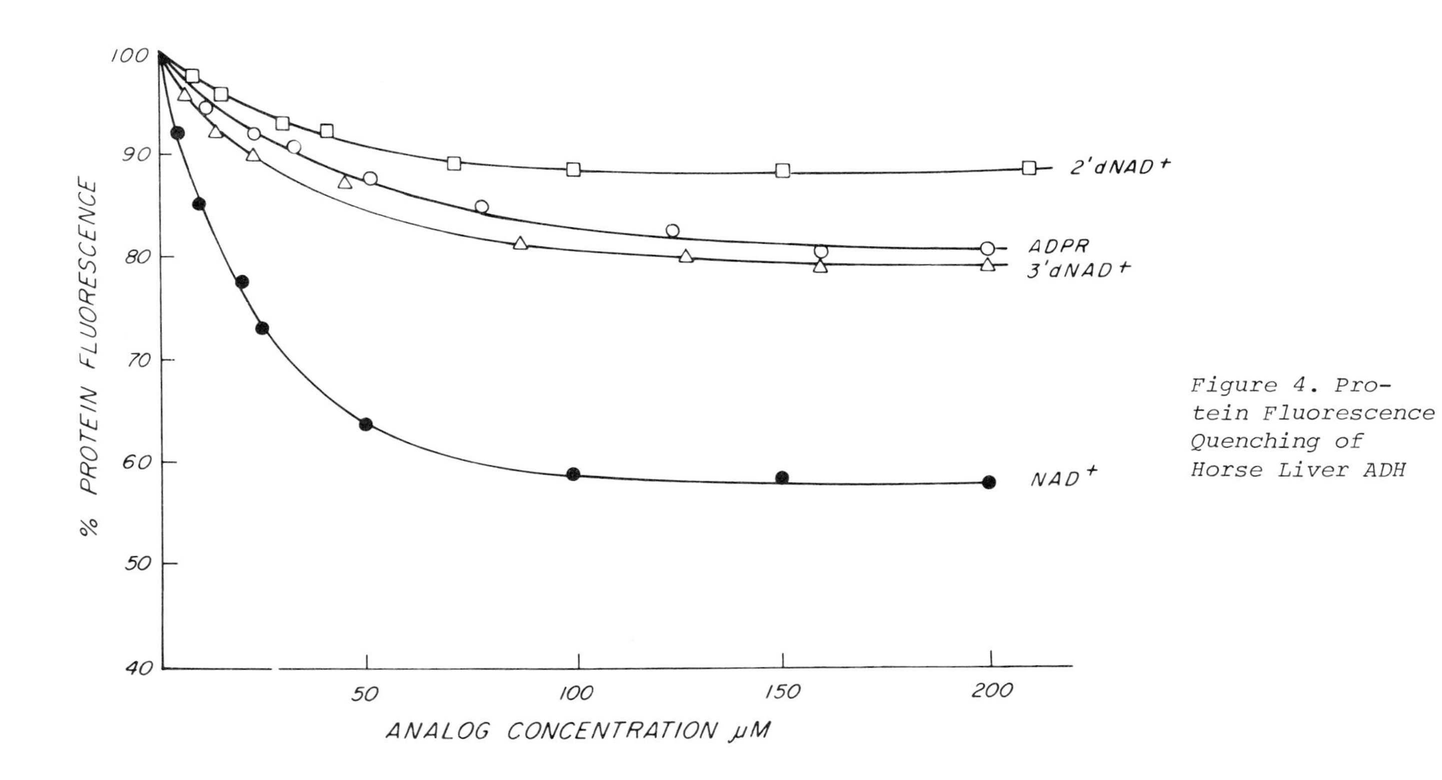

Figure 4. Protein Fluorescence Quenching of Horse Liver ADH

The data obtained here with 2'dNAD$^+$ and 3'dNAD$^+$ show that the contribution of the 2' and 3'-hydroxyl groups of NAD$^+$ for hydrogen bonding is very small and their removal does not change the affinity for the enzyme; however, these 2' and 3'-hydroxyl groups are essential for productive complex formation and subsequent efficient catalysis. In addition, the absence of the 2' or 3'-hydroxyl group slows k_1, which reflects the decrease in binding interaction. With 2'dNAD$^+$ and 3'dNAD$^+$, assuming that the NAD$^+$ analogs do not alter the rate determining step, there is also a slower decomposition of E·2'dNADH or E·3'dNADH (Tables I and II). With horse liver ADH with 2'dNAD$^+$, the K_D changes from 19.8 to 38 (Fig. 2). This calculates to a 2'-hydroxyl-carboxylate aspartate hydrogen bond of only 0.42 kcal/mole which is about 10% of the 3-5 kcal/mol for the same type bond in a hydrophobic pocket (17).

A possible explanation for the equal binding affinity of 2'dNAD$^+$ with NAD$^+$ is that the β-carboxyl group of the conserved aspartate residue interacts with N_3 of the adenine ring; this produces an abnormal binding of 2'dNAD$^+$ which results in the formation of non-optimal ternary complex. This in turn results in decreased catalysis as evidenced by the slight decrease in fluorescence quenching with horse liver ADH, the marked decrease in the V_{max} and the slow down in the rate of addition of the conenzyme analog to the enzyme. Therefore, the 2' and 3'-hydroxyl groups, although not necessary for hydrogen bonding, are essential for proper orientation of NAD$^+$ and the subsequent formation of a more productive complex.

<u>Quenching of protein fluorescence and binding sites for NAD$^+$.</u>

The 42% quenching of fluorescence observed with NAD$^+$ could be due to an interaction with any three of the four tryptophan residues of horse liver ADH, since all of these groups are within the prescribed limits for energy transfer (18). It is probably tryptophan 314 which is the species primarily quenched by NAD$^+$ (1). The 14 and 21% protein fluorescence quenching observed with 2'dNAD$^+$ and 3'dNAD$^+$ indicate that the critical amino acids are not in an optimal position to allow the nicotinamide moiety of NAD$^+$ to be in the proper conformation to catalyze hydrogen transfer with its substrate (Fig. 4). This correlates well with the lowered V_{max} for 2'dNAD$^+$ and 3'dNAD$^+$ for the four dehydrogenases studied. The fluorescence quenching for NTuD$^+$ and NID$^+$ are similar to NAD$^+$ (Table I).

<u>Importance of the Adenine Ring of NAD$^+$.</u> To understand the selectivity of the adenine ring for the coenzyme domain of the dehydrogenases, K_m, K_D, and V_{max} data were obtained in which the adenine of NAD$^+$ was either modified (NTuD$^+$, NFD$^+$, or NID$^+$)

suggest that the value for K_D should not be altered significantly. Experimentally, this was observed (Fig. 3 and Table I). It is possible that the 0.7 V_{max} for horse liver ADH with 2'dNAD$^+$ compared to 3.5 for NAD$^+$ reflects a decrease in k_5 or other steps; this possibility could be tested by measuring kinetic deuterium isotope effects. In the case of NAD$^+$, V_H/V_D = 1 (13). These experiments are currently under investigation.

With yeast ADH, V_{max} contains k_5 and k_9; therefore, with this enzyme we are not looking solely at k_9. The change in V_{max} could be due to an affect on k_5 since k_5 is partially rate determining (15). With yeast ADH, the reduction of the V_{max} is not clearly defined; the interpretation of the V_{max} data in Tables I and II is done with these reservations in mind. From Table II, the greatest changes in V_{max}/K_m when the adenine ring is modified occurs with yeast ADH; therefore, yeast ADH must contain a more restricted pocket. With NTuD$^+$, NFD$^+$, 2'dNAD$^+$ and 3'dNAD$^+$ there is a decrease in k_1 and V_{max} but little or no change in K_m. In contrast, Plapp *et al.* (12,16) reported that alkylation of the active site by picolinimidylation increased activity and decreased the binding of conenzyme to horse liver ADH. Based on the fluorescence quenching, K_m, and K_D's with 2' dNAD$^+$ and 3'dNAD$^+$, it appears that the lack of conformational changes of either the enzyme of the lack of proper positioning of the nicotinamide ring results in marked reduction of available interaction with the substrate that is needed for maximum catalysis.

Importance of the 2' and 3'-hydroxyl groups of the adenine ribose moiety of NAD$^+$. An important feature of NAD$^+$ binding in the coenzyme domain is the proper orientation and hydrogen bonding between the 2'-hydroxyl group of the adenine ribose and the β-carboxyl of the conserved aspartate residue and the 3'-hydroxyl group with the peptide backbone for productive complex formation. The x-ray diffraction data, with NAD$^+$ bound at the catalytic site, clearly show that the β-carboxyl group of the conserved aspartate residue of all the dehydrogenases hydrogen bonds to the 2'-hydroxyl group; with the 3'-hydroxyl, hydrogen bonding to the polypeptide chain occurs only with GAPDH and LDH (6, 9). With the dehydrogenases studied here, when the 2' or 3'-hydroxyl group of NAD$^+$ is removed, there are small changes in the affinity (Fig. 3 and Table I). However, even though the affinity for 2'dNAD$^+$ and 3'dNAD$^+$ is the same as NAD$^+$, the proper positioning of the nicotinamide ring with the substrate and subsequent formation of a productive complex for catalysis is markedly decreased as evidenced by the decrease in the V_{max}, the small decrease in the percent fluorescence quenching (Fig. 4) and the decrease in the V_{max}/K_m relative to NAD$^+$ (Table II).

IV. REFERENCES

(1). Boyer, P., The Enzymes, Vol. XI, Academic Press, New York (1975).

(2). Fawcett, C.P., and Kaplan, N.O., J. Biol. Chem. 237, 1709-1715 (1962).

(3). Greenfield, J.C., Leonard, N.J., and Gumport, R.I., Biochemistry 14, 698-706 (1975).

(4). Woenckhaus, D., and Jeck, R., Hoppe-Seyler's Z. Physiol. Chem. 352, 1417-1423 (1971).

(5). Suhadolnik, R.J., Lennon, M.B., Uematsu, T., Monahan, J.E., and Baur, R., J. Biol. Chem., in press (1976).

(6). Eklund, H., Branden, C.-J., and Jornvall, H., J. Mol. Biol. 102, 61-73 (1976).

(7). Fox, J.B.,Jr., and Dandliker, W.B., J. Biol. Chem. 221, 1005-1017 (1955).

(8). Moras, D., Olsen, K.W., Sabesan, M.N., Buehner, M., Ford, G.C., and Rossmann, M.G. J. Biol. Chem. 250, 9137-9162 (1975).

(9). Liljas, A., and Rossmann, M.G., Annu. Rev. Biochem.43, 475-507 (1974).

(10). Dalziel, K., in:"The Enzymes" (Boyer, P., Ed.) Vol. XI, p. 1-60, Academic Press, New York (1975).

(11). Wratten, C.C., and Cleland, W.W., Biochemistry 2, 935-941 (1963).

(12). Plapp, B.V., Brooks, R.L., and Shore, J.D., J. Biol. Chem.. 248, 3470-3475 (1973).

(13). Klinman, J.P., in:"Steenbock Symposium on Isotope Effects in Enzymology" (W.W. Cleland, Ed.) University Park Press, in press, (1976).

(14). Vestling, C.S., and Kunsch, U., Arch. Biochem. Biophys. 127, 568-575 (1968).

(15). Klinman, J.D., J. Biol. Chem. 250, 2569-2573 (1975).

(16). Plapp, B.V., J. Biol. Chem. 245, 1727-1735 (1970).

(17). Jencks, W.P., in:"Advances in Enzymology" (A.Meister, Ed.) Vol. 43, 219-410 (1975).

(18). Luisi, P.L., Favilla, R., Eur. J. Biochem. 17, 91-94 (1970).

(19). Pfleiderer, G., Sann, E., and Ortanderl, F., Biochim.Biophys. Acta 73, 39-49 (1963).

(20). Honjo, M., Furukawa, Y., Moriyama, H., and Tanaka, K., Chem. Pharm. Bull. 11, 712-720 (1963).

MODIFICATION OF TRYPTOPHAN IN LIVER ALCOHOL DEHYDROGENASE

Edyth L. Malin and J. Maitland Young

Bryn Mawr College, Pa.

Modification of tryptophan in liver alcohol dehydrogenase (LADH) by 2-hydroxy-5-nitrobenzyl bromide (HNB) resulted in the incorporation of 1.5 to 5 moles of HNB per mole of LADH. When LADH was denatured in urea before modification, 7 or 8 HNB groups were incorporated. Ultra-violet, visible and fluorescence spectra suggest: (1) all tryptophan residues are accessible to HNB only when LADH is denatured; (2) accessible tryptophans are located on the exterior surface and can be doubly labeled by HNB. These conclusions agree with reports describing two classes of tryptophan residues in LADH. HNB-LADH was subjected to tryptic digestion, gel filtration, and high-voltage electrophoresis. An HNB-containing peptide was isolated and identified by dansyl-Edman sequencing as the octapeptide containing tryptophan-15. No evidence of altered structure was observed in HNB-LADH by circular dichroism and analytical gel filtration studies. Essential cysteine groups in HNB-LADH were as readily alkylated by iodoacetate as those in LADH. Activity loss of modified LADH correlated with the number of HNB groups incorporated, but the failure of modification to abolish all activity indicates that tryptophan-15 cannot be involved in hydride transfer, in agreement with the evidence from X-ray studies (10,11).

I. INTRODUCTION

Tryptophan residues of liver alcohol dehydrogenase (LADH) were chemically modified by the use of 2-hydroxy-5-nitrobenzyl bromide (HNB), a reagent specific for this amino acid (1,2). HNB is an extremely versatile reagent because of the pH dependence of its ultraviolet and visible spectra, its speed of reaction, and the relative ease with which its protein derivatives can be manipulated.

The chemistry of HNB reactions with model compounds and with proteins has been extensively reported (3-6). Following initial reaction of HNB at position 3 or the indole ring, internal cyclizations may occur via attack at position 2 by the α -amino nitrogen (N-cyclization) or by the phenoxide group of HNB (O-cyclization) (3). In addition, double labeling of tryptophan may occur by incorporation of a second HNB molecule at ring position 2 (7) or at the indole nitrogen (5).

II. INCORPORATION OF HNB

Modification reactions were conducted in 0.1 M phosphate buffer (ph 7.0) using conditions described previously (8). The results show that labeling of tryptophan residues was only partially dependent on the molar excess of HNB in the incubation mixture. At low molar ratios of reagent to protein tryptophan (10:1), the incorporation observed was ∿ 1.5 moles HNB/mole LADH. With a greater excess of HNB (25:1) present during the reaction, the incorporation increased to 3-5 moles HNB/mole LADH. No further increase in the amount of HNB incorporation was observed in the presence of still greater excess of HNB. However, when LADH was denatured in 8 M urea (pH 3.0) before modification, the incorporation increased to 7 or 8 moles HNB/ mole LADH.

Two possible explanations of these results are: (1) all four tryptophan residues per dimer of LADH can incorporate only one HNB moiety when the reaction is conducted in phosphate buffer, while all four residues are doubly labeled when the protein is first denatures in 8 M urea; (2) all four tryptophans are accessible to HNB only when the enzyme is denatured. The latter explanation implies further that in the absence of urea, two of the tryptophan residues can be doubly labeled. This possibility is particularly attractive in view of previous spectroscopic (9) and x-ray crystallographic evidence (10,11), for two classes of tryptophan residues in LADH. Other explanations for the wide range of HNB incorporation observed in phosphate buffer cannot be totally excluded, but they would be of minor signficance.

Amino acid analyses (Table I) of chemically modified LADH are consistent with the second explanation. The amounts of unreacted tryptophan recovered upon amino acid analyses of alkaline hydrolysates of HNB-LADH were greater than expected if all four tryptophan residues were available for incorporation of only one HNB group per residue. Only in samples containing 7 or 8 moles HNB/mole LADH (modified in urea) was there no recovery of unreacted tryptophan. Therefore, some tryptophan residues must be doubly labeled both in the presence and in the absence of urea. Furthermore, all of the tryptophans have been modified only when the protein is first denatured.

TABLE 1
Recovery of Tryptophan from Hydrolysates of HNB-LADH after Amino Acid Analysis

HNB/LADH	Recovered Tryptophan, Moles/Moles LADH	
	Observed	Theoretical
0.0	4.4	4.0
2.8	2.1	1.0
3.0	1.5	1.0
8.4	0.0	0.0

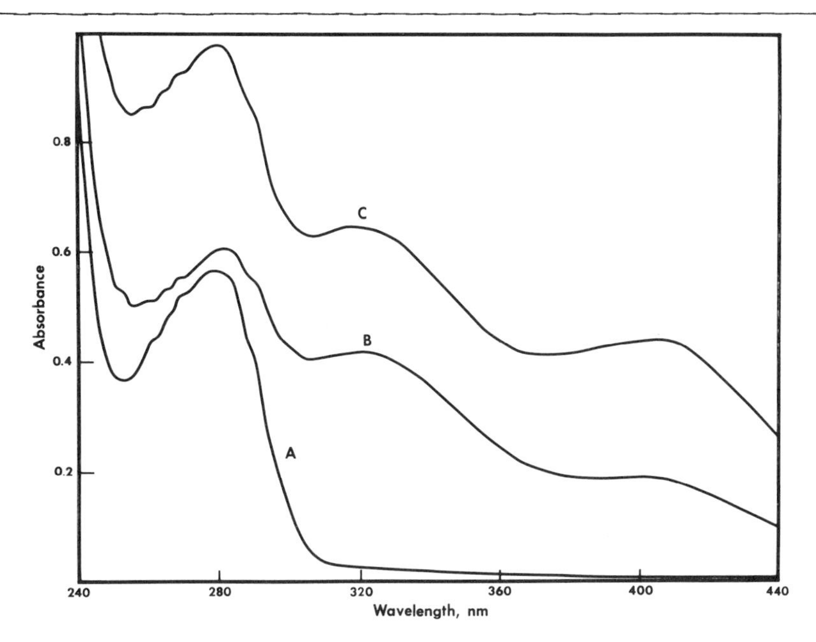

Figure 1. Ultraviolet-Visible Spectra of LADH and HNB-LADH. Samples in 0.1 M phosphate buffer, pH 7.0, normalized to 15 μM A. LADH B. HNB (1.5)-LADH C. HNB(3.0)-LADH

Further investigation revealed no evidence of reaction of HNB with amino acid residues other than tryptophan. Both HNB-LADH and unmodified LADH were equally reactive toward ^{14}C-iodoacetate, as demonstrated by the amount of ^{14}C incorporation and by isolation of S-^{14}C-carboxymethylcysteine after hydrolysis of the alkylated proteins. In addition, modification by HNB occurred to the same extent both with native LADH and with S-^{14}C-carboxymethyl-LADH. Thus, neither modification of tryptophan by HNB nor alkylation of cysteine by iodoacetate precludes reaction at the other amino acid residue.

III. SPECTROSCOPIC STUDIES OF MODIFIED ENZYME

Ultraviolet and visible spectra were compared with spectra of model compounds (12) which are typical of uncyclized, N-cyclizedc and O-cyclized HNB-tryptophan:Uncyclized HNB chromophore absorbs at 410 nm at pH 7.0, with a distinct shoulder beginning at 320 nm. At pH 5.0 the 410 nm peak disappears and the maximum absorption occurs at 320 nm.

At pH 7.0, N-cyclized product has a weak absorption at ∿ 415 nm and a broad shoulder from 280 to 375 nm. At pH 3.0 only the shoulder is present.

The O-cyclized form absorbs weakly at ∿ 320 nm at pH 7.0 and pH 3.0.

The observed spectra of HNB-LADH at pH 7.0 (Figure 1) and pH 3.0 (not shown) suggest that each sample contained a mixture of O-cyclized and uncyclized products, with only a small proportion of N-cyclized HNB-tryptophan. Double labeling of one class of tryptophan is consistent with the presence of cyclized and uncyclized forms.

Fluorescence spectra of HNB-LADH provide further evidence that only one class of tryptophan residues was modified at neutral pH under non-denaturating conditions and identify that class as those residues partially exposed to solvent. First, the fluorescence of LADH was not totally quenched in any samples of HNB-LADH prepared under non-denaturating conditions. Second, the greater quenching of fluorescence on the long wavelength side of the emission peak indicates that the modified tryptophan is probably exposed to a polar environment on the protein surface. Previous spectroscopic studies (9) indicated that there are two classes of tryptophan residues, one of which is exposed to solvent and the other is in a hydrophobic but polarizable environment. The tertiary structure of LADH (10,11) shows that tryptophan-15 is, indeed, on the exterior surface, partially exposed to solvent. In contrast, tryptophan-314 is located in the coenzyme binding domain, a non-polar region at the interface between dimers. These considerations lead to the conclusion that tryptophan-15 was modified by HNB under non-denaturing conditons.

IV. PEPTIDE SEQUENCE STUDIES

Confirmation that tryptophan-15 was modified by HNB is provided by sequence studies. HNB-LADH was digested with trypsin; pooled fractions of peptides eluted from a column of Sephadex G-50 were subjected to high-voltage electrophoresis. One pool contained a peptide which reacted positively with ninhydrin, with Ehrlich reagent, and with ammonia vapor. This peptide was separated by preparative electrophoresis and analyzed by dansyl-Edman sequencing procedures.

The tryptic peptides containing tryptophan which were reported in the primary structure of LADH (13) are:

11-18	ala-ala-val-leu-trp-glu-glu-lys[1]
313-315	thr-trp-lys

The first three amino acids determined in the sequence of the HNB-peptide were ala, val, and leu.[2] No subsequent Edman degradation or dansylation was observed. An aliquot of the peptide was hydrolyzed and the mixture of amino acids reacted with dansyl chloride. Thin-layer chromatography of this mixture and dansyl-amino acid standards revealed the presence of all amino acids in the octapeptide 11-18, as well as dansyl-HNB-tryptophan. Thus, the sequence of the peptide is:

ala-val-leu-(HNB-trp,glu)-lys

V. EFFECTS OF HNB MODIFICATION ON STRUCTURE AND REACTIVITY

Circular dichroism spectra of native LADH, HNB-LADH, LADH in the presence of ethanol, and LADH in urea solution were examined. The differences in the spectra of LADH and HNB-LADH are small compared to changes produced by the addition of ethanol or urea to LADH. Therefore, HNB modification has minimal effects on the over-all secondary structure of LADH.

Samples of LADH and HNB-LADH were examined by analytical gel filtration to determine the effect of modification on the quaternary structure of the enzyme. Co-elution of LADH and HNB-LADH indicated that HNB modification does not interfere with subunit association. The recent x-ray crystallographic data (11) show that the two tryptophan-314 residues are hydrogen-bonded to each other at the dimer interface. Modification of these amino acids under non-denaturing conditions would likely disrupt this connection between subunits. Such a disruption into subunits was observed when glutamate dehydrogenase was modified by HNB (15). The results reported here provide further support for the conclusion that tryptophan-314 is not modified by HNB under the non-denaturating conditions employed. Modification of only tryptophan-15 is again indicated.

[1] In the steroid-active subunit of LADH the sequence in this region is: ala-ala-val-leu-trp-glu-gln-lys (14).

[2] Trypsin preparations treated with tosylphenylalaninecheloromethyl ketone (TPCK) have been found to cleave ala-ala linkages (H. Jörnvall, personal communication). This is a possible explanation for our observation of only one ala residue.

Enzymatic activity decreased linearly with increasing extent of HNB incorporation. However, even samples having the greatest incorporation of HNB under non-denaturing conditons exhibited greater than 60% of the activity of the native enzyme. Such partial decreases in activity are not sufficient for tryptophan to be essential for catalytic activity, as was proposed by Schellenberg (16) for all dehydrogenases.

The lower activity observed cannot be due to modification of cysteine-46 by HNB, since alkylation of that residue by ^{14}C-iodoacetate occurred to the same extent in both HNB-LADH and unmodified LADH, as described above. Decreased activity most likely reflects a perturbation of the local environment when tryptophan-15 is modified. This local perturbation could be distributed through the tertiary structure of LADH and affect other catalytic parameters, resulting in a partial loss of enzymatic activity.

VI. ACKNOWLEDGEMENTS

This work was supported by Grant GU 3181 from the National Science Foundation. We wish to express our appreciation to Dr. David J. Prescott for helpful suggestions and discussions throughout the course of this work.

VII. REFERENCES

(1). Koshland, D.E.,jr., Karkhanis, Y.D., and Latham, H.G., J.Am.Chem.Soc. 86, 1448 (1964).
(2). Horton, H.R., and Koshland, D.E., jr., Ibid. 87, 1126 (1965).
(3). Loudon, G.M., and Koshland, D.E.,jr., J.Biol.Chem. 245, 2247 (1970).
(4). Means, G.E., and Feeny, R.E., "Chemical Modifications of Proteins" Holden-Day, San Francisco, (1971).
(5). Spande, T.F., Wilchek, M., and Witkop, B., J.Am.Chem.Soc. 90, 3256 (1968).
(6). Loudon, G.M., Portsmouth, D., Lukton, A., and Koshland, D.E.,jr., Ibid. 91, 2792 (1969).
(7). Robinson, G.W., J.Biol.Chem. 245, 4832 (1970).
(8). Malin, E.L., Ph.D.Dissertation, Bryn Mawr College, (1975).
(9). Purkey, R.M., and Galley, W.C., Biochemistry 9, 3569 (1970).
(10). Eklund, H., Norström, B., Zeppezauer, E., Söderlund, G., Chlsson, I., Boiwe, T., and Brändén, C.I., FEBS Letters 44, 200 (1974).
(11). Eklund, H., Nordström, B., Zeppezauer, E., Söderlund, G., Ohlsson, I., Boiwe, T., Soderberg, B.-O., Tapia, O., Brändén, C.I., and Akeson, A., J. Mol. Biol. 102,27 (1976).
(12). Naik, V.R., and Horton, H.R., J. Biol. Chem. 248,6709 (1973).

(13). Jörnvall, H., Eur. J. Biochem.16, 25 (1970)
(14). Ibid., p. 41.
(15). Witzemann, V., Koberstein, Sund, H., Rasched, I., Jornvall, H., and Noack, K., Ibid. 43, 319 (1974).
(16). Schellenberg, K.A., J. Biol. Chem..242, 1815 (1967).

STRUCTURAL AND FUNCTIONAL CHANGES IN DIFFERENT ALCOHOL DEHYDROGENASES DURING EVOLUTION

Hans Jörnvall

Karolinska Institutet

Different alcohol dehydrogenases show large variations in subunit size, quaternary structure, genetic organization, multiplicity of forms, substrate specificity and metabolic function. The primary structure of the yeast enzyme was determined and permits more detailed comparisons. The yeast and horse enzymes have an overall positional identity of only 25% but regions of maximal similarity (40% identity) can be correlated with functionally important structures and the most variable regions (random similarities) with differences in quaternary structure and zinc binding. These aspects establish a connection between the enzymes but show that extensive changes may obscure possibly still more distant relationships within dehydrogenases. The present variabilities support previous evolutionary conclusions but also suggest that additional relationships may exist and the possibility of a repetitive and variable ancestral building block of widespread occurrence cannot be excluded.

I. INTRODUCTION

Alcohol dehydrogenase was the first pyridine nucleotide-dependent dehydrogenase to be purified in crystalline form, first from yeast and then from horse liver (*cf.* 1). It has later been obtained from several other species, and is now known to be of widespread occurrence in nature (2). Early studies on catalytic mechanisms and molecular properties revealed many similarities between the enzymes from different species but also established important differences (1). These observations raised two general questions: First, regarding the structural explanations to all similarities and differences; secondly, regarding the evolutionary relationships between different alcohol dehydrogenases and, possibly, the connections between different types of dehydrogenases (*cf.* 3-6), especially since a few other well-studied dehydrogenases apparently exhibited less species differences. Further insight has now been obtained into both these questions by different types of progress in structural studies. Thus, after the determination of the primary and tertiary structures of horse liver alcohol dehydrogenase (2), and the same advances for other dehydrogenases (7,8), further correlations within alcohol dehydrogenases (9-11) and between different dehydrogenases (12) were possible. Recently, the determination of the amino acid sequence of yeast alcohol dehydrogenase (13) and the structural studies of Drosophila alcohol dehydrogenase (14), made it possible to judge additional structural, functional and evolutionary questions (15). In the present study, these aspects will be considered and available knowledge summarized.

II. MATERIALS AND METHODS

The determination of the primary structure of yeast alcohol dehydrogenase was largely based on sequence analysis of small fragments by the dansyl-Edman method. Peptides were prepared from eight different types of mixtures derived from the ^{14}C-carboxymethylated protein. Proteolytic treatments used were digestions with trypsin, chymotrypsin, pepsin, thermolysin, a staphylococcal protease or a myxobacterial protease, and cleavages with CNBr. Peptide purifications were centered on Sephadex chromatography, paper electrophoresis at different pH values and paper chromatography, with some supplementation of DEAE-Biogel chromatography. Dansyl amino acids were identified by thin-layer chromatography on polyamide sheets. Full descriptions of the methodologies are given in Ref. 13. In another study on Drosophila alcohol dehydrogenase, about half of all amino acids in the enzyme were determined in tryptic peptides, and one mutational difference between two enzyme forms was characterized (14). For the sequence comparisons, the original references to all structures involved and methods employed are described elsewhere (15).

III. RESULTS AND DISCUSSION

A. Comparisons of Molecular Properties of Different Alcohol Dehydrogenases

An extensive variation is noticed when superficial properties of alcohol dehydrogenases are compared. For example, the characterized enzymes differ in size from about 24,000 to about 40,000 for the subunits, and from dimers to tetramers for the whole molecules. Substrate specificities, catalytic efficiencies and metabolic functions of the enzymes are also different in different species (2). Furthermore, zinc binding is apparently variable (*cf.* 15) and the presence of many thiol groups and at least one reactive cysteine residue, which are characteristics of the mammalian and yeast enzymes (3), are not general properties, as revealed by the studied bacterial (16) and insect (14) enzymes. Some of these differences are summarized in Table 1. Another type of variability concerns

TABLE 1
Variations and Multiplicities of Alcohol Dehydrogenases

Enzyme source	*Subunit size*	*Quaternary structure*	*Different genes*	*Allelic variations*	*Chemically derived subgroups*	*Substrate specificity*	*Metabolic functions*
Mammals	40,000	Dimer				Broad	Bile acid formation; fatty acid ω-oxidation; removal of intestinally produced alcohols.
Horse			2	+[a]	+		
Man			3	+	+		
Rat			1	-	+		
Yeast (Saccharomyces cerevisiae)	35,000	Tetramer	3	+[b]	(+)[b]	Strict	Fermentation (one form); gluconeogenesis (another form).
Insect (Drosophila melanogaster)	24,000	Dimer	1	+	+		Nutrition
Bacterium (Bacillus stearothermophilus)	35,000	Tetramer					

Statements are based on presently available estimates and additional variations or multiplicities are not excluded.

Original references are given elsewhere (2), except where indicated by footnotes. [a]from (17). [b]from (13).

the multiplicity of forms of alcohol dehydrogenases in a single species. Isozymes frequently occur and may have different organ distributions (*cf.* 18). Different gene loci, allelic variability and secondarily modified forms are also known (Table 1).

These observations have several consequences. First, the divergence of forms suggests that the metabolic function(s) of alcohol dehydrogenases may vary between species and between organs. This is also known is some cases (Table 1) and has been specially considered in the case of steroid binding (19). The multiplicities may perhaps also contribute an explanation to the peculiar fact that mammalian alcohol dehydrogenases are abundant (and well-studied) enzymes but still without a common, generally recognized main physiological role.

Secondly, the variations suggest that the genetic organizations and the gene architecture of alcohol dehydrogenases have been subject to changes. Not only are there multiple gene loci, indicating possible ancestral duplications, but also, the number of loci is apparently different even between closely related species (Table 1), indicating frequent and separate duplications. Furthermore, the differences in subunit size indicate quite extensive genetic alterations, and the variable organ distributions of isozymes shows the presence of different control mechanisms for related genes.

Thirdly, the variation may be interpreted in structural and functional relationships (10). In spite of the differences, a common enzymatic function is conserved as is apparently also the general folding of the protein (11). The divergencies and similarities may be correlated with other properties (13) to distinguish different molecular influences. Finally, from evolutionary aspects, the large variations in alcohol dehydrogenases to some extent resemble variations between different proteins rather than just within one protein type. Alcohol dehydrogenases may therefore conceivably further illustrate some suggested inter-dehydrogenase relationships. With the knowledge of the primary structure of yeast alcohol dehydrogenase, all comparisons and conclusions may be extended.

B. The Primary Structure of Yeast Alcohol Dehydrogenase

The amino acid sequence of the subunit is schematically given in Table 2, and was determined by analysis of small peptides from the ^{14}C-carboxymethylated protein. Complete data for all peptides necessary to deduce the structure are

TABLE 2
Schematic Sequence Comparisons between Yeast and Mammalian Alcohol Dehydrogenases

```
    3
Acyl------Lys-----Glu-----------Val--Pro Lys Ala-Glu--Ile-----
    1       10           20                    30           40

                    0   0
-Gly-Cys--Asp-His---Gly-----Thr-Leu Pro---Gly His Glu-Ala Gly-
                  50             60                          70

                                    0
Val---Gly Glu-Val-----Gly Asp-----------Cys--Cys--Cys---------
             80                90              100          110

   1           21
Cys---Asp Leu Ser--------Phe-Gln Tyr---Asp----Ala-Ile-----Leu--
                 120            130               140

                 1                                     0
Val----Cys---Thr-Tyr - Ala-Lys-Ala----Gly--/-Ala--Gly--Gly Gly
  150             160                  170             180

                          1
-Gly-------Lys Ala-Gly-Arg--Gly-Asp----Lys-------Gly----------
        190                 200            210           220

              1                                  1
Lys-----Val Leu---Asn Gly Gly------Val--------------Ala-Gly--
   230                         240        250    260

                                     4
Val-Val-Val Gly - Pro------Ser--------/---Ile-Gly------------
               270            280        (290)         (300)

          2                                     0       0  0
Asp Phe-Ala--Leu------Val-------Glu-------Gly----Arg--------.
         (310)        (320)      (330)        (340)  (347)
```

The continuous signs give all positions in the yeast enzyme and the vertical bars indicate tentative assignments. Abbreviated amino acids show the residues at all positions where the horse E-type enzyme is identical - dashes show positions where residues are different. Superpositioned figures denote extra and zeros missing residues in mammalian proteins.

reported elsewhere (13). Experimental support for the structure of most regions is abundant but microheterogeneities, desamidations, insoluble peptides, repetitive sequence elements and other factors complicated the determination at other positions (13). Therefore, some assignments rely on just minimal data and the direct connection between positions 286 and (287) is still tentative (*cf.* Table 2). In spite of the difficulties, however, the known peptides appear consistent, and the reliability of the structure is considered to be accurate enough for the present, comparative purposes.

Some general features of the protein chain are noticeable. The N-terminus is acylated. The active-site cysteine residues are somewhat shifted in comparison with their positions in the corresponding mammalian enzymes (positions 43 and 153 versus 46 and 174, respectively). Unexpected distributions are found for tryptophan, histidine, proline, systeine, phenylalanine and arginine which occur solely or preferentially in one-half of the protein chain (the former four in the N-terminal half, the latter two in the C-terminal half). Branched-chain residues, especially valine, are often associated. Repetitive elements occur in the structure, both tripeptide segments of exact repetitions (13) and long regions of incomplete sequence coincidences (Fig. 1, below).

Ordinary enzyme preparations apparently contain more than one type of subunit. This finding may be correlated with the large population from which a single enzyme preparation originates, and has also been encountered in the case of another yeast dehydrogenase (20). One microheterogeneity in yeast alcohol dehydrogenase is strongly indicated to be a threonine → isoleucine exchange at position 235 (13). Regarding desamidations, most of those determined (13) are derived from treatments of peptides but some labile asparagine structures may probably also be sensitive in the native protein (13).

C. Structural and Functional Aspects

The amino acid sequences of yeast and mammalian alcohol dehydrogenases are compared in Table 2. Due to the low degree of similarity in some places, gaps and larger shifts are not always uniquely defined, but the alignment shown in Table 2 is considered to be the best (15) and is compatible with considerations of the tertiary structure (21). The overall positional identity between the yeast and horse liver proteins is only 25%, and along the protein chains there is a variation in the similarity from about 40% identity to random values below 10% identity (15). Each of these extremes may be correlated with different aspects of the structure.

Regions of comparatively large similarity are centered around the ligands to the catalytic zinc atom in the horse protein (22), corresponding to positions 43, 66 and 153 in Table 2. The finding of a major structural conservance in these segments suggests that the enzymes have similar active sites and catalytic mechanisms. In spite of the sequence differences, a generally unaltered folding is suggested by more detailed comparisons of all amino acid substitutions in relation to the known tertiary structure of the horse protein. Many compensating exchanges and similarities/differences that may be interpreted in relation to structural and functional properties are then revealed (11,21).

One region of dissimilarity is a segment around position 100, where the identity is low and the horse protein is internally longer (Table 2). This segment corresponds in the tertiary structure of the horse protein (22) to two different loops, one around the second zinc atom and contributing to some subunit interactions, the other at the surface of the

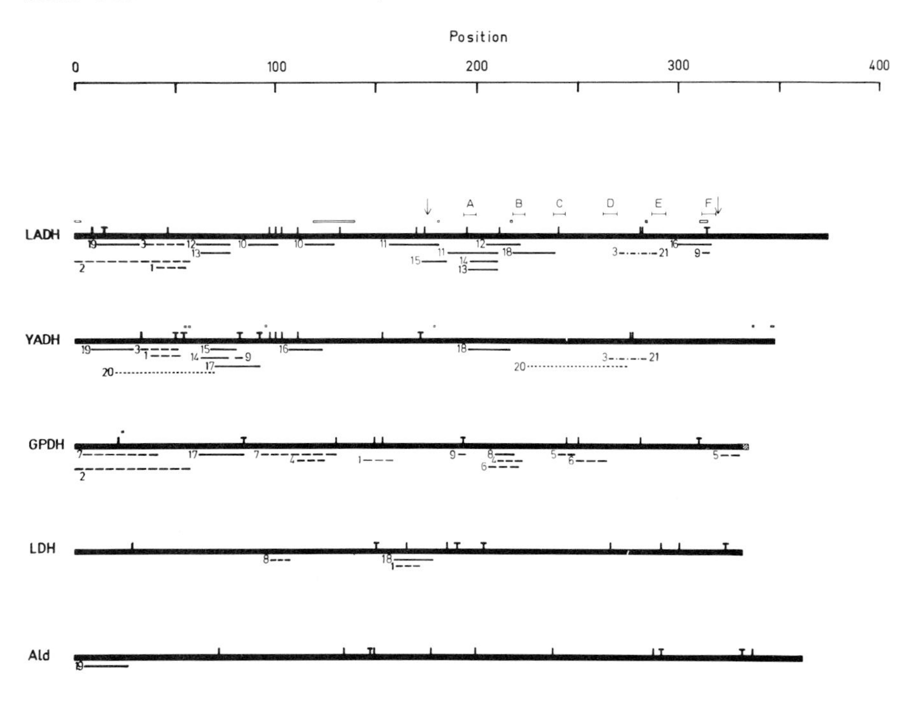

Fig. 1. Representation of Sequence Coincidences in Some Enzymes. LADH, YADH, GPDH, LDH *and* Ald *denote, respectively, horse liver alcohol dehydrogenase, yeast alcohol dehydrogenase, glyceraldehyde 3-phosphate dehydrogenase, lactate dehydrogenase and aldolase. Arrows indicate domain borders and lines A-F pleated sheet structures in* LADH *(22). Non-filled boxes above one alcohol dehydrogenase show a region absent in the other alcohol dehydrogenase. All positions of cysteine residues are indicated by and all those of tryptophan residues by . Sequence coincidences 1-8 (----), are from Refs. (3,6,23,24), 9-19 (——) from Ref. (15) and 20-21 (·····) are unpublished.*

catalytic domain (15), but neither is associated with specific, known functions (22). It is conceivable, therefore, that this dissimilarity may be accommodated in the enzyme structures without general changes but with alterations in zinc binding, surface properties and subunit interactions. Another region of dissimilarity consists of large segments in the C-terminal halves of the molecules, especially one between positions 270 and 300 (Table 2). This region corresponds to the structures contributing the main subunit contacts in the horse protein (22) and therefore supports the conclusion of alterations in subunit interactions. Combined, the dissimilar regions may probably explain the different quaternary structures of the enzymes (Table 1).

D. Evolutionary Aspects

The overall similarities between yeast and mammalian alcohol dehydrogenases establish an ancestral connection between the enzymes while the differences show that evolutionary changes have been extensive. Both conclusions are compatible with the superficial differences between the enzymes (Table 1) and with previous sequence comparisons when only the more closely related rat and horse proteins were known (10). Thus, intra-mammalian comparisons of alcohol dehydrogenases also show several changes and the two most unequal regions between the rat and horse proteins were found around residues 115 and 280 (10), corresponding closely to the dissimilar regions between the yeast and mammalian enzymes. The fact that parallel variabilities are noticed when both distantly and closely related alcohol dehydrogenases are compared suggests that evolutionary changes have not been limited to particular time periods or to different regions at different periods. Therefore, the changes behind the differences in quaternary structures, surface properties and subunit sizes need not be isolated alterations but may correspond to frequent or stepwise occurrences. This conclusion is supported by the variability shown in Table 1. In particular, successive changes appear most probably in the case of the binding of the non-catalytic zinc atom (*cf.* 15).

The alcohol dehydrogenase comparisons furthermore reveal that changes in the corresponding genes have not only involved base exchanges but also frequent deletions/insertions of single codons or longer regions. In total, 41 residues in the yeast or horse proteins are without equivalence in the other sequence (Table 2). The occurrence of gaps is also compatible with intra-mammalian variations. Thus, in comparison with the horse enzyme, an insertion of a cysteine residue (after position 111) has been noticed in the rat enzyme (10) and extra structures have been suggested in the human enzyme (23). In summary, the alcohol dehydrogenase comparisons show that evolutionary changes have been extensive, that the identities of some related regions may be hardly distinguishable from random similarities, and that internal deletions/insertions may shift corresponding structures.

E. Possible Dehydrogenase Relationships

The evolutionary conclusions may also be considered in relation to possible inter-dehydrogenase connections. Sequence similarities among some dehydrogenases were indicated early (3,6,24,25) but were later found to be apparently inconsistent with tertiary structures (12). An evolutionary connection based on a common ancestral mononucleotide binding unit for the coenzyme-binding domains of dehydrogenases and kinases has been suggested from the foldings, but does not apply to other regions of the subunits (12).

The variability in alcohol cehydrogenases demonstrates that single types of dehydrogenases need not be as fixed structures as might be judged from species variations in another well-studied dehydrogenase, glyceraldehyde 3-phosphate dehydrogenase, which is both metabolically and structurally more constant (8). It is conceivable that alcohol and glyceraldehyde 3-phosphate dehydrogenases illustrate different degrees of divergence and that the species variability in the former more closely resembles the type of possible changes between different dehydrogenases.

Another point of interest from the present variations regards the previously suggested mononucleotide binding unit (12). From the comparisons in Table 2 and Ref. (15) it is clear that all parts of alcohol dehydrogenases contain variations, and there are no immediate details that evolutionary distinguish the mononucleotide binding regions from the rest of the molecules. It is possible that common functional requirements in coenzyme binding may contribute an explanation to the increased similarities in tertiary structures, but this need not apply to other possible ancestral connections.

These considerations, as well as the presence of short repetitive elements complicating the structural work (13) on alcohol dehydrogenases, provide the basis for further comparisons of known amino acid sequences. The comparative results have been shown in detail (15) and reveal a number of sequence coincidences between functionally and positionally non-equivalent regions in several enzymes. This is schematically shown in Fig. 1. The representations in the figure do not imply any necessary significance but give a summary of all presently or previously noticed coincidences. Statistical analyses of sequence distributions are complicated but some estimates have been given (15). None of the coincidences is statistically fully established, and some may even mutually exclude each other. Nevertheless, the species variations in alcohol dehydrogenases as revealed in Tables 1 and 2, show that similarities between even definitely related structures need not be extensive or continuous since deletions/insertions and point mutations occur. Therefore, the possibility cannot be excluded that some of the coincidences in Fig. 1 represent repetitions and partial variations of a common old ancestral unit relating the entire protein chains of many enzymes. The unit need not be limited to mononucleotide binding and could have a size corresponding to about one fourth of the monomers. This possibility is compatible with all known facts and it would extend accepted connections. In particular, the spread of subunit sizes in both alcohol dehydrogenases and dehydrogenases in general is compatible with different unit-multiplicities (15).

IV. GENERAL CONCLUSIONS

The characterized alcohol dehydrogenases show different properties in many respects. The functions range from several suggested roles for the mammalian enzymes with broad substrate specificities to established roles for the more specific yeast enzymes. A corresponding divergence in genes and in control mechanisms is also apparent. The known primary structures show sequence variations larger than those for another well-studied dehydrogenase and may illustrate the type of changes that have occurred in dehydrogenase evolution. From these conclusions, accepted evolutionary connections are supported, but additional relationships may exist. The presence of a widespread, repetitive and variable old ancestral unit relating many oligomeric enzymes cannot be excluded. The alcohol dehydrogenase structural variations are of great interest and indicate possibilities for distant relationships, but the knowledge of additional structures is necessary for further distinctions.

V. ACKNOWLEDGEMENT

This work was supported by a grant from the Swedish Medical Research Council (project 13X-3532).

VI. REFERENCES

(1). Sund, H., and Theorell, H., "The Enzymes," 2nd ed., 7, 25 (1963).

(2). Brändén, C.-I., Jörnvall, H., Eklund, H., and Furugren, B., "The Enzymes," 3rd ed., 11, 103 (1975).

(3). Harris, I., Nature 203, 30 (1964).

(4). Fondy, T.P., and Holohan, P.D., J. Theor. Biol. 31, 229 (1971).

(5). Holbrook, J.J., Pfeiderer, G., Mella, K., Voltz, M., Leskowac, W., and Jeckel, R., Eur. J. Biochem. 1, 476 (1967).

(6). Jörnvall, H., Eur. J. Biochem. 14, 521 (1970).

(7). Holbrook, J.J., Liljas, A., Steindel, S.J., and Rossmann, M.G., "The Enzymes," 3rd ed., 11, 191 (1975).

(8). Harris, J.I., and Waters, M., "The Enzymes," 3rd ed., 13, 1 (1976).

(9). Jörnvall, H., Proc. Nat. Acad. Sci. USA 70, 2295 (1973).

(10). Jörnvall, H., in "Alcohol and Aldehyde Metabolizing Systems," p. 23. Academic Press, New York, 1974.

(11). Eklund, H., Brändén, C.-I., and Jörnvall, H., J. Mol. Biol. 102, 61 (1976).

(12). Rossmann, M.G., Liljas, A., Brändén, C.-I., and Banaszak, L.J., "The Enzymes," 3rd ed., 11, 61 (1975).

(13). Jörnvall, H., Eur. J. Biochem. (1976), in press.

(14). Schwartz, M.F., and Jörnvall, H., Eur. J. Biochem. 68, 159 (1976).

(15). Jörnvall, H., Eur. J. Biochem. (1976), in press.

(16). Bridgen, J., Kolb, E., and Harris, J.I., FEBS Letters 33, 1 (1973).

(17). Pietruszko, R., Biochem. Biophys. Res. Commun. 60, 687 (1974).

(18). Smith, M., Hopkinson, D.A., and Harris, H., Ann Hum. Genet. 34, 251 (1971).

(19). Jörnvall, H., Biochem. Soc. Transact. (1977), in press.

(20). Jones, G.M.T., and Harris, J.I., FEBS Letters 22, 185 (1972).

(21). Jornvall, H., Eklund, H., and Bränden, C.-I., to be published.

(22). Eklund, H., Nordström, B., Zeppezauer, E., Söderlund, G., Ohlsson, I., Boiwe, T., Söderberg, B.-O., Tapia, O., Brändén, C.-I., and Åkeson, Å., J. Mol. Biol. 102, 27 (1976).

(23). Berger, D., Berger, M., and von Wartburg, J.P., Eur. J. Biochem. 50, 215 (1974).

(24). Engel, P.C., FEBS Letters 33, 151 (1973).

(25). Bennett, C.D., Nature 248, 67 (1974).

METAL ION EFFECTORS OF HORSE
LIVER ALDEHYDE DEHYDROGENASES

R. Venteicher, L. Mope and T. Yonetani

University of Pennsylvania

Salts of Ma^{2+}, Ca^{2+}, Y^{3+}, Mn^{2+}, Fe^{3+}, Co^{2+}, Ni^{2+}, Cu^{2+}, Zn^{2+}, La^{3+}, Ce^{3+}, Pr^{3+}, Nd^{3+}, Sm^{3+}, Eu^{3+}, Gd^{3+}, Tb^{3+}, Dy^{3+}, Ho^{3+}, Er^{3+}, Tm^{3+}, Yb^{3+} and Lu^{3+} were investigated as effectors of the horse liver aldehyde dehydrogenase isoenzymes, F1 and F2. While Ca^{2+} and Mg^{2+} were relatively insensitive effectors (requiring > 200 μM concentrations), most remaining metal ions produced dramatic activation (0.1-10μM) and inhibition (5-40μM) with the F1 isoenzyme and large activation but weak inhibition with the F2 isoenzyme. The representative pattern of activation and inhibition of F1 with Zn^{2+} possessed a maximal 200% activation (0.5μM) followed by 90% inhibition (2.0 μM Zn^{2+}) with 1 mM acetaldehyde and 0.1 mM NAD^+. The stoichiometry of NADH binding established by fluorescence titration with F1 in the presence of inhibitory levels of Zn^{2+} was shown to be two moles of NADH per mole of enzyme. Tigher NADH binding to the enzyme was noted. The tighter binding of NADH could be the mechanism of inhibition where the dehydrogenase activity is moderated by the rate limiting NADH dissociation step. Because NAD^+ is not required, the effect of metal ions upon the esterase activity discriminates as to whether the metal ions are probing the substrate or coenzyme binding sites. Zinc was the only metal ion tested which showed equivalent esterase and dehydrogenase inhibition. This indicates that the dehydrogenase activity is controlled through the binding of metal ions to the cofactor, NAD^+, and their corresponding influence upon slower release of NADH.

I. INTRODUCTION

The purified isozymes of horse liver aldehyde dehydrogenase, F1 and F2 are catalytically active enzymes whose analysis did not indicate the presence of incorporated metal ions (1,2). However, the ability to artifically incorporate metal ions into these enzymes has in some cases produced activation and in most cases produced inhibition behaviour towards acetaldehyde as well as the ability to initiate paramagnetic structural probe studies for spectroscopies such as electron paramagnetic resonance (epr), paramagnetic nuclear magnetic resonance (nmr), water proton relaxation rate (PRR) and fluorescence enhancement or quenching studies.

Previous studies of the effect of metal ion chelating agents such as EDTA and 1,10-phenanthroline (OP) towards the related aldehyde dehydrogenase enzymes, partially purified bovine liver (3), human liver (4), and yeast aldehyde dehydrogenase (5), have demonstrated the ability of these reagents to inhibit the enzymes. With human liver aldehyde dehydrogenase, OP was shown to be a competitive inhibitor of NAD^+ and not to be binding to a catalytic metal ion as was implicated in the previous studies with the bovine liver and yeast aldehyde dehydrogenase enzymes. Alcohol dehydrogenases from horse liver, rat liver, and yeast are known to contain essential zinc ions which are sensitive to such chelating agents as 1,10-phenanthroline and 2,2'-dipyridyl by tight complex formation with the active site zinc ions. Our investigations were undertaken to ascertain whether the purified horse liver aldehyde dehydrogenase isozymes, F1 and F2 were metallo-enzymes and, additionally, upon artifical incorporation of metal ions into the enzyme systems, by which mechanism does the observed activation and inhibition occur.

Therefore, we have studied the inhibition properties of OP and the related compound, 2,9-Dimethyl-1, 10-phenanthroline, whose complexing ability to zinc is reduced 1000-fold due to the steric hinderance of the methyl substituents. Our investigations have covered a wide variety of metal ions whose selections were based primarily with regards to metal ions of possible physiological importance, such as Zn^{2+}, Cu^{2+}, Ca^{2+}, Mg^{2+}, Fe^{2+} and Fe^{3+}; those metal ions possessing potent epr and pRR probe abilities, Gd^{3+} and Mn^{2+}; and those metal ions previously demonstrating powerful nmr structural probe abilities as nmr shift reagents, i.e. the lanthanide ions (6). While the conccentration requirements of the metal ions varied for activation and inhibition behaviour, similar trends were noticed for both the F1 and F2 isoenzymes. Because of limited solubility of metal ions in phosphate and pyrophosphate buffers, the dehydrogenase activity studies were performed predominantly in 0.04M Tris-Cl or 0.04M Imidazole-Cl (pH 7.0). At higher pH's, the lanthanide ions formed hydroxides and were non-interactive towards the enzyme systems.

A. Experimental Procedures

Fresh horse livers were obtained from a local slaughterhouse within 30 minutes of removal and were immediately perfused with cold 0.15M sodium chloride. The purification procedures established by Eckfeldt, *et al.* (1), were followed to purify the two isozymes, F1 and F2. The enzyme solutions were stored at pH 5.5 and 5.8, respectively in 10.0mM sodium phosphate buffer, 0.25% mercaptoethanol, and 1.0mM EDTA. However, previous to use with metal ions, the enzyme solutions were extensively dialyzed against nitrogen-saturated 0.04M Tris-Cl (pH 7.0) buffer to free the enzyme of mercaptoethanol, EDTA, and phosphate. Enzyme solutions were used within two days after dialyzing into Tris-Cl buffer. The metal chlorides were prepared as 2.0mM stock solutions in de-ionized water. The metal ion concentrations were standardized by a semi-micro complexiometric titration with Eriochrome Black T (Fisher Chemical Co., Pittsburgh, Pa.) as indicator of where the end-point was observed by monitoring the absorbance at 665 nm. 2,9-Dimethyl-1, 10-phenanthroline was obtained from the G. Frederick Chemical Co. (Columbus, Ohio), and 1,10-phenanthroline was obtained from Eastman Organic Chemicals (Rochester, N.Y.). β-NAD^+ was obtained from Sigma Chemical Co. (St. Louis, Mo.) and acetaldehyde was purchased from the J.T. Baker Chemical Co., (Phillipsburg, N.J.

Dehydrogenase activity was determined spectrophotometrically at 25° by measuring the formation of NADH at 340 nm in 0.04M Tris-Cl or Imidazole-Cl (pH 7.0) with 1.0mM acetaldehyde and 0.1mM NAD^+. For the inhibition titrations, the appropriate volume of 2.0mM metal ion solution was pipetted into the assay mixture followed by an approximate 30 sec incubation period before the initiation of the NADH production upon addition of the substrate. Esterase activity was measured by the rate of formation of p-nitrophenol from p-nitrophenylacetate by monitoring the absorbance at 400 nm. The buffer was $\mu = 0.1$ sodium phosphate (pH 7.0) containing 1% acetone at 25°.

II. RESULTS AND DISCUSSION

A. The Influence of 1,10-Phenanthroline and 2,9-Dimethyl-1,10-Phenanthroline on F2.

Under the experimental conditions in which the purified enzymes, F1 and F2 are obtained, they do not appear to possess a metallo catalytic site. The enzymes are purified and stored at pH 5.5 and 5.8, respectively, with a metal ion complexing agent, EDTA. These are conditions known to labilize the metal-to-enzyme binding for both the catalytic and non-catalytic zinc sites of horse liver alcohol dehydrogenase. It is possible that in case the enzymes originally contained metal ions *in vivo*, that during the experimental purification procedures, the removal of the metal ions could have occurred. Our studies of the purified enzymes did not indicate the presence of metal ions as determined by metal ion analysis nor was an observable epr spectrum obtained for concentrated samples of approximately 50 μM. Additional evidence was obtained by the inhibition studies with 1,10-phenanthroline and 2,9-Dimethyl-1,10- phenanthroline.

As demonstrated with human liver aldehyde dehydrogenase by Sidhu and Blair (4), metal ion chelating agents can inhibit an enzyme other than by direct metal ion coordination, in this case, by being a competitive inhibitor of NAD^+ suggesting that the NAD^+ binding site has an affinity for aromatic ring structures based on hydrophobic interactions. Eckfeldt and Yonetani (1,2,7) have previously demonstrated many physical and kinetic similarities between F2 and the one human isozyme that has been studied (8,9). For F2, double reciprocal plots gave inhibition constants (K_i's) for 1,10-phenanthroline and 2,9-Dimethyl-1, 10-phenanthroline of 0.23 mM and 0.20 mM, respectively, as compared to the values of 0.13 and 0.14 mM found for human liver aldehyde dehydrogenase (4). The similar megnitudes of the inhibition constants suggest that the inhibitory effect of these two reagents occurs through the hydrophobic interaction at the coenzyme binding site and not through a metal binding site. If a metal ion site was effected by these chelating agents, a greater interaction would be expected for 1,10-phenanthroline versus 2,9-Dimethyl-1,10-phenanthroline because of the 1000-fold greater stability of 1,10-phenanthroline towards zinc upon complex formation (10).

B. Influence of Metal Chlorides on the Acetaldehyde Dehydrogenase Activity of F1 and F2 Isozymes

The incremental additions of metal chlorides to the bioassay mixtures produced dramatic activation and inhibition behavior for both F1 and F2. The data presented in Figure 1 demonstrate typical results obtained in Tris-Cl buffer for a representative sample of transition metal ions. The lanthanide ions produced similar effects and showed less variation in behavior from metal-to-metal than did the first row transition metal ions. The behavior for the GD^{3+} ion (Figure 2) demonstrates the behavior characteristic of the entire lanthanide series. This fact probably reflects the very similar chemical and coordination properties shared by the lanthanide ions. Activation of the enzymes was also more prevalent for the lanthanide ions and is probably related to the greater affinity for higher coordination numbers and the extra charge normally portrayed by lanthanide ions relative to the first row transition metal ions. For example, in thermolysin, when lanthanide ions were replaced into Ca^{2+} sites, a perturbation of the protein's structure was observed and found to be due to the occupancy of two adjacent calcium sites by a single lanthanide ion, quite clearly motivated

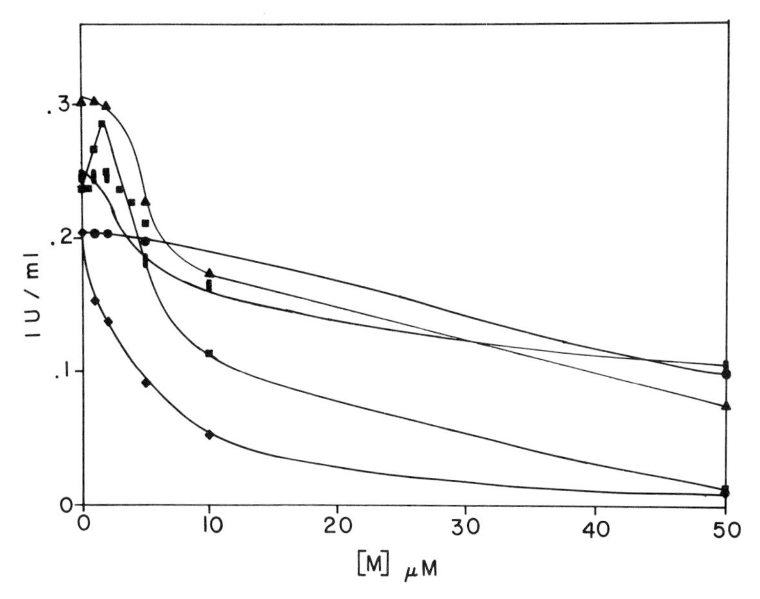

Fig. 1. The influence of metal chlorides upon the aldehyde dehydrogenase activity of the F1 isozyme of horse liver aldehyde dehydrogenase towards acetaldehyde. The activity was measured in 0.04M Tris-Cl (pH 7.0) in 1.0 mM acetaldehyde and 0.1 mM NAD^+ with varying concentrations of Zn^{2+} (■), Mn^{2+} (◆), Ca^{2+} (●), Ni^{2+} (●), and Co^{2+} (▲) chlorides. $MgCl_2$ produced an equivalent effect as that of $CaCl_2$.

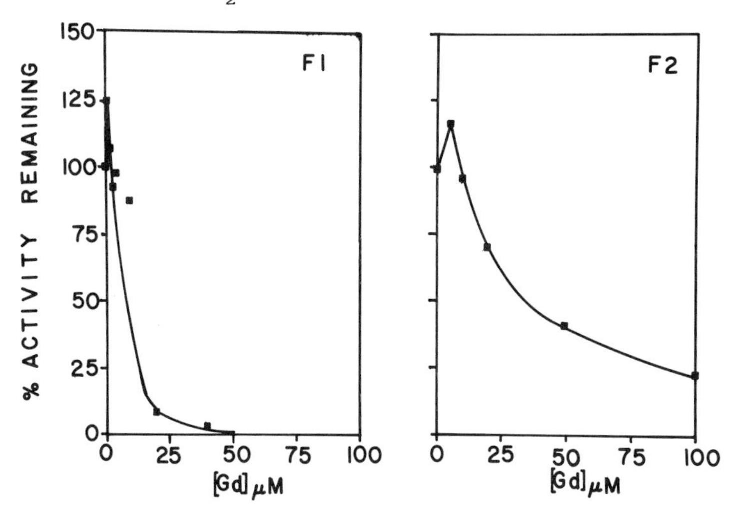

Fig. 2. Activation and inhibition of F1 and F2 isozymes by Gd^{3+} in 0.04M Tris-Cl buffer (pH 7.0). The enzyme solutions were previously extensively dialyzed to remove the protective agents, EDTA and mercaptoethanol.

by the larger ionic size and higher coordination affinity of the lanthanide ion (11). Reasonably, such distortions in coordination polyhedra about an active site or coenzyme binding site could account for the enhancement in activity.

Clearly, F1 shows greater sensitivity towards metal ion than does F2. The K_m's of NAD^+ towards F1. and F2 are 3 μM and 30 μM, respectively (1) and the 10-fold greater requirement of F2 for binding of NAD^+ parallels the nearly 4-fold greater requirement of lanthanide ion required to produce 50% inhibition of F2. This correlation indicates that the metal ions probably effect the enzymes' activity through binding with the coenzyme. In general, the relative activity of both purified enzymes was observed as the decreasing order according to buffer system; pyrophosphate > phosphate > Imidazole-Cl > Tris-Cl. This dependence on buffer system must be considered when comparing the relative activity enhancement factors. For example, the activation of F1. by Zn^{2+} was not always reproducible in Tris-Cl, but was usually observed when performed in Imidazole-Cl buffer. A maximal increase of 25% in activity in Tris-Cl has been observed as a 200% increase in Imidazole-Cl. The precise reason for this behavior cannot yet be explained but probably is related to imidazole's ability to coordinate metal ions compared to the relatively weak coordination affinity of Tris.

C. The Influence of Metal Chlorides on the Esterase Activity of F1 Towards p-Nitrophenylacetate

Coenzyme is not required for esterase activity and a metal ion's ability to effect this function, thereby, serves as one criterion as to the mode of metal ion-enzyme interaction. Removal of activity by a metal ion indicates either binding at the active site, thereby stopping substrate interaction; an exterior interaction of metal ion and substrate preventing the enzyme's interaction with substrate, or possibly an interaction of metal ion and the enzyme at an alternate site capable of sterically blocking the substrate from passing into an active site passageway should, indeed, such a passageway exist. A sulfydrul active site was indicated by p-chloromercuribenzoate and phenylmethylsulfonyl fluoride inhibition studies of the enzymes (1). The existence of zinc metalloenzymes where the catalytic metal atom is coordinated to sulfydryl or nitrogenous ligands has been previously well established. Therefore, activation or inhibition through metal ion binding with a sulfydryl active site could be expected. Zinc was the only metal ion tested that showed equivalent esterase and acetaldehyde inhibition whereas the remaining metal ions effected substantially only aldehyde activity (Fig. 3). The failure of the remaining metal ions towards effecting esterase activity indicates that

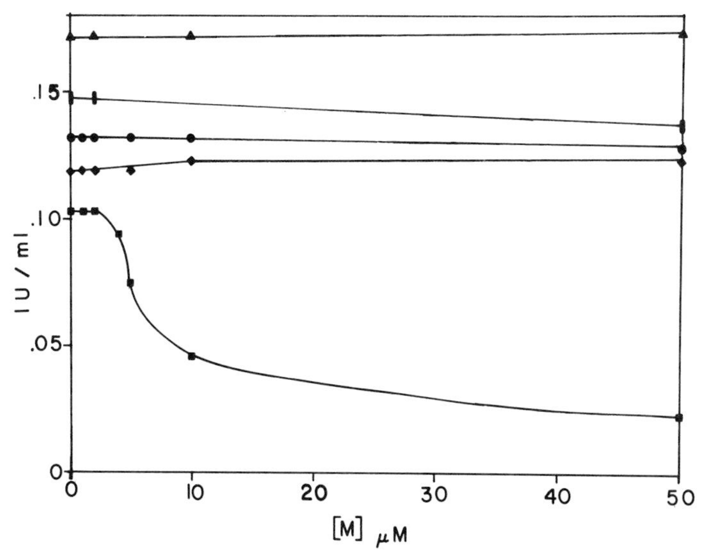

Fig. 3. Esterase activity of the F1 isozyme of horse liver aldehyde dehydrogenase toward p-nitrophenylacetate in the presence of varying concentrations of metal chlorides. The activity was measured in 0.04M Tris-Cl (pH 7.0) in 0.22mM p-nitrophenylacetate and 1.0% acetone at 25°. The initial tangent of the absorbance at 400 nm was taken upon addition of the enzyme. The Chlorides of Zn^{2+} (■), Mn^{2+} (◆), Ca^{2+} (●), Ni^{2+} (▮), and Co^{2+} (▲) were used as inhibitors.

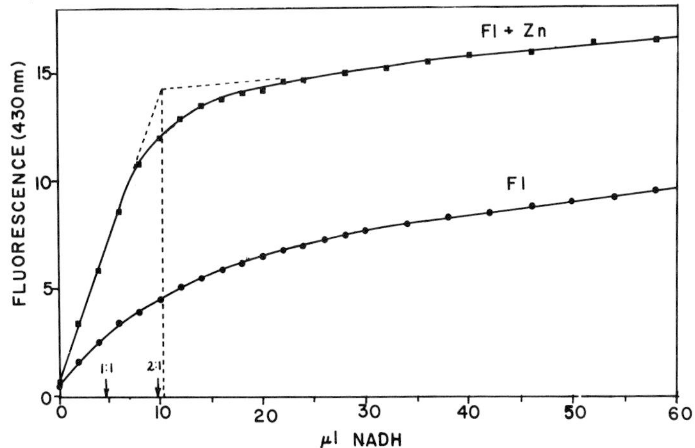

Fig. 4. Fluorescence enhancement titration curve for the F1 isozyme of horse liver aldehyde dehydrogenase upon addition of 100 μM $ZnCl_2$ with NADH. The experiment was performed at 25° in 0.04M Tris-Cl (pH 7.0). The enzyme concentration was 0.8 μM as determined by an extinction coefficient of 0.95 mg/ml cm at 280 nm. The concentration of the titrant, NADH, was 0.2mM and the experiment was performed on a Perkin-Elmer fluorescence spectrometer (Model MPF-2).

their acetaldehyde inhibition mechanism probably proceeds through a site other than the enzyme's substrate binding site.

D. Fluorescence Enhancement Titration of F1 with NADH in the Presence of Metal Chlorides

When inhibitory levels of diamagnetic metal chlorides were added to the buffer medium, an enhancement in the fluorescence emission spectrum of NADH was observed during titration against F1. The fluorescence enhancement reflects tighter NADH binding to the enzyme in the presence of metal ions and allows one to determine the stoichiometry of binding. As demonstrated in a typical titration curve (Fig. 4), for 100 μM Zn^{2+}, the molar ratio of 2:1 NADH molecules per mole of F1 isozyme was obtained which agrees with previously determined stop flow results (2). The use of inhibitory levels of metal ions, in this case zinc, allows a convenient method of determining the enzyme's concentration, similar to the isobutyramide inhibited NADH titration of horse liver alcohol dehydrogenase. In the presence of paramagnetic ions, such as Co^{2+}, Ni^{2+}, Fe^{3+}, or Mn^{2+}, the fluorescence emission spectrum of NADH was diminished due to paramagnetic quenching effects.

E. Water PRR Titrations of F1 and F2 Isozymes of Aldehyde Dehydrogenases with 100 μM $MnCl_2$

The formation of binary complexes of Mn^{2+} with F1 and F2 isoenzymes and subsequent formation of ternary complexes upon addition of NAD^+ or NADH was demonstrated by water proton relaxation rate titrations (Fig. 5). By measuring the spin lattice relaxation rate ($1/T_1$) of the solvent water protons both in the presence and absence of paramagnetic ions, in this case Mn^{2+} and applying the theory of nuclear relaxation by paramagnetic ions as developed by Bloembergen and Solomon (12,13), one is able to determine the mechanism by which the metal ion inhibition occurs. The relaxation enhancement factor, ε, is defined as follows:

$$\varepsilon = (1T_1^*)/(1/T_1)$$

where T_1 is the spin-lattice relaxation time of the water protons of the paramagnetic ion solution and T_1^* denotes the relaxation time when the paramagnetic ion is bound to the macromolecule. For the case in which the enhancement factor for the ternary complex is greater than that of the binary complex, the metal ion binds through the substrate (coenzyme) and not directly to the enzyme. Therefore, although it appears that both F1 and F2 may have a very weak binding Mn^{2+} site for binary complex formation, upon coenzyme addition, a tight binding occurs through

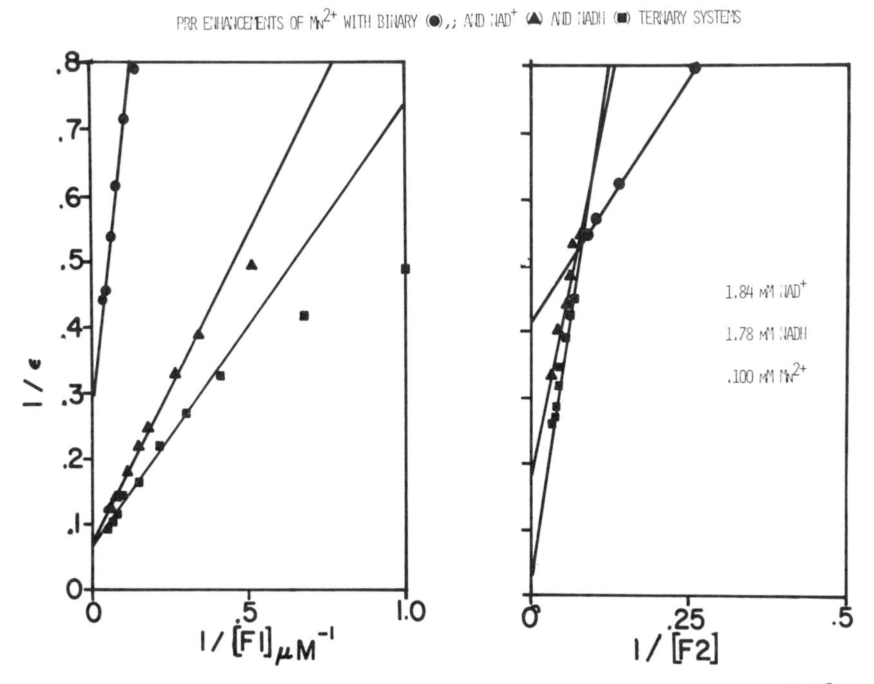

Fig. 5. Water proton relaxation rate titrations of F1 and F2 with 100 μM $MnCl_2$ and formation of ternary complexes with NAD and NADH. Data was obtained at 23° and 24.3Mhz in 0.04 M Tris-Cl buffer containing 0.25% mercaptoethanol.

the coenzyme binding to the macromolecule forming a ternary complex with Mn^{2+}, i.e. the catalytically inhibited complex.

III. CONCLUSIONS

1, 10-Phenanthroline and 2,9-Dimethyl-1,10-phenanthroline equivalently inhibit the F2 isozyme of horse liver aldehyde dehydrogenase by competitive inhibition with NAD^+. This property was also noted with human liver aldehyde dehydrogenase (4) and further promotes F2 as a good model enzyme for the analogous human enzyme. Both horse liver and human aldehyde dehydrogenase do not possess a functional metal ion capable of coordination by such chelating agents. Horse liver aldehyde dehydrogenase displayed activation and inhibition of activity upon incorporation of metal ions into the bioassay buffers. While the precise mechanism of activation cannot yet be explained explicitly, at higher concentrations of metal ions, the formation of a ternary metal ion complex occurs between the enzyme, metal ion, and coenzyme as indicated by fluorescence enhancement, bioassay and PRR titration data. The ternary complex appears to be a system in which the enzyme binds metal ions primarily through the coenzyme. The stoichiometry of two moles of coenzyme per mole of enzyme, Fl, was established by fluorescence enhancement titrations and provides a convenient method of enzyme concentration determination.

IV. ACKNOWLEDGEMENTS

We wish to acknowledge Dr. M. Cohn for use of the pulsed nmr spectrometer for obtaining PRR data. Supported by NIH , AA-00292-05.

V. REFERENCES

(1). Eckfeldt, J., Mope, L., Takio, K., and Yonetani, T., J. Biol. Chem. 251, 236 (1976).
(2). Eckfeldt, J., Dissertation in Biophysics, University of Pennsylvania, Philadelphia (1975).
(3). Schwarcz, M.N., and Stoppani, A.O.M., Biochim.Biophys. Acta 39, 383 (1960).
(4). Sidhu, R.S., and Blair, A.H., Biochem. J. 151, 443 (1975).
(5). Stoppani, A.O.M., Schwarcz, M.N., and Freda, C.E., Arch. Biochem. Biophys.113, 464 (1966).
(6). LaMar, G.N., Horrocks, W. DeW., Jr., Holm, R.H., "NMR of Paramagnetic Molecules" p. 479, Academic Press, New York (1973).
(7). Eckfeldt, J.H., and Yonetani, T., Arch. Biochem. Biophys. 175, 717 (1976).
(8). Blair, A., and Bodley, F., Canad. J. Biochem. 47,265 (1969).
(9). Kraemer, R., and Deitrich, R., J.Biol.Chem. 243, 6402 (1968).
(10). Martell, A.E., and Sillen, L.G., Spec. Publ. Chem. Soc. No. 17: "Stability Constants", p.665 (1964).
(11). Colman, P.M., Weaver, L.H., Matthews, B.W., Biochem. Biophys. Res. Comm. 46, 1999 (1972).
(12). Bloembergen, N., Purcell, E.M., and Pound, R.L., Phys. Rev. 73, 679 (1948).
(13). Solomon, I., Phys. Rev. 99, 559 (1955).

INTERACTION OF DISULFIRAM WITH HORSE LIVER ALDEHYDE DEHYDROGENASE

Charles G. Sanny and Henry Weiner

Purdue University, Indiana

Disulfiram was shown to inactivate the pI 5 isozyme of horse liver aldehyde dehydrogenase at different rates depending upon pH and the presence of various salts, sulfhydryl containing compounds, NAD and substrate. The inactivation rates were biphasic only in the absence of NAD and substrate suggesting differences in the reactive sites toward disulfiram. NAD was not a competitive inhibitor of inactivation by disulfiram, though it decreased the rate of inactivation.

I. INTRODUCTION

To deter people from drinking ethanol, the drug antabuse (disulfiram; tetraethylthiuram disulfide) is often prescribed. The physiological consequence of ingesting ethanol after taking disulfiram is an increase in acetaldehyde levels (1). The drug has been shown to inhibit aldehyde dehydrogenase (ALDH), the enzyme responsible for acetaldehyde oxidation.

Aldehyde dehydrogenases have been isolated from different mammalian livers including sheep (2), bovine (3), horse (4,5) and human (6,7). The interaction of disulfiram with these enzymes has been extensively studied with the sheep (8) and the bovine enzymes (9).

Two major forms of horse liver aldehyde dehydrogenase have been isolated (4,5) which react irreversibly with disulfiram *in vitro* at different rates (5). The pI 5 isozyme, presumably a mitochondrial enzyme (10), is inactivated by disulfiram at much slower rates than is the cytosolic, pI 6, isoenzyme (5).

When studying the inactivation of the pI 5 isozyme by disulfiram, it was found that the inactivation rate varied considerably depending upon experimental conditions. Therefore, an investigation was undertaken to obtain a better understanding of factors affecting ALDH inactivation by disulfiram *in vitro* as part of a long-range objective to ultimately better understand inactivation *in vivo*.

A. Experimental

Homogenous horse liver ALDH, pI 5 (4) was dialysed against 50 mM sodium phosphate, pH 7.5 at 4° to remove sulfhydryl containing compounds used to stabilize the stored enzyme. The enzyme was assayed at 25° by following the production of NADH at 340 nm. Disulfiram was dissolved in 95% ethanol, and the final inactivation mixtures contained 10% ethanol (v/v). Inactivation was followed as a function of time by one of two methods: A.) discontinuous--a sample of ALDH and disulfiram was assayed at time, t, by a 100-fold dilution into 50 mM sodium phosphate, pH 7.5, containing 0.4 mM NAD and 0.14 mM propionaldehyde; B.) enzyme was continuously added to a cuvette containing buffer, NAD, propionaldehyde and disulfiram at t = 0, and the production of NADH was followed by 340 nm. The rate of inactivation was calculated from the change in slope as a function of time.

III. RESULTS AND DISCUSSION

The inactivation rate of ALDH by disulfiram, as measured by discontinuous assay, appears to be biphasic as shown in Fig. 1. Inactivation measured by continuous assay in the presence of NAD and substrate follows first-order kinetics. The reason for biphasic inactivation is not known, though the data suggest possible differences in the reactivity of sites in the native enzyme. Equilibrium dialysis (11) and pre-steady state burst studies (12) show two NAD bind per tetrameric enzyme and two NADH are formed before the rate limiting step. This "half-of-the sites" reactivity might be responsible for the biphasic disulfiram inactivation rates. The presence of NAD and substrate could modify the active sites such that first-order inactivation kinetics are obtained.

Though the detailed mechanism of inactivation of LADH by disulfiram is not known, disulfiram probably irreversibly binds to a cysteine of the enzyme forming a mixed disulfide (13). If an enzyme-inactivator complex forms prior to irreversable binding, represented by the general equation:

$$E + I \underset{k_{-1}}{\overset{k_1}{\rightleftarrows}} E \cdot L \xrightarrow{k_2} E\text{-}I$$

where E represents free enzyme, I is the inactivator (disulfiram), E·I is an enzyme-inactivator complex, and E-I is the irreversibly

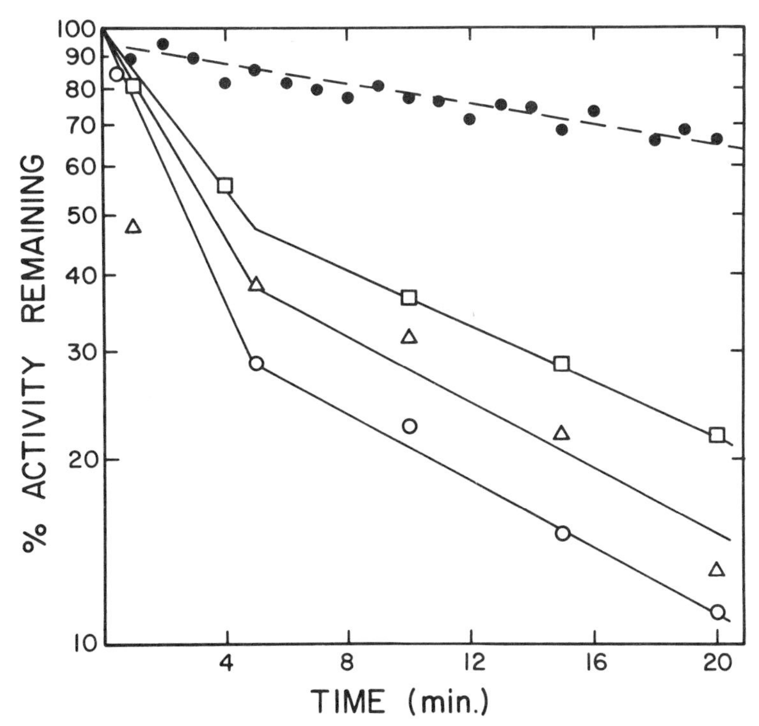

Figure 1. Inactivation of ALDH by disulfiram at pH 7.5, 25°. Inactivation was followed using the discontinuous method (see Experimental) at 100 μM (O), 50 μM (Δ), and 25 μM (□) disulfiram. The dashed line shows inactivation using the continuous method at 100 μM disulfiram in the presence of 0.4 mM NAD and 0.14 mM propionaldehyde.

inactivated enzyme, then a linear inactivation rate expression can be derived:

$$t_{\frac{1}{2}} = \frac{1}{[1]} (TK_{inac}) + T$$

where $t_{\frac{1}{2}}$ is the inactivation half-time [1] is the concentration of inactivator, T is the minimum inactivation half-time, and K_{inac} is $(k_{-1} + k_2)/k_1$ (14,15). Figure 2 presents a plot of $t_{\frac{1}{2}}$ *vs.* 1/[1] for both the initial rate of inactivation and the post-5 minute inactivation rate. K_{inac} is 25 μM and 13 μM, calculated from the initial and post-5 minute inactivation rates, respectively, with corresponding first-order rate constants of 0.31 min^{-1} and 0.07 min^{-1} calculated from T. The fact that $t_{\frac{1}{2}}$ is finite at high inhibitor concentrations means that a complex does form prior to covalent binding.

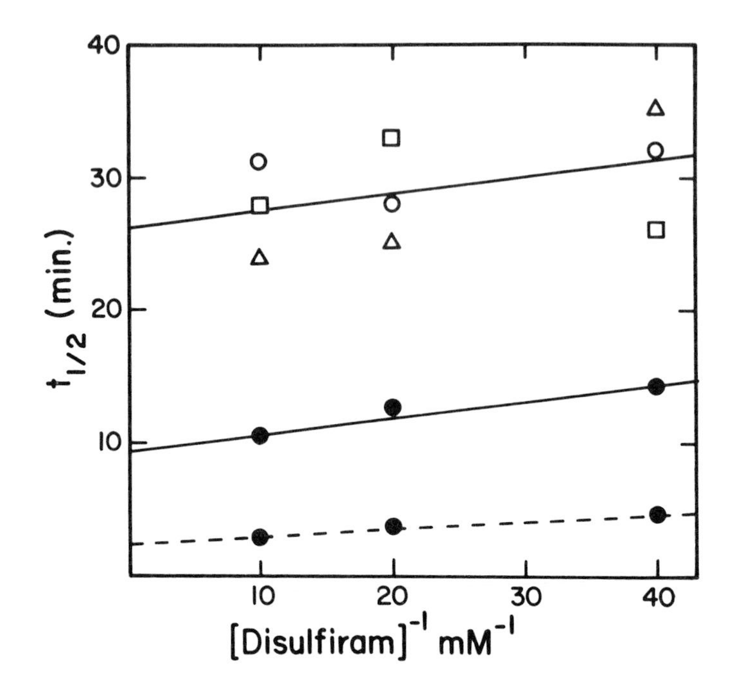

Figure 2. Inactivation half-time as a function of the reciprocal of disulfiram concentration at various NAD concentrations. In the absence of NAD, $t_{1/2}$ was calculated from the initial inactivation rates (● -- ●) and from the post 5-min inactivation rates (●——●). $t_{1/2}$ was also calculated from the post-5 min inactivation rates at 85 μM (Δ), 170 μM (□) and 340 μM (O) NAD. Initial inactivation rates in the presence of NAD gave similar results.

NAD has been reported to be a competitive inhibitor of disulfiram inactivation (9). If this were so, the following rate expression could be derived (14,15):

$$t_{1/2} = \frac{1}{[1]} T(K_{inact} + \frac{K_{inact}[NAD]}{K_d} + T$$

where K_d is the dissociation constant of an enzyme-NAD complex. K_d for NAD and ALDH has been calculated by equalibrium dialysis to be *ca.* 10 μM (11). A plot of $t_{1/2}$ *vs.* 1/[1] at various NAD concentrations did not show NAD to be a competitive inhibitor of disulfiram inactivation (Figure 2).

Disulfiram as well as diethyldithiocarbamate can inactivate aldehyde dehydrogenase *in vivo*, whereas only disulfiram

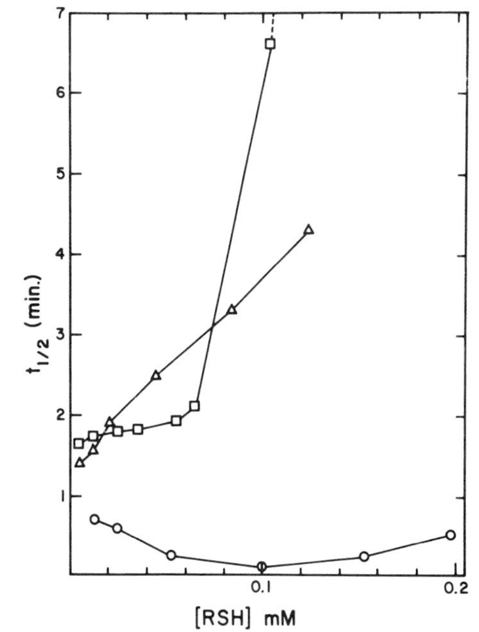

Figure 3. Effect of sulfhydryl containing compounds on the inactivation half-time. Disulfiram (0.1 mM) was preincubated with 2-mercaptoethanol (O), reduced glutathione (Δ) or dithiothrietol (□) in 50 mM sodium phosphatem pH 7.5, containing 10% ethanol, 0.4 mM NAD, and 0.14 mM propionaldehyde before the addition of enzyme. Inactivation was followed using the continuous method (see Experimental)

is effective *in vitro* (16). To be an inhibitor of ALDH a disulfide bond must be present in the compound. Kitson has shown that a mixed disulfide of the form $(Et)_2NCS_2$-SR was an inhibitor of the sheep liver enzymes (8). Since disulfiram is rapidly reduced *in vivo* to diethyldithiocarbamate (17), this compound must subsequently form a disulfide bond prior to being an inhibitor of ALDH. To determine the effects of mixed disulfide derivatives of diethyldithiocarbonate, disulfiram was preincubated with various -SH containing compounds prior to the addition of enzyme. The results of these inactivations are presented in Fig. 3.

Reduced glutathione linearly decreased the rate of inactivation of the enzyme while 2-mercaptoethanol increased the rate of inactivation up to equimolar concentrations of disulfiram. Kitson (8) found similar results with 2-mercaptoethanol and disulfiram inactivation of sheep mitochondrial ALDH. Dithiothreitol had little effect on the inactivation rate up to equimolar ratios of -SH to disulfiram, after which marked protection from inactivation was observed.

TABLE 1

The Effects of Salts on the Specific Activity of ALDH and on the Rate of Inactivation by Disulfiram at pH 7.5[a]

Salt[b]	Relative Specific Activity	Inactivation Half-time (min)
$(CH_3)_4NCl$	0.9	26
NaCl	1.0	13
NaCl+ $MgCl_2$ (10^{-5}M)	1.2	15
KCl	0.5	44

[a]*Assay conditions: 50 mM imidazole-HCl, pH 7.5, 100 μM disulfiram, 10% ethanol, 0.4 mM NAD, 0.14 mM propionaldehyde.*

[b]*All salt concentrations are 0.1 M unless indicated otherwise.*

[c]*Specific activity in NaCl as 1.0*

Inactivation of ALDH by disulfiram *in vivo* could also be influences by ions in the cell as well as cofactors, substrates, and -SH groups. The effect of pH on inactivation resembles the pH effect on activity with inactivation being greater as the pH is increased. The relative inactivation rates at pH 7, 8 and 9 are 1, 3 and 6 respectively. The effects of various salts on the inactivation rate are shown in Table 1. Major differences are observed between K^+ and Na^+ with respect to both the specific activity of ALDH and the inactivation rates. Small amounts of Mg^{+2} (10^{-5}M) also affected the specific activity (18) and the inactivation rate.

IV. CONCLUSIONS

The rate of reaction of disulfiram with ALDH may be dependent upon the microenvironment in the cell. Small changes in potassium or magnesium levels would alter the reactivity as would pH. The physiological inhibitor may not be disulfiram, by may be a mixed disulfide of diethyldithiocarbamate and some -SH containing compound. The compound may provide specificity toward different isozymes. It was observed that after feeding a rat disulfiram only some liver isozymes of ALDH were inhibited (19) but *in vitro* incubations showed that all were inhibited (unpublished observations); thus, *in vitro* determinations of rates of inactivation of ALDH may not be comparable to those in the intact cell.

It is known that NAD binds to ALDH prior to aldehyde and that NAD enhances the esterolytic reaction catalysed by the enzyme (20). Since disulfiram binds to presumably the active site of the enzyme, NAD could enhance its rate of inactivation analogous to the esterase effect or, if the binding sites overlapped, protect against inactivation. With beef liver ALDH, NAD was found to be a competitive inhibitor against disulfiram inactivation (9). It is not a competitive, but an un- or non-competitive inhibitor with the horse liver enzyme. Since the K_d of NAD is 10 μM, physiologically disulfiram must inacticate an enzyme-NAD complex and if NAD were truly a competitive inhibitor of inactivation, the drug would not be an effective inhibitor *in vivo*.

V. ACKNOWLEDGEMENTS

This work was supported in part by Grants NIAAA 01395 and BSF BMS 75-03926. Journal Paper Number 6510 from the Purdue University Agricultural Experiment Station.

VI. REFERENCES

(1). Hald, J., and Jacobsen, E., Acta Pharmacol. Toxicol. 4, 305, (1948).
(2). Crow, K.E., Kitson, T.M., MacGibbon, A.K.H., and Batt, R. D., Biochem. Biophys. Acta 350, 121-128 (1974).
(3). Deitrich, R.A., Hellerman, L., and Wein, J., J. Biol.Chem. 237, 560-564 (1962).
(4). Feldman, R.I., and Weiner, H., J. Biol. Chem. 247, 260-266, (1972).
(5). Eckfeldt, J., Mope, L., Takio, K., and Yonetani, T., J. Biol. Chem. 251, 236-240 (1976).
(6). Blair, A.H., and Bodley, F., Can.J. Biochem. 47, 265-272 (1969).
(7). Pietruszko, R., in:"Alcohol and Aldehyde Metabolizing Systems" (R.G. Thurman, J.R. Williamson, H. Drott and B. Chance, Eds.) Vol. 2, Academic Press, New York (1977).
(8). Kitson, T.M., Biochem. 151, 407-412 (1975).
(9). Deitrich, R.A., and Hellerman, L., J. Biol. Chem. 238, 1683-1698 (1963).
(10). Eckfeldt, J.H., and Yonetani, T., Arch. Biochem. Biophys. 175 717-722 (1976).
(11). Weiner, H., King, P., Hu, J.H.J., and Bensch, W.R., in: "Alcohol and Aldehyde Metabolizing Systems" (R.G. Thurman, T. Yonetani, J.R. Williamson, and B. Chance, Eds.) Vol. 1, p. 101-115 Academic Press, New York (1974).
(12). Weiner, H.,, Hu, J.H.J., and Sanny, C.G., J. Biol. Chem. 251, 3853-3855 (1976).
(13). Deitrich, R.A., and Hellerman, L., J. Biol. Chem. 238, 1683-1689 (1963).

(14). Kitz, R., and Wilson, I.B., J. Biol. Chem. 237,3245-3249 (1962).
(15). Meloche, H.P., Biochemistry 6, 2273-2280 (1967).
(16). Deitrich, R.A., and Erwin, V.G., Mol. Pharmacol. 7, 301-307 (1971).
(17). Fairman, M.D., Dodd, D.E., Nolan, R.J., and Hanzlik, R., in:"Alcohol and Aldehyde Metabolizing Systems" (R.G. Thurman, H. Drott, J.R. Williamson, and B. Chance, Eds.) Vol.2, Academic Press, New York (1977).
(18). Weiner, H., Brown, C.S., Sanny, C.G., and Hu, J.H.J., Fed. Proc. 35, 1499 (1976).
(19). Berger, D., and Weiner H., Biochem. Pharmacol., in press.
(20). Feldman, R,I., and Weiner, H., J. Biol. Chem. 247, 267-272 (1972).

INTRACELLULAR LOCATION OF 3,4-DIHYDROXYPHENYLACETALDEHYDE AND ACETALDEHYDE OXIDATION IN RAT LIVER[1]

A.W. Tank[2] and H. Weiner

Purdue University

In rat liver acetaldehyde does not inhibit the oxidation of the aldehyde derived from dopamine, 3,4-dihydroxyphenylacetaldehyde (DOPAL). It was found that both aldehydes are oxidized primarily in the matrix of the mitochondria. Two isozymes of aldehyde dehydrogenase were found to exist in the matrix, both possessing low Km's for acetaldehyde and DOPAL. Disulfiram (Antabuse), administered to the animal, inhibits primarily one matrix isozyme of aldehyde dehydrogenase as well as acetaldehyde oxidation. However DOPAL oxidation is only slightly retarded. Thus the lack of inhibition of DOPAL oxidation by acetaldehyde can be rationalized by postulating that the disulfiram sensitive isozyme is primarily responsible for acetaldehyde oxidation while the other matrix is required for DOPAL oxidation.

I. INTRODUCTION

The catabolism of dopamine involves two major steps: oxidative deamination catalyzed by monoamine oxidase[1] producing an aldehyde (DOPAL) followed by either oxidation to the corresponding acid (DOPAC) or reduciton to the alcohol derivative (DOPET). Ethanol has been shown to alter the catabolism of dopamine such that DOPET, not DOPAC, is the major product (1,2). Two explanations have been presented for this alteration. The first is based on the fact that the oxidation of ethanol produces NADH causing the cell to become reductive (3). Hence, the biogenic aldehyde (DOPAL) would be reduced rather than be oxidized. The second explanation presented

---

[1]Abbreviations: MAO, monoamine oxidase; DOPAL, 3,4-dihydroxyphenacetaldehyde; DOPAC, 3,4-dihydroxyphenyl acetic acid; DOPET, 3,4-dihydroxyphenylethanol. [2]David Ross predoctoral Fellow.

is that the acetaldehyde produced during ethanol metabolism prevents DOPAL from being oxidized; hence DOPAL accumulates and is ultimately reduced (4,1). We recently showed that neither explanation is totally correct and presented a third hypothesis which involves alcohol dehydrogenase being converted from an E-NAD complex to an E-NADH complex only during ethanol metabolism and this complex then reduces DOPAL (2). Increasing NADH by alternative means did not mimic the effect of ethanol. It was most surprising to find that acetaldehyde did not inhibit DOPAL oxidation, since both aldehydes are substrates for aldehyde dehydrogenase.

It has been established that acetaldehyde is primarily oxidized in the mitochondria (5-7). Many isozymes of rat liver aldehyde dehydrogenase exist in different regions of the cell as well as within the same region of the cell (8-11) and the physiological role of each isozyme is not known. Thus it is possible that different isozymes located in the same or in different subcellular organelles could be responsible for the different oxidations. Hence we undertook an investigation to determine the specificity of the isozymes of rat liver aldehyde dehydrogenase and to determine the subcellular localization of DOPAL oxidation.

Experimental Procedures

Isolation and quantification of the products obtained from dopamine metabolism was performed as described elsewhere (2). Incubations of dopamine with various subcellular liver fractions were carried out at 37° in 1 ml of 10 mM sodium phosphate pH 7.4 buffer containing 10 mM $MgCl_2$ and either 0.1 mM or 1.0 mM ADP. Coenzymes when added were at a concentration of 0.55 mM NAD and NADPH, 0.55 μM NADH and NADP. ^{14}C-Dopamine (2 mM, 0.2 μCi) was added to all assay mixtures in order to initiate the reaction. The reaction was terminated after a 15 min incubation period by the addition of two drops of 0.2 M HCl, and the particulate material removed by centrifugation at 40,000 × g for 10 min. The recovery of radioactivity was *ca.* 85%. Monoamine oxidase, prepared from rat liver (12), was added to the mixtures that did not contain mitochondria.

Subcellular fractions of rat liver were obtained as outlined by Hogeboom (13) by first homogenizing the livers with a tissue grinder in the presence of 0.25 M sucrose-10 mM tris-HCl buffer, pH 7.4. The material was subjected to centrifugation at 900 × g for 15 min to remove the nuclear fraction. The supernatant was recentrifuged at 9000 × g for 15 min to obtain the mitochondria. This pellet was washed twice in the same buffer followed by centrifugations. Centrifugation of the post-mitochondrial supernatant at 12,000 x g for 15 min was performed to separate the lysosomal pellet. The post-lysosomal was centrifuged at 104,000 x g for 60 min to isolate the microsomal pellet. The final supernatant was designated as the cytosol fraction. Marker enzymes were assayed in order to determine the degree of cross-contamination of the various fractions (7).

Aldehyde dehydrogenase was assayed by measuring the increase in fluorescence of NADH with an Aminco microfluorometer. Assays were performed in a 0.1 M pH 7.4 sodium phosphate buffer containing 0.5 mM NAD and 10 mM acetaldehyde.

II. RESULTS AND DISCUSSION

Incubating dopamine with liver slices show that approximately 60% of the deamined products is isolated as DOPAC, and only 6% as the alcohol derivative, DOPET; 21% of the deaminated products are isolated as the aldehyde, DOPAL (Table 1).

TABLE 1
Subcellular localization of DOPAL oxidation

	ADP (mM)	*Percent Deaminated Products*		
		DOPAL	*DOPET*	*DOPAC*
Liver Slice		21.5 5.0	6.6 3.1	59.9 8.7
Mitochondria	1.0	8.0 2.0	1.9 0.9	82.7 4.2
Cytosol	1.0	50.4 5.2	30.6 6.0	6.5 0.6
Cytosol	0.1	33.1 1.8	10.0 0.4	38.4 1.5
Microsomes	1 or 0.1	47.3 3.6	10.4 1.0	22.6 2.4
Cytosol + Microsomes	0.1	28.2 2.5	8.9 3.4	49.5 5.7

MAO was added to incubation not containing mitochondria. Coenzymes were added to all incubations, except for the liver slice. Protein concentrations added were 4.3 mg/ml cytosolic, 3 mg/ml mitochondrial and 3.6 mg/ml microsomal. A 50 mg liver slice was employed.

The remainder of the products are tetrahydropapaveroline and homovanillic acid. However for the sake of presentation these metabolites are not included. Incubating dopamine with liver mitochondria produces results which are similar to those found with the whole liver slice except that the percent DOPET is substantially lower as is the percent DOPAL formed. Thus it seems that the aldehyde dehydrogenase responsible for oxidizing DOPAL is located in the mitochondria. However, it was necessary to incubate dopamine with other subcellular organelles to determine if they were capable of oxidizing the aldehyde.

The cytosolic fraction in the presence of low levels of ADP (0.1 mM) is capable of oxidizing DOPAL to its acid. However, with more physiological levels of ADP (1.0 mM) the cytosol no longer is capable of oxidizing DOPAL; large concentrations of DOPAL accumulates and more DOPET is formed. The microsomal fraction is also capable of oxidizing DOPAL but it is much less efficient than is the mitochondrial fraction. ADP does not have an inhibitory effect on the microsomal activity. Incubating cytosol and microsomes together produce results similar to the liver slice only if the ADP level is low. Thus it seems that all subcellular fractions of the liver can oxidize DOPAL. However, with physiological concentrations of ADP it appears that the mitochondria most closely parallels the results found from liver slice incubations.

It is not known why ADP inhibits DOPAL oxidation in the cytosol. It is possible that it acts as an allosteric modulator of aldehyde dehydrogenase.

Aldehyde dehydrogenase has been shown to be located in both the matrix (8-11) and intramembrane space (9) of the mitochondria. Incubations of intact mitochondria in the presence of rotenone were performed in order to determine if the matrix enzymes were responsible for the bulk of the activity. Since the intramembrane space enzymes are sensitive to added coenzymes (14) the reactions were run in the presence or absence of added coenzymes, and the results presented in Table 2 (Reduced coenzymes were added to mimic non-mitochondria incubations). Since the matrix enzymes are insensitive to added coenzymes the results in the presence and absence of rotenone but in the absence of coenzymes give an indication of the capacity of the matrix space to oxidize DOPAL. It was observed that the bulk of the oxidation is inhibited when rotenone is present as indicated by the large increase in DOPAL concentration. In the absence of rotenone, adding coenzymes to allow the intramembrane space enzymes to operate results in no statisticaLly significant increase in DOPAC or decrease in DOPAL percentages. However, in the presence of rotenone and coenzyme there is an increase in DOPAC formation and a decrease in DOPAL formation compared to the run with rotenone and no coenzymes. These results suggest that the enzymes in the matrix

TABLE 2
Localization of DOPAL oxidation within the liver mitochondria.

*Coenzymes**	*Rotenone*	*DOPAL*		*DOPET*		*DOPAC*	
-	-	9.3	2.9	2.2	0.5	80.3	2.6
-	+	53.7	7.4	8.5	2.4	22.9	5.7
+	-	8.0	2.0	1.9	0.9	82.7	4.2
+	+	36.9	4.1	7.0	3.5	42.0	6.8

**NAD and NADPH = 0.55 mM; NADH and NADP = 0.55 μM. Rotenone concentration was 2 μM. Mitochondria protein was 3 mg/ml. Incubation contained 1.0 mM ADP.*

are mainly responsible for DOPAL oxidation. When the matrix enzymes are not functioning then the intramembrane space enzymes are capable of performing the reaction, though not as efficiently as the matrix enzymes as indicated by the higher levels of DOPAL that accumulate.

Since the data in Table 1 show that each subcellular frac- is capable of oxidizing DOPAL the oxidations were performed in the presence of mitochondria and cytosol or microsomes in the presence and absence of rotenone. Even though the cytosol is capable of oxidizing DOPAL, in incubations containing cytosol and mitochondria rotenone causes a large increase in the percent DOPAL as compared to incubations of mitochondria alone (Table 3). In addition, a decrease in DOPAC formation is obtained. These data show that mitochondria will preferentially oxidize DOPAL when cytosol is present. The same conclusions are reached when cytosol, microsomes and mitochondria are used: in the absence of rotenone the oxidation can be attributed to the mitochondria.

These observations lead us to conclude that the matrix of the mitochondria is the major site of DOPAL oxidation. Other experiments performed in this laboratory (7) and elsewhere (5,6) show that the matrix is the site of acetaldehyde oxidation. Thus, the dilemma remains that both acetaldehyde and DOPAL are oxidized in the same subcellular region, yet acetaldehyde does not impede the oxidation of DOPAL (2). The isozymes of aldehyde dehydrogenase were isolated from the various subcellular organelles in order to determine if there were more than one isozyme in each and to determine the substrate specificity with respect to the two aldehydes in question. We subjected the cytosolic and mitochondrial material to column isoelectric focusing to separate isozymes. The apparent K_m of

TABLE 3

Metabolism of DOPAL in subcellular fractions of rat liver in the presence of mitochondria.

	Rotenone	Percent Deaminated Products					
		DOPAL		DOPET		DOPAC	
Liver Slice	-	21.5	5.0	6.6	3.1	59.9	8.7
Mitochondria	-	8.0	2.0	1.9	0.9	82.7	4.2
	+	36.9	4.1	7.0	3.5	42.0	6.8
Cytosol + Mitochondria	-	26.6	6.5	8.4	2.2	53.0	6.9
	+	54.4	14	14.9	4.9	20.1	3.8
Cytosol + Microsomes + Mitochondria	-	23.9	8.9	6.9	4.4	57.9	6.9
	+	45.5	9.7	12.2	9.5	32.0	14

Incubation contained added coenzymes and 1.0 mM ADP. Rotenone was 0.2 μM. Protein concentrations added were 4.3 mg/ml cytosolic 3 mg/ml mitochondrial and 3.6 mg/ml microsomal. A 50 mg liver slice was employed.

the various isozymes from the cytosol, mitochondria and microsome are presented in Table 4. The microsomes were not subjected to isoelectric focusing and were used as an impure preparation. Low K_m enzymes for each substrate exist in the cytosol and mitochondria, and the same pI form of the enzymes have low K_m's for both substrates. The enzymes at pI = 5.4 and 5.6 isolated from the mitochondria have the lowest K_m for the substrates and were shown to be located in the matrix (7). Thus, the two matrix space enzymes have low K_m's for the two aldehydes. These data alone do not allow us to conclude why acetaldehyde does not prevent DOPAL oxidation. In a separate study we investigated the inhibition of liver aldehyde dehydrogenase produced by administering disulfiram to rats (15). It was observed that only 16% of DOPAL oxidation was inhibited, compared to a drastic inhibition of acetaldehyde oxidation (16). When investigating the isozyme pattern before and after administering the drug it was found that only one of the matrix enzymes (pI = 5.4) was inhibited (17). Thus, it appears that even though both enzymes are capable *in vitro* of oxidizing acetaldehyde and DOPAL, only one may be involved in acetaldehyde oxidation *in vivo*. The other matrix isozyme must be involed in DOPAL oxidation or more inhibition of DOPAL oxidation would have been observed in disulfiram treated animals. If *in vivo* each matrix isozyme possessed unique specificity then the lack of inhibition of DOPAL oxidation in the presence of acetaldehyde can be explained. It is not possible to explain why under *in vivo* conditions the two enzymes may possess unique substrate specificity while in the *in vitro* assays they have the same specificity. It is possible that yet to be discovered modulators may be changing the *in vivo* specificity of the enzyme.

III. CONCLUSIONS

Even though the matrix of the mitochondria is the major location of DOPAL and acetaldehyde oxidation the two aldehydes do not mutually inhibit one another's oxidation. The most likely explanation is that *in vivo* the two different isozymes located in the matrix space have unique substrate specificity. Neither the physiological role of the isozymes located in the cytosol or microsomes nor why the oxidation is so ADP-sensitive is yet known. It is possible that these isozymes are involved in the oxidation of aldehydes derived from steroids or other lipid material. The observations obtained for rat liver cannot be extended to all other livers for it has been shown that in horse (18) or in man (19) there are only two isozymes of aldehyde dehydrogenase. Similarly the isozyme pattern of ALDH in the brain of rat appears to be much less complex than that of the liver (unpublished observations).

TABLE 4
Rat liver aldehyde dehydrogenase isozymes

	pI	*Apparent K_m (mM)*	
		DOPAL	*Acetaldehyde*
Cytosol	7.4	2.5×10^{-1}	13
	6.6	2.2	9.4×10^{-1}
	6.15	*	1.0×10^{-2}
	6.05	5.1×10^{-3}	2.2×10^{-2}
	5.8	*	4.5×10^{-3}
Mitochondria	6.9	1.8×10^{-1}	1.3
	5.4	1.7×10^{-3}	10^{-3}
	5.6	3.0×10^{-4}	1.3×10^{-4}
Microsomes		4.5×10^{-1}	4.0

Assays were performed at pH 7.4. The isoelectric points were determined by column isoelectric focusing.

IV. ACKNOWLEDGEMENTS

The work was supported in part by Grants NIAAA 01395 and NSF BMS75-03926. Journal paper number 6504 from the Purdue University Agriculture Experiment Station.

V. REFERENCES

(1). Davis, V.E., Walsh, M.J. and Yamanaka, Y., J. Pharmacol. Exp. Ther. 174, 401-412 (1970).
(2). Tank, A.W., Weiner, H. and Thurman, J.A., Ann. N.Y. Acad. Sci. 273, 219-226 (1976).
(3). Feldstein, A., Hoagland, H., Freeman, H., and Williamson, O., Life Science 6, 53-61 (1967).
(4). Lahti, R.A., and Majchrowicz, E., Biochem. Pharmacol. 18, 535-538 (1969).
(5). Marjenon, L., Biochem. J. 127, 633-639 (1972).
(6). Grunnet, N., Quistorff, B., and Thieden, H.I.D. in "Alcohol and Aldehyde Metabolizing Systems" (R.G. Thurman, T. Yonetani, T.R. Williamson, B. Chance, Eds.), p. 137-146. Academic Press, N.Y., 1974.
(7). Tank, A.W., Ph.D. Thesis, Purdue University (1976).
(8). Tottmar, S.O.C., Pettersson, H., and Kiessling, K.H., Biochemical J. 135, 577-586 (1973).
(9). Deitrich, R.A., and Siew, C., in "Alcohol and Aldehyde Metabolizing Systems" (R.G. Thurman, T. Yonetani, J.R. Williamson and B. Chance, Eds.), p. 125-136. Academic Press, N.Y., 1974.
(10). Weiner, H., King, P., Hu, J.H.J., and Bensch, W.R., in "Alcohol and Aldehyde Metabolizing Systems" (R.G. Thurman, T. Yonetani, J.R. Williamson and B. Chance, Eds.), p. 125-136. Academic Press, N.Y., 1974.
(11). Koivula, T. and Koivusalo, M., Biochem. Biophys. Acta 397, 9-23 (1975).
(12). Greenawalt, J.W., in Methods in Enzymology 31A, 310-323 (1974).
(13). Hogeboom, G.H., Methods of Enzymology 1, 16-19 (1955).
(14). Ernster, L., and Kuylenstierna, in "Membranes of Mitochondria and Chloroplasts (E. Racker, Ed.), p. 172-212. Van Nostrand, N.Y., 1970.
(15). Berger, D., and Weiner, H., Biochem. Pharmacol.,in press.
(16). Truitt, E.B., and Walsh, M.J., in "The Biology of Alcoholism" I (B. Kissin and H. Begleiter, Eds.), p. 161-195 Plenum Press, N.Y., 1971.
(17). Berger, D., and Weiner, H., Alcoholism: Clinical and Experimental Research, in press.

(18). Eckfeldt, J. Mope, L., Takio, K., and Yonetani, Y., J. Biol. Chem. 251, 236-240 (1976).

(19). Pietruszko, R., in "Alcohol and Aldehyde Metabolizing Systems" III (R.G. Thurman, J.R. Williamson, H. Drott, and B. Chance, Eds.), p. Academic Press, N.Y., 1977.

LOCALIZATION AND PROPERTIES OF ALDEHYDE REDUCTASE

J.P. von Wartburg, Margret M. Ris
and B. Wermuth

University of Berne

Two aldehyde reductases have been isolated from rat brain (Eur.J.Biochem. 37, 69, 1973). Their coenzyme, substrate and inhibitor specificity was used to localize the major isoenzyme 1 in cytosol and the minor isoenzyme 2 in mitochondria. An additional isoenzyme 3 is detected in microsomes. A similar isoenzyme distribution is found in isolated synaptosomes. An aldehyde reductase corresponding to the cytosolic brain enzyme was purified to homogeneity from human liver. A molecular weight of 37,000, and S-value of 2.9 and an IEP of 5.3 were determined. A time-dependent irreversible inactivation was observed by p-mercuribenzoate and N-ethylmaleimide but not by iodoacetamide. NADPH or $NADP^+$, but not NADH, NAD^+ or aldehydes protected against inactivation. The amino acid composition was dissimilar from the ones observed for oligomeric dehydrogenases and characterized by a high proline content. Only 10% α -helicity was detected by circular dichroism spectra.

I. INTRODUCTION

In brain biogenic aldehydes, the products of oxidative deamination of biogenic amines by monoamine oxidase are further metabolized either by oxidation via aldehyde dehydrogenase (E.C. 1.2.1.3) or by reduction via aldehyde reductase (E.C. 1.1.1.2) (1). In general it seems that catecholamines without β-hydroxyl group are mainly oxidized to their acids, while the β-hydroxylated amines are reduced to the alcohol products (2). While these pathways are well documented, relatively little information is available on the isoenzymes of aldehyde reductase, their function and subcellular localization. Two NADPH-dependent aldehyde reductases, differing in their pH optima, Michaelis constants for substrates and their inhibitor sensitivity have been discerned in pig brain (3). An NADPH and NADH-dependent enzyme have been isolated from ox brain (4,5). Two isoenzymes have also been found in rat brain and four different aldehyde reductases could be distinguished in human brain (6). The NADPH-dependent enzymes from bovine, pig and rat brain are mainly localized in the cytosol. In rat brain the major isoenzyme is NADPH-dependent and is strongly inhibited by anticonvulsive drugs (7). For the minor form NADH serves also as coenzyme and the inhibition by the anticonvulsants is less pronounced. These characteristics served to study the distribution of both isoenzymes.

In addition, NADPH-dependent aldehyde reductases have been isolated from various other mammalian tissues (8-12). Judged from their substrate and conenzyme specificity and their inhibitor sensitivity they correspond to the major isoenzyme in brain. A monomeric structure and a molecular weight of 30,000 to 40,000 have been reported by several authors for these enzymes (6-12). However, no detailed physico-chemical studies have been carried out, except for the enzyme from pig kidney cortex (9). Considering a recently proposed model of an evolutionary tree of pyridine nucleotide dependent dehydrogenases (13) it seems possible that the monomeric aldehyde reductase represents a link between a hypothetical ancestral dehydrogenase and today's diversified oligomeric dehydrogenases. Structural and functional comparisons between the monomeric aldehyde reductase and the functionally similar oligomeric alcohol dehydrogenase might give evidence for such relations. In earlier publications we have described the heterogeneity and properties of alcohol dehydrogenase from human liver (14,15). In this paper we report on the purification and the physico-chemical properties of human liver aldehyde reductase.

A. Subcellular and Subsynaptosomal Distribution of Aldehyde Reductase Isoenzymes from Rat Brain

Subcellular fractions from rat brain obtained by differential centrifugation were characterized using lactate dehydrogenase, glutamate dehydrogenase and NADPH-linked cytochrome c reductase as markers for cytosol, mitochondria and microsomes, respectively. Total homgenate (withouth nuclei and cell debris) contained 323 mIU/g tissue NADPH- and 114 mIU/g tissue NADH-dependent aldehyde reductase activity. As shown in Table I, the NADPH-dependent activity is distributed over all subcellular fractions, most of it in the cytosol. The NADH-linked aldehyde reductase activity is exclusively detected in mitochondria. Further purification of the crude mitochondrial fraction by a 3-step sucrose gradient allowed up to 85% of the NADH-linked activity found in the homgenate to be recovered whereas the contaminating lactate dehydrogenase decreased. Microsomes only contain NADPH-dependent activity. The characteristics of the major isoenzyme 1 and the minor isoenzyme 2 with respect to substrate and coenzyme specificity and inhibitor sensitivity are summarized in Table II for comparison with the activities in the subcellular fractions. The cytosolic aldehyde reductase activity is relatively low with p-chlorbenzaldehyde and butyraldehyde, strongly inhibited by phenobarbital and exclusively NADPH-dependent.Thus, the cytosolic enzyme corresponds to the isoenzyme 1. The results obtained with the mitochondrial fraction correlate best with the isoenzyme 2, especially with regard to the activity observed with NADH. The microsomal aldehyde reductase shows a remarkably different activity and inhibition pattern: only p-nitrobenzaldehyde and p-nitroacetophenone serve as substrates and glutethimide does not inhibit this enzyme. These results suggest that microsomes contain an additional NADPH-dependent aldehyde reductase which was not isolated previously (6).

TABLE 1
Subcellular Localization of Aldehyde Reductase in Rat Brain

Fraction	Enzyme % of Total Activity in Homogenate				
	NADPH-dependent Aldehyde Reductase	NADH-dependent Aldehyde Reductase	Lactate Dehydrogenase	Glutamate Dehydrogenase	NADPH-dependent Cytochrome c Reductase
Cytosol	44	--	43	1	26
Mitochondria	21	31	26	70	5
Microsomes	10	--	8	4	72

Assay conditions are described in reference 6 and 7. p-nitrobenzaldehyde served as substrate for aldehyde reductase; (--) no detectable activity.

In order to study the distribution of aldehyde reductase isoenzymes in synaptosomes, the crude mitochondrial fraction was subfractioned into myelin (A), synaptosomes (B) and "glial" mitochondria (C). Subsequently, synaptosomes were disrupted by osmotic shock and further fractioned into cytosol (O), vesicles (D), and membrane fragments (E,F,G) incompletely disrupted synaptosomes (H) and synaptosomal mitochondria (I). Characterization of these fractions was achieved by determination of the following marker enzymes: glutamate dehydrogenase for mitochondria, acetylcholinesterase for cholinergic synaptosomal membranes and Mg^{2+} - and Na^{+}, K^{+} - stimulated adenosine triphosphates for membranes in general. Figure 1 shows a typical distribution

TABLE II

Characterization of Aldehyde Reductase in Subcellular Fractions of Rat Brain

Substrate-Aldehyde	Coenzyme	Inhibitor	Relative Activities				
			Cytosol	Mitochondria	Microsomes	Iso-enzyme I	Iso-enzyme II
p-Nitrobenz-	NADPH	-	100	100	100	100	100
p-Chlorbenz-	NADPH	-	21	44	0	13	66
p-Nitroaceto-phenone	NADPH	-	10	13	36	0	3
Butyr-	NADPH	-	17	31	0	5	34
p-Nitrobenz-	NADH	-	0	47	0	6	64
p-Nitrobenz-	NADPH	Phenobarbital	18	41	51	5	66
p-Nitrobenz-	NADPH	Glutethimide	56	73	100	38	61
p-Nitrobenz-	NADPH	Pyrazole	100	100	87	86	89

All assay conditions are described in references 6 and 7.

pattern of aldehyde reductase and marker enzymes. The synaptosomal fractions B and H contain both NADPH- and NADH- linked activity and intermediary values for the inhibition by phenobarbital of 60 and 40% are observed, respectively. Aldehyde reductase of cytosol entrapped in synaptosomes (O) is exclusively NADPH-dependent and strongly inhibited by phenobarbital. NADH is as good a conenzyme as NADPH for the aldehyde reductase in the "glial" (C) and the synaptosomal mitochondria (I). These results indicate that both the cytosolic isoenzyme I and the mitochondrial isoenzyme 2 are present in nerve endings.

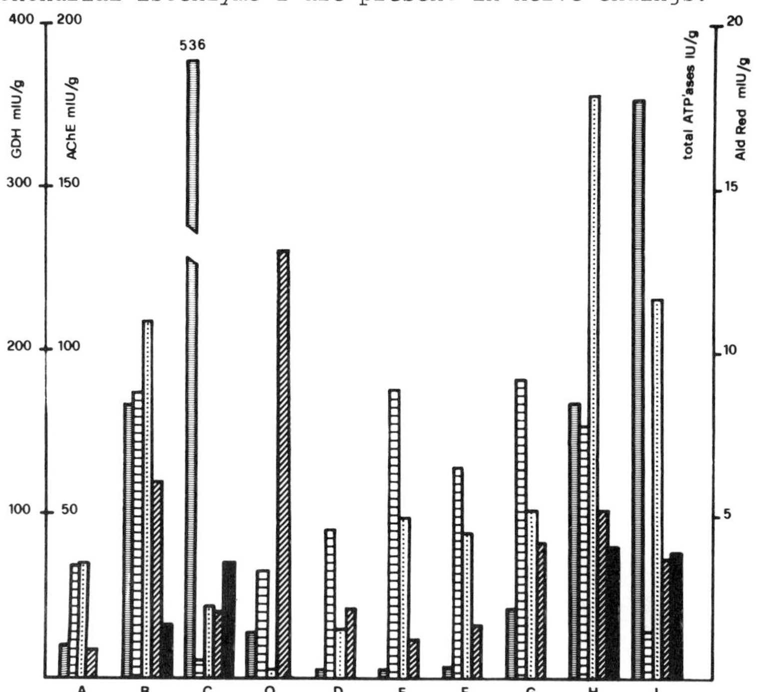

Fig.1. Distribution of Aldehyde Reductase and Some Marker Enzymes in Subsynaptosomal Fractions from Rat Brain. For designation of fractions see text. For reasons of better comparison only one tenth of the real activities are drawn for fractions A, B and C. ▤ *represent glutamate dehydrogenase (GDH),* ⊟ *acetylcholinesterase (AChE),* ⬚ *total ATP'ases,* ▨ *NADPH-dependent aldehyde reductase and* ■ *NADH-dependent aldehyde reductase (AldRed.).*

B. Purifaction and Properties of Human Liver Aldehyde Reductase

Purification of human liver aldehyde reductase was achieved by passing crude liver homogenate over the following chromatography resins: DEAE-cellulose, Sephadex G-100, hydroxyl-apatite and DEAE-Sephadex A-50, which all previously had been equilibrated against 10 mM Na-phosphate buffer pH 7.0. The Sephadex G-100 and hydroxylapatite columns were developed with the same buffer; from the ion exchange resins the enzyme was eluted by a gradient 10 to 100 mM Na-phosphate pH 7.0. Table III summarizes the results of a typical purifiaction of aldehyde reductase. The most highly purified enzyme from the DEAE-Sephadex column was enriched approximately 2400 fold and was stable for several months when stored at 4°C.

TABLE III
Purification of Aldehyde Reductase from Human Liver

Fraction	Volume	Protein	Activity	Specific Activity	Recovery
	ml	mg	units[1)]	units/mg	%
Extract	1970	190,000	1230	0.0065	100
DEAE-Cellulose	560	3,600	327	0.09	26.5
Sephadex G-100	154	169	288	1.7	23.4
Hydroxylapatite	160	26	187	7.2	15,2
DEAE-Sephadex	105	10	157	15,7	12.8

1) One unit of activity is the amount of enzyme required to catalyze the oxidation, by 0.5 mM p-NO_2-benzaldehyde, of 1µmol of NADPH per minute.

The purity of the isolated enzyme was established by polyacrylamide gel electrophoresis. A single protein and activity band was obtained when the electrophoresis was carried out in the presence of thiol reagents. On the other hand, the presence of oxidizing substances such as oxygen or o-iodosobenzoate favored the formation of multiple bands, indicating that the enzyme can form disulfide bonds. On SDS polyacrylamide gels again the reduced enzyme gave a single protein band and the oxidized enzyme showed multiple faster moving bands. Since this method separates proteins according to their molecular weight we can conclude that the enzyme forms intramolecular disulfide bridges which retain the enzyme in a partially folded conformation giving rise to a lower apparent molecular weight.

In order to compare human liver aldehyde reductase with the enzyme from other tissues and species its substrate and coenzyme specificity as well as inhibitor sensitivity were determined. NADPH was solely used as coenzyme (Km=5μM) but a relatively low substrate specificity was observed. Para-substituted benzaldehydes, methylglyoxal and glyceraldehyde as well as medium chain aliphatic aldehydes were reduced. The biogenic aldehydes derived from octopamine and norepinephrine were also found to be good substrates. No activity was detected with acetaldehyde, pyridoxal and its phosphate and hexoses. The two anticonvulsive drugs phenobarbital and 5-ethyl, 5-phenylhydantoin strongly inhibited the reduction of aldehydes by aldehyde reductase in the presence of NADPH. By these criteria aldehyde reductase from human liver is similar to the main isoenzyme from pig (3), bovine (4) human and rat brain (6) and the enzymes isolated from pig kidney (8,9) sheep (10) or bovine heart (11) and rabbit muscle (12). Thus, it seems that aldehyde reductase is an ubiquitous mammalian enzyme. Its physiological function, however, remains to be elucidated.

Furthermore, the physico-chemical properties of aldehyde reductase were investigated. A summary of the findings is given in Table IV. Three independent methods were used to determine the molecular weight: SDS-gel electrophoresis (MW = 39,400), ultracentrifugation (36,300) and gel filtration on Sephadex G-100 (33,400). The values obtained by SDS-gel electrophoresis and gel filtration differ significantly, a finding which had already been observed for the pig kidney enzyme. An unspecific adsorption to dextran could be the cause for the lower molecular weight found by gel filtration. Such a phenomenon was excluded by the use of Bio-Gel, which gave the same results as Sephadex. The stokes radius was obtained by gel filtration on a calibrated Sephadex G-100 column and the sedimentation coefficient by sedimentation velocity experiments in an analytical ultracentrifuge. From these experimentally measured parameters the values for the diffusion constant and the frictional ratio were calculated .

Circular dichroism spectra were recorded to get some information on the secondary structure of the enzyme and an α-helix content of 7-18% was computed from these recordings. Thus, a monomeric structure of near globular shape is the main characteristic of aldehyde reductase and contrasts with the oligomeric structure of most other pyridine nucleotide dependent dehydrogenases.

The amino acid composition was determined to get a rough estimate of homology with dehydrogenases of known amino acid compositions. Table V depicts the amino acid compositions of

TABLE IV
Physico-Chemical Properties of Human Liver Aldehyde Reductase

Property	Value
Molecular Weight	36,200
Sedimentation Coefficient	2.95
Stokes Radius	2.6 nm
Isoelectric Point	5.3
Extinction Coefficient	54,300 $M^{-1}cm^{-1}$
α - Helix Content	$\sim$ 10%
Diffusion Constant	7.5 · $10^{-7} cm^2 s^{-1}$
Frictional Ratio	1.18

aldehyde reductase from human liver. Only 3.4 carboxymethylcysteine residues were found when native aldehyde reductase was analysed but 6.4 residues were obtained when the enzyme had been denatured in 8M urea prior to the carboxymethylation. No similarities were detected between the amino acid compositions of aldehyde reductase and alcohol dehydrogenase from human liver as well as the ones of yeast alcohol dehydrogenase, lactate dehydrogenase, glyceraldehyde-3-P dehydrogenase or supernatant malate dehydrogenase. An obvious similarity, however, was detected between the aldehyde reductase from human liver and pig kidney. A further homology seems to exist between the liver reductase and octopine dehydrogenase, a monomeric pyridine nucleotide-dependent dehydrogenase occurring in mollusc muscle (16). Most oligomeric dehydrogenases contain an active cysteine residue in or close to the catalytic site and modification of this residue usually results in the reduction or loss of enzymatic activity. We have, therefore, tested the influence on aldehyde reductase activity of several thiol modifying reagents. Figure 2 depicts the effect of p-mercuribenzoate on the catalytic activity depending on time and the reagent concentration. A time dependent inactivation occurs, which can be prevented by NADPH. Concentrations of $NADP^+$ more than ten times higher were needed to obtain the same protective effect and NADH, NAD^+ or p-NO_2-benzaldehyde did not prevent loss of catalytic activity. Inactivation of the enzyme was also obtained by N-ethylmaleimide but only at concentrations above 1mM. Iodoacetamide, even at concentrations of 50mM and overnight did not inactivate the enzyme. This finding and the results from the amino acid analyses showing that not all cysteine residues are accessible to iodoacetamide in the native enzyme indicate that

TABLE V
Amino Acid Composition of Human Liver Aldehyde Reductase [1]

ASP	28	ILE	14
THR[2]	12	LEU	35
SER[2]	16	TYR	13
GLU	35	PHE	8
PRO	27	HIS	9
GLY	21	LYS	20
ALA	29	ARG	15
VAL	21	CYS[3]	6
MET	4	TRP[4]	7

1) Mean values of four hydrolysates
2) Extrapolated to zero hour
3) As CM-cysteine and cysteic acid
4) Spectrophotometrie according to Edelhoch (17)

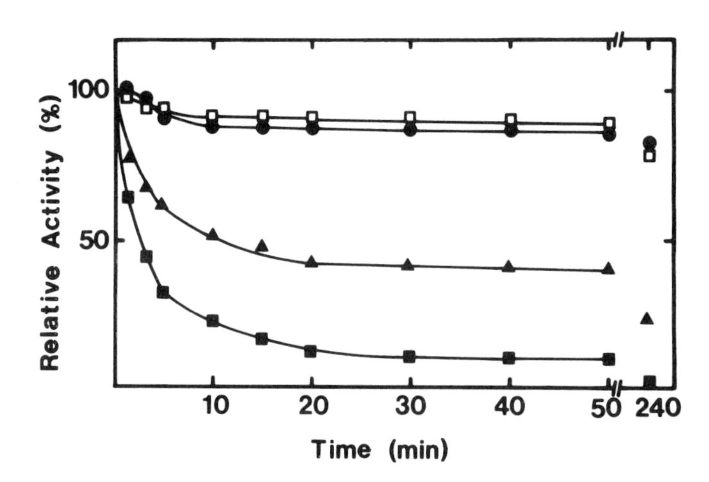

Fig. 2. Inhibition of Human Liver Aldehyde Reductase by p-Mercuri-Benzoate (pMB). Aldehyde Reductase (9.6 μM) was incubated with pMB in 10mM Na-phosphate, pH 7.0. Aliquots were taken for activity determinations after specified time intervals. Assay conditions were: 100 mM Na-phosphate, pH 7.0, 0.5mM p-NO_2-benzaldehyde and 80 μM NADPH.

the active cysteine residue is buried at the bottom of a largely hydrophobic catalytic cleft. This could also explain the preference of aldehyde reductase for aromatic and hydrophobic aldehydes. Thus, it may be concluded that aldehyde reductase and alcohol dehydrogenase differ considerably in their structural properties despite their functional similarities.

III. REFERENCES

(1). Rutledge, C.O., and Jonason, J., J. Pharmacol. Exp. Ther. 157, 493-502 (1967).
(2). Breese, G.R., Chase, T.N., and Kopin, I.J., Biochem.Pharmacol. 18, 863-869 (1969).
(3). Turner, A.J., and Tipton, K.F., Biochem. J. 130, 765-772 (1972).
(4). Tabakoff, B., and Erwin, V.G., J. Biol. Chem. 245, 3263-3268 (1970).
(5). Erwin, V.G., Heston, W.P.W., and Tabakoff, B., J. Neurochem. 19, 2269-2278 (1972).
(6). Ris, M.M., and von Wartburg, J.P., Eur. J. Biochem. 37, 69-77 (1973).
(7). Ris, M.M., Deitrich, R.A., and von Wartburg, J.P. Biochem. Pharmacol.24, 1865-1869 (1975).
(8). Bosron, W.R., and Prairie, R.L., J. Biol. Chem. 247,4480-4485 (1972).
(9). Flynn, T.G., Shires, J., and Walton, D.J. J. Biol. Chem. 250, 2933-2940 (1975).
(10). Smolen, A., and Anderson, A.D., Biochem. Pharmacol. 25, 317-323 (1976).
(11). Kawalek, J.C., and Gilbertson, J.R., Arch. Biochem. Biophys. 173, 649-657 (1976).
(12). Kormann, A.W. Hurst, R.O., and Flynn, T.G., Biochem. Biophys. Acta 258, 40-55 (1972).
(13). Rossmann, M.G., Moras, D., and Olsen, K.W., Nature 250, 194-199 (1974).
(14). Lutstorf, U.M., Schürch, P.M., and von Wartburg, J.P., Eur. J. Biochem. 17,497-508 (1970).
(15). Berger, D., Berger, M., and von Wartburg, J.P., Eur. J. Biochem. 50, 215-225 (1974).
(16). Olomucki, A., Huc, C., Lefebure, F., and Thoai, N., Eur. J. Biochem. 28, 261-268 (1972).
(17). Edelhoch, H., Biochemistry 6, 1948-1954 (1967).

HUMAN LIVER ALDEHYDE DEHYDROGENASE

R. Pietruszko, N.J. Greenfield
and C.R. Edson

Rutgers University

Two aldehyde dehydrogenases have been purified from human liver to apparent homogeneity by employing techniques of ion exchange and affinity chromatography. One of the enzymes, E_2 appears to be identical with that partially purified previously by Kramer and Deitrich (1968) and Blair and Bodley (1969). The other enzyme, E_1 has been isolated from the human liver for the first time. The enzymes differ greatly in kinetic properties with acetaldehyde as substrate and in susceptibility to inhibition to disulfiram (an inhibitor used to produce alcohol aversion in humans). Even though human autopsy livers are unsuitable for subcellular fractionation, it appears possible that E_2 is a mitochondrial enzyme. Aldehyde dehydrogenase isolated previously were reported to differ in molecular weight. Our results show that molecular weights of all aldehyde dehydrogenases isolated so far from human liver are in the region of 230 000 daltons. Previously reported differences appear to be due to interactions with the Sephadex columns. The data presented also indicate that there are more than two aldehyde dehydrogenases in the crude extracts of the human liver.

I. INTRODUCTION

Aldehyde dehydrogenases have been purified from several sources (1,2,3,4,5,6,7). In 1968 Kramer and Deitrich (8) reported purification from human liver, Blair and Bodley (1969) (9) also purified the same enzyme by an independent procedure. With the exception of molecular weight, reported to be 90,000 by the first group of investigators and 200,000 by the second group, the properties of the isolated enzyme were similar. The enzyme had a low Km for acetaldehyde and other aldehydes and showed relatively low sensitivity to disulfiram, an inhibitor used to induce alcohol aversion in humans.

Horse (7), rat (10) and sheep (4) aldehyde dehydrogenase activity has been shown to be distributed among various subcellular components. In the rat, the activity has been found in the cytosolic, mitochondrial and microsomal fractions of the liver. The isolation of two homogenous aldehyde dehydrogenases (7) from horse liver, one of cytosolic and the other of mitochondrial origin, and partial purification of two similar enzymes from sheep liver cytosol and mitochondria (4) are evidence supporting distribution of multiple molecular forms among different subcellular organelles.

1. Subcellular Distribution of Aldehyde Dehydrogenase

When samples of human liver, obtained from autopsy 10 - 24 hours after death, are used for subcellular fractionation (11) and subsequent quantitation by assay, the majority of aldehyde dehydrogenase activity appears to be localized in the cytoplasm. Mitochondria contain only a few percent of total activity. In Table 1 the result of fractionation of human liver sample is compared with that of (a) fresh rat liver, (b) frozen to -70° immediately after decapitation and (c) stored at room temperature overnight. The latter was done to simulate temperature conditions prevalent before autopsy in the slow cooling human body. The results show that levels of aldehyde dehydrogenase activity extractable from mitochondria and microsomes are profoundly affected by storage at room temperature, total activity being considerably more affected than the specific activity. The levels of aldehyde dehydrogenase activity in the cytosolic fraction are unaffected by either storage or freezing. In the control liver sample about 17% of total activity is present in the cytosolic fraction. The relative contribution of the cytosolic fraction to the total activity increases on freezing or storage to ca. 40% due to the loss of activity from the mitochondria and the microsomes. The virtual absence of significant activity in the human microsomes and low mitochondrial levels may therefore be due to conditions prevalent before autopsy.

It is of interest to note that at 0.068 mM propionaldehyde the level of aldehyde dehydrogenase activity per gram of human liver is about twice as high as that in the rat (Table 1).

2. Purification of Human Liver Aldehyde Dehydrogenase

Two aldehyde dehydrogenases have been purified to homogeneity employing a sequence of procedures in conditions described in Table 2. The most important step is affinity chromatography on 5'AMP-Sepharose 4B, the use of which allows for a relatively simple purification with good yield (12). Two aldehyde dehydrogenases co-purify and are separated in the last step of the

TABLE 1

Subcellular distribution of aldehyde dehydrogenase activity in human autopsy liver; effect of freezing or storage overnight at room temperature on apparent distribution of aldehyde dehydrogenase in rat liver.

Assay	*Tissue*	*Subcellular Fractions*								*Sum of Activity of Fractions*	
Propionaldehyde (mM)		*Mitochondria Extract*		*Mitochondria Membranes*		*Microsomes*		*Supernatant*			*%*
		A	*SA*	*A*	*SA*	*A*	*SA*	*A*	*SA*	*A*	*of Control*
	Rat liver										
0.068	control	168	10.1	128	2.7	192	2.6	91	0.7	579	100
0.068	frozen	106	8.7	14	0.5	135	2.0	--	--	-	-
0.068	stored	48	23.1	76	1.6	22	1.0	119	1.3	265	46
13.5	control	593	35.6	1012	21.2	1417	18.8	618	4.7	3640	100
13.5	frozen	270	22.0	680	21.9	2084	30.5	556		3590	99
13.5	stored	53	25.4	511	11.1	282	13.7	438	4.8	1284	35
	Human Liver										
0.068	autopsy	59		0		0		1004		1063	

The assay system contained: pyrophosphate buffer, 0.09M pH 9.0; pyrazole, 0.1M; NAD, 0.45mM; at propionaldehyde concentrations listed in the table. The reaction was started by addition of enzyme and followed on a Varian 635 spectrophotometer at 340nm and 25°. A=activity in nmoles of NADH formed/min/organelles from 1g of liver. SA= Specific activity in nmoles NADH/mg protein/min. Protein was determined by a Lowry procedure, employing bovine serum albumin as a primary standard. Control = fractionation done immediately after decapitation; frozen = liver frozen to -70° immediately after decapitation; stored = liver left on bench overnight.

procedure on Whatman DE-32 microgranular cellulose.

TABLE 2
Purification of Human Liver Aldehyde Dehydrogenases E_1 and E_2.

Step	Buffer	Total Activity (μmoles/min)	Specific Activity (μmoles/min)
Centrifuged Dialysed Homogenate	Buffer 1	376	0.02
CM-Sephadex	applied and eluted in Buffer 1	220	0.03
DE-Sephadex	applied in Buffer 2, eluted by NaCl gradient in Buffer 2	167	0.06
5'AMP-Sepharose	applied in Buffer 1, eluted with Buffer 3	135	0.64
Sephadex G-100	applied and eluted with Buffer 1	88	0.7
Separation of E,from E_2,DE-52	applied in Buffer 4, eluted in Buffer 1		
	E_1	12	0.58
	E_2	51	1.0

Buffer 1 = 30 mM sodium phosphate pH 6.0; Buffer 2 = 30 mM sodium phosphate pH 6.8; Buffer 3 = 100 mM sodium phosphate pH 8.0; Buffer 4 = 25 mM sodium phosphate pH 6.0. All buffers contained: 0.1% 2-mercaptoethanol and 1 mM EDTA. Assay system as in reference (5).

The isolated homogenous enzymes, E_1 and E_2 have relatively low specific activity indicating that human liver contains a large amount of aldehyde dehydrogenase protein. Taking an average specific activity as 0.7 μmoles/min/mg it can be calculated that human liver contains about 1.8 g of aldehyde dehydrogenase per Kg.

TABLE 3
Catalytic Properties of Homogenous E_1 and E_2 Enzymes

Substrate	E_1 Km (μM)		E_2 Km (μM)	
	pH 7.0	pH 9.0	pH 7.0	pH 9.0
Acetaldehyde	2500	125	3.0	2.4
Propionaldehyde	1000	11	0.7	1.2
NAD	40	8	70	70

Km values for acetaldehyde and propionaldehyde were determined in 0.1 M sodium phosphate buffer at pH 7.0 and in 0.09 M sodium pyrophosphate buffer pH 9.0 at 0.46 mM NAD and 25°. The Km values for NAD were determined at 13.5 mM propionaldehyde for E_1 and 0.068 propionaldehyde for E_2. Measurements were made in a Varian 635 spectrophotometer at 340 nm, the reaction was started by addition of enzyme.

3. Some Properties of the Homogenous Enzymes

On electrophoresis on starch, polyacrylamide and cellulose acetate between pH 5.5 and 9.0 both enzymes migrate anodally and are readily but not widely separable; E_1 migrates slower than E_2. On ployacrylamide gradient gel (13) E_1 also migrates slower than E_2 (molecular weight calculated by this method = 245,000 and 225,000 ± 10% respectively). On polyacrylamide SDS gels (14) the molecular weights of subunits were 54,000 and 53,000 respectively. Both enzymes gave similar extinction coefficients at 280 nm (0.96 for E_1 and 1.0 for E_2), indicating similar tryptophan content.

Michaelis constants for acetaldehyde and propionaldehyde with NAD as coenzyme and for NAD at constant propionaldehyde are listed in Table 3. The constants were determined at pH 7.0 as well as at pH 9.0. E_1 has lower affinity for acetaldehyde and propionaldehyde, which is much more apparent when the Km is determined at pH 7.0 Although at pH 9.0 E_1 has higher affinity for NAD this difference is considerably less at pH 7.0.

Since the Km value for E_1 with acetaldehyde at pH 7.0 is in the millimolar range (lethal to a mammalian organism) it is surprising to find that this enzyme, and not E_2 is highly susceptible to inhibition by disulfiram (Ki = 0.01 μM), an inhibitor of aldehyde dehydrogenase *in vivo*. E_2, which has a low Km with acetaldehyde is not inhibited by disulfiram concentrations below 40 μM. E_2 has physical and catalytic properties of the enzyme isolated previously from human liver (8,9). E_1 was never isolated previously from human liver, probably due to its relative lability in the absence of stabilizing agents, is similar in properties to the F_1 enzyme isolated from horse liver by Eckfeldt *et al.* (7).

4. Comparison of E_1 and E_2 with Previous Purifications

Aldehyde dehydrogenase isolated from human liver by Kramer and Deitrich (8) and by Blair and Bodley (9) were reported to have molecular weights of 90,000 and 200,000, respectively. In order to investigate this, aldehyde dehydrogenase was prepared in our laboratory from the same human liver, by exact repetition of the methods described by Kramer and Deitrich (A) and Blair and Bodley (B). Neither preparation was homogeneous. The preparation (B) had higher specific activity and less electrophoretically identifiable protein impurities than preparation (A). Both preparations were compared with homogeneous enzymes E_1 and E_2.

The molecular weights of (A), (B), E_1 and E_2 were determined by employing Sephadex G-200 column in 5.0 mM sodium phosphate buffer - conditions employed by Kraemer and Dietrich (8). Aldehyde dehydrogenase activity of (B), E_1 and E_2 were eluted at the same volume as lactate dehydrogenase marker, giving apparent molecular weight of *ca.* 145,000. Aldehyde dehydrogenase activity of (A) was eluted at larger volume giving apparent molecular weight of 215,000 daltons. On electrophoresis on starch gel and cellulose acetate however, (A) and (B) migrated identically with the homogeneous E_2 enzyme. On polyacrylamide gradient gel both (A) and (B) also migrated in the way identical to E_2. Both (A) and (B) therefore are the same as E_2 and have identical molecular weights in the region of 230,000 daltons.

Aldehyde dehydrogenases from human liver appear to interact with Sephadex making gel filtration methods on Sephadex unsuitable for the molecular weight determination. At low ionic strengths human liver aldehyde dehydrogenases are retarded by a Sephadex G-200 column and this retardation is decreased but not eliminated by increasing the ionic strength. This may also apply to aldehyde dehydrogenases from sources other than human liver.

5. Use of E_1 and E_2 to Identify Aldehyde Dehydrogenases in the Human Liver

In 5.0 mM phosphate buffer at pH 7.0 electrophoresis of crude human liver extracts yields two bands staining with

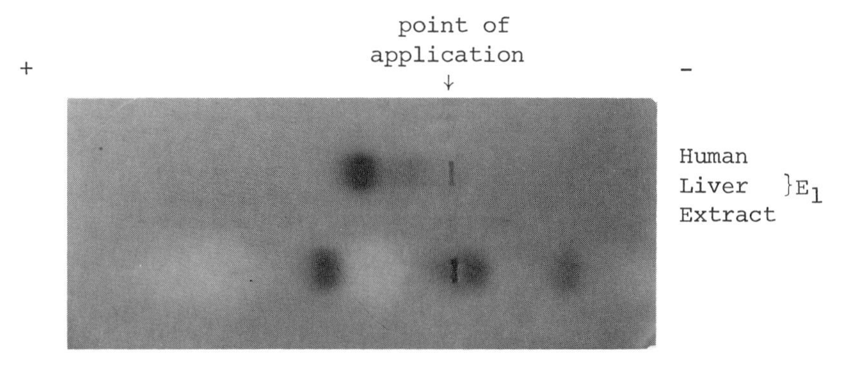

Fig. 1. Electrophoresis of human aldehyde dehydrogenase on starch gel in citrate-tris buffer pH 5.5. The buffer concentration in the gel was 10 mM and 50 mM in the electrode vessels. The gel was run overnight at 150 V and developed by incubation with 13.5 mM propionaldehyde, 1 mM NAD, 0.1 mM phenazine methosulphate and 0.5 mM nitroblue tetrazolium in 0.1 M pyrophosphate buffer, pH 8.5. In a separate experiment human liver alcohol dehydrogenase was seen to migrate in this system towards the cathode faster than any of the bands seen in the figure.

aldehyde substrates, in the presence of NAD, phenazine methosulphate and nitroblue tetrazolium. The more anodal band corresponds in electrophoretic mobility to the E_2 enzyme and the less anodal band to the E_1 enzyme. In these conditions a third weakly staining band can be detected whose mobility towards anode is less than of E_1.

At pH 5.5 in citrate-tris buffer aldehyde dehydrogenase migrates in well defined compact bands (Fig. 1). In the crud-homogenates three bands staining with propionaldehyde as substrate can be readily visualised. The most anodal of these corresponds to E_2 and also to preparations (A) and (B). There is however a bleached area (possibly due to superoxide dismutase) where E_1 should be, as demonstrated in Fig. 1 by employing homogeneous E_1 as control. Presence of E_1 in the crude liver extracts cannot be for this reason demonstrated by electrophoresis in this system. This electrophoresis however readily demonstrates demonstrates existence in the crude homogenate of more than two aldehyde dehydrogenase enzymes. Crude human liver extracts appear to contain at least four protein bands detectable on starch gel with propionaldehyde in the presence of NAD, phenazine methosulphate and nitroblue tetrazolium.

Because of the conditions of availability of the human tissues the samples of human liver are unsuitable for the study of subcellular localization of the enzymes. We have been able, however, to demonstrate the presence of the E_2 enzyme by electrophoresis of a mitochondrial extract of human liver alongside E_2 control. It appears likely therefore that E_2 occurs in the human mitochondria.

II. REFERENCES

(1). Racker, E., J. Biol. Chem. 177, 883 (1949).
(2). Maxwell, E.S. and Topper, Y.J., J. Biol. Chem. 236, 1032 (1961).
(3). Erwin, V.G. and Deitrich, R.A., J. Biol. Chem. 241, 3533 (1966).
(4). Crow, K.E., Kitson, T.M., McGibbon, A.K., and Batt, R.D., Biochem. Biophys. Acta. 350, 121 (1974).
(5). Feldman, R.I., and Weiner, H., J. Biol. Chem. 247, 260 (1972).
(6). Shum, J.G. and Blair, A.H., Canad. J. Biochem. 50, 741 (1972).
(7). Eckfeldt, J., Mope, L., Takio, K., and Yonetani, T., J. Biol. Chem. 251, 236 (1976).
(8). Kramer, R.J., and Deitrich, R.A., J. Biol. Chem. 243, 6402 (1968).
(9). Blair, A.H., and Bodley, F.H., Canad. J. Biochem. 47, 265 (1969).
(10). Tottmar, S.O.C., Peterson, H., and Klessling, K.H., Biochem. J. 135, 577 (1973).
(11). Hogeboom, G.H., Methods in Enzymology 1, 16 (1957).
(12). Greenfield, N.J., and Pietruszko, R., (submitted to Biochim. Biophys. Acta).
(13). Andersson, L.O., Borg, H., and Mikaelsson, M., FEBS Letters 20, 199 (1972).
(14). King, T. and Laemli, M., J. Mol. Biol. 62, 465 (1971).

III. ACKNOWLEDGEMENT

Financial support of UPHS NIAAA Grant 00AA186 is gratefully acknowledged.

INHIBITION OF RAT-LIVER ALDEHYDE DEHYDROGENASES *IN VITRO* AND *IN VIVO* BY DISULFIRAM, CYANAMIDE AND THE ALCOHOL-SENSITIZING COMPOUND COPRINE

O. Tottmar, H. Marchner and P. Lindberg

Institute of Zoophysiology and the Lund Institute of Technology

Coprine (N^5-(1-hydroxycyclopropyl)-L-glutamine), the disulfiram-like constituent of Coprinus atramentarius, is an inhibitor of aldehyde dehydrogenase (ALDH) in vivo. It has a rapid onset of action like cyanamide and demonstrates highest inhibition after 2-12 h with a long duration of action like disulfiram with measurable inhibition after 144 h. All three inhibitors decreased the activity of the low-K_m ALDH strongly in vivo, but only cyanamide affected the high-K_m ALDH. Coprine did not inhibit ALDH in vitro, but one of its probable metabolites, 1-aminocyclopropanol hydrochloride, was effective both in vitro and in vivo. Disulfiram was 4-5 times more potent in vitro than the two other inhibitors. The inhibition produced by the inhibitors appeared to be irreversible. Acetaldehyde protected the low-K_m ALDH from inhibition both in vitro and in vivo. An ALDH-inhibitor, calcium cyanamide, was found in calcinated bonemeal, which is widely used in animal diets. Rats fed such diets showed an apparent low sensitivity to disulfiram treatment, and in alcohol feeding experiments, a two-fold increase in the ALDH-activity and a 30% increase in the rate of ethanol elimination were observed. These diets may cause serious misinterpretations of the results obtained in alcohol studies.

I. INTRODUCTION

Disulfiram (antabuse) and calcium cyanamide (Temposil) have been used as adjuncts in alcohol therapy for many years (1,2). These drugs are well-known aldehyde-dehydrogenase inhibitors and cause an accumulation of acetaldehyde in the body during ethanol metabolism. Several other compounds are known to interfere with the metabolism of acetaldehyde and to evoke a hypersensitivity to alcohol (1,2). Recently, the disulfiram-like constituent (coprine) of the inky cap mushroom, *Coprinus atramentarius*, has been isolated and identified as a cyclopropanone derivative (N^5-(hydroxycyclopropyl)-L-glutamine) (3,4). Coprine and several analogues have been synthesized (3). Coprine inhibits the ALDH-activity in rat and mouse liver *in vivo* but not *in vitro* (4,5). It has been suggested that coprine *in vivo* acts *via* the hydrolytic product, 1-aminocyclopropanol (ACP), which is a potent ALDH-inhibitor *in vitro* (5).

A disulfiram-like compound has also been reported to be present in animal charcoal (6,7). Recent studies have shown that animal charcoal and calcinated bonemeal contain a cyanamide derivative (probably calcium cyanamide) (8-10). Calcinated bonemeal is widely used in animal diets as a source of calcium and phosphorus.

The aims of the present study were 1) to compare the effects of disulfiram, cyanamide, coprine and 1-aminocyclopropanol hydrochloride on the metabolism of acetaldehyde *in vitro* and *in vivo* and 2) to discuss the effects caused by feeding animals on diets containing calcinated bonemeal on the experimental results in alcohol studies.

II. MATERIALS AND METHODS

Coprine (N^5-(1-hydroxycyclopropyl)-L-glutamine) and 1-aminocyclopropanol hydrochloride (ACP) were synthetized as described by Lindberg *et al.* (3). Details about other chemicals used in this work have been given elsewhere (8-12).

Female Sprague-Dawley rats, weighing 200-250 g, obtained from Anticimex, Sollentuna, Sweden, were used in all experiments. In the long-term alcohol-feeding experiments, the rats were fed on a commercial standard diet (diet 1) contaminated with calcium cyanamide (for details, see refs. 8-10). In all other experiments, a standard diet (control diet) which did not contain calcium cyanamide was used.

In the alcohol-feeding experiments, the rats were given a 15% (v/v) solution of ethanol as the sole drinking fluid for 3 months. Pair-fed controls were given an amount of diet 1 corresponding to the total caloric intake in the alcohol-treated rats.

The low- and high-K_m ALDH were assayed as described

previously (8-12). The inhibition studies *in vitro* were performed on a semipurified preparation of the low-K_m ALDH isolated by ammonium sulphate fractionation of a matrix + intermembrane fraction isolated from rat liver mitochondria. Details about the inhibition experiments are given in the legends for the figures and tables.

Drugs were administered by a stomach tube as suspensions in 5% (w/v) gum arabicum at the doses indicated. Controls received a corresponding volume of 5% gum arabicum. Ethanol was given intraperitoneally as a 7.8% (w/v) solution in saline at a dose of 1.5 g/kg. Blood samples were taken from the tip of the tail. Ethanol and acetaldehyde were determined enzymatically as described previously (12).

III. RESULTS AND DISCUSSION

A. Inhibition of the ALDH-activity *in vitro*

Coprine at concentrations up to 3 mM did not inhibit the low-K_m ALDH, whereas 1-aminocyclopropanol hydrochloride (ACP) was found to be a potent inhibitor. A progressive decline in the enzyme activity was observed when the enzyme was preincubated with NAD^+ and ACP in the absence of acetaldehyde, and similar results were obtained in experiments with disulfiram and cyanamide (Fig. 1A). At an inhibitor concentration of 20 μM, the inhibition was 50% after an incubation period of 1-2 min with disulfiram, and after 5 min with ACP and cyanamide. The inhibition at different concentrations of the inhibitor after an incubation time of 5 min is shown in Figure 1B. The enzyme was inhibited by 50% at 5 μM disulfiram and at 20 μM ACP or cyanamide.

Cyanamide and ACP at concentrations below 0.2 mM did not inhibit the enzyme when they were added to the complete reaction mixture containing acetaldehyde (25 μM). However, at higher concentrations (0.2 - 1 mM), a gradual decrease in the activity was observed during the course of the reaction. Similar results were obtained in experiments with disulfiram, in agreement with previous findings (13). It appears that a similar protection by acetaldehyde occurred *in vivo* (see below). The effects of different concentrations of NAD^+ or acetaldehyde on the inhibition were not studied with this crude enzyme preparation. It is quite clear, however, that the inhibition by disulfiram is influenced by the concentration of NAD^+ and substrate. The time of observation is also an important parameter (13,14).

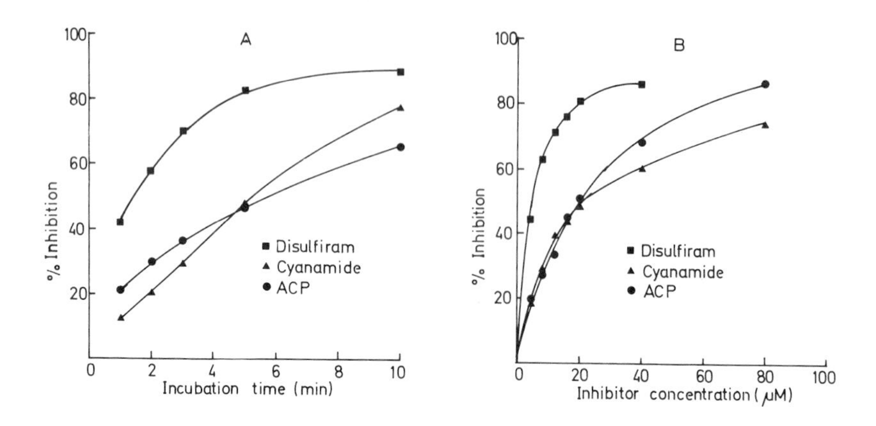

Fig. 1. Inhibition of low-K_m ALDH *in vitro by disulfiram, cyanamide and 1-aminocyclopropanol hydrochloride (ACP). The enzyme preparation (0.15 mg of protein in a final assay volume of 2.1 ml) was preincubated for various times at 23° in 50 mM sodium pyrophosphate buffer (pH 8.8) containing 0.5 mM* NAD *and inhibitor. Acetaldehyde (25 μM) was added to start the reaction. The reaction rates were linear for at least 5 min, and in the absence of the inhibitors, no loss in activity was observed during the preincubation. A) Enzyme activity after a preincubation time of 1-10 min at an inhibitor concentration of 20 μM. B) Enzyme activity at different inhibitor concentrations after a preincubation time of 5 min.*

Disulfiram appears to inhibit the aldehyde dehydrogenase in the liver irreversibly both *in vitro* and *in vivo* (14,15), and it has been suggested that disulfiram blocks SH-groups of the enzymes through a mixed-disulfide formation (14,15).

In agreement with previous reports (14,15), the inhibition by disulfiram *in vitro* could be prevented, but not reversed, by SH-reagents. This protection can probably be explained by a reduction of disulfiram to diethyldithiocarbamate by the SH-reagents. Diethyldithiocarbamate is not an ALDH-inhibitor *in vitro* (14,15). The inhibition by cyanamide and ACP could not be prevented, nor reversed, in the presence of 0.1 - 1 mM of 2-mercaptoethanol or dithiothreitol. Attempts to restore the enzyme activity by gel-filtration on Sephadex G-25 were unsuccessful in experiments with all three inhibitors. These results suggest that cyanamide and ACP, like disulfiram, cause an irreversible inhibition of the low-K_m ALDH *in vitro*. It has been reported, however, that cyanamide is a reversible inhibitor of ALDH (16).

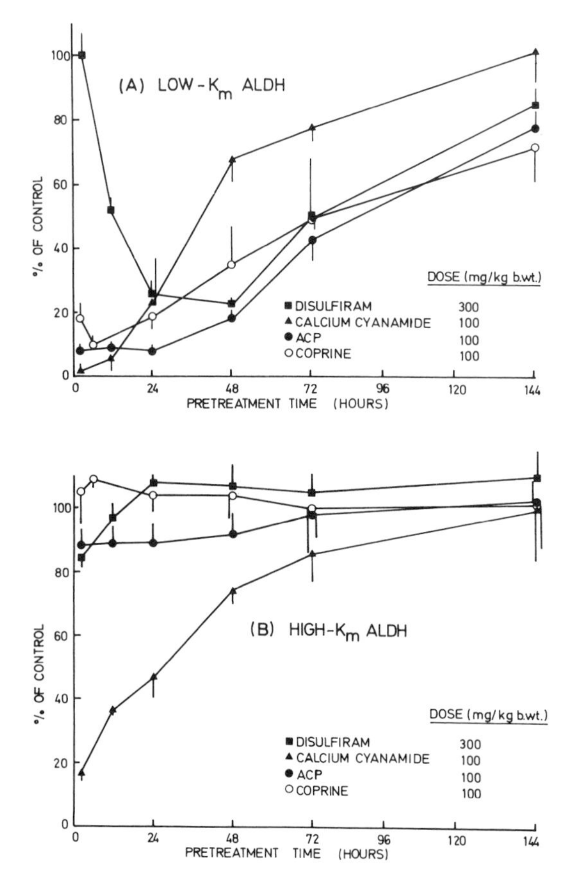

Fig. 2. Time-course of the inhibition of the low-K_m ALDH *(Fig. 2A) and the high K_m* ALDH *(Fig. 2B) in rat liver in vivo after administration of disulfiram, calcium cyanamide, coprine and 1-aminocyclopropanol. The* ALDH-*activity in liver homogenates was measured as previously described (8-11). The values are the means ± S.D. from experiments on 3 rats. The activity of the low- and high K_m* ALDH *in control rats were 6.2 ± 0.6 and 14.3 ± 2.6 nmole of* NADH*/min/mg of protein, respectively (mean ± S.D. from 6 rats).*

B. Inhibition of the ALDH-activity *in vivo*

The time-course of the inhibition of the low- and high-K_m ALDH in rat liver is shown in Fig. 2. In agreement with our previous findings, disulfiram inhibited the low-K_m ALDH markedly, whereas the total activity of the high-K_m enzyme was largely unaffected. The onset of the inhibition was slow and the highest inhibition was observed after 24 - 48 h. The activity of the low-K_m ALDH had not returned to normal levels even after 144 h. Similar results have previously been reported by Deitrich and Erwin (15).

Calcium cyanamide, coprine and ACP caused a rapid decline in the activity of the low-K_m ALDH. The inhibition was 82-98% after 2-12 h and was long-lasting in rats given coprine or ACP. In rats given calcium cyanamide, the activity returned to normal values much faster, which is consistent with the short duration of action of this drug observed in clinical situations (17) but is not consistent with all three substances being irreversible inhibitors of equal potency.

The activity of the high-K_m ALDH in the liver homogenate was markedly decreased in rats treated with calcium cyanamide, whereas no significant effects were found in rats treated with disulfiram coprine or ACP. In some preliminary experiments performed *in vitro* on the microsomal high-K_m ALDH, it was found that high concentrations of disulfiram and ACP inhibited this enzyme, whereas cyanamide at concentrations as high as 5 mM had no effect. The discrepant results obtained with cyanamide cannot be explained. It is possible, however, that a metabolic product of cyanamide potentiates the inhibition *in vivo* (see ref. 18).

C. Effects of Coprine and 1-Aminocyclopropanol (ACP) on Blood Acetaldehyde Levels during Ethanol Metabolism

The blood acetaldehyde level during ethanol metabolism was approximately 20 times higher in rats treated with coprine or ACP for 24 h at a dose of 27 mg/kg, and the rate of ethanol elimination was decreased by 20-30% (324 ± 30, 293 ± 13 and 410 ± 18 mg ethanol/h/kg for coprine-treated, ACP-treated, and control rats, respectively. Mean ± S.D. for 5 rats) (Fig. 3). Similar results were obtained in rats pretreated for 2 h. The time-course of the inhibition of acetaldehyde oxidation was followed at different doses of coprine (9, 27 and 81 mg/kg). High acetaldehyde levels were found 72 h after administration of coprine, and at the highest dose, an increased level was measurable after 144 h. The results obtained with ACP were almost identical.

The fact that coprine inhibits the low-K_m ALDH *in vivo* but not *in vitro* suggests that an active metabolite of coprine is formed *in vivo*. Acid hydrolysis of coprine yields glutamic acid and 1-aminocyclopropanol hydrochloride (ACP), and it is likely that the latter compound is responsible for the inhibition caused by coprine *in vivo*.

Work is now in progress to evaluate the possible use of coprine and some related compounds for alcohol therapy.

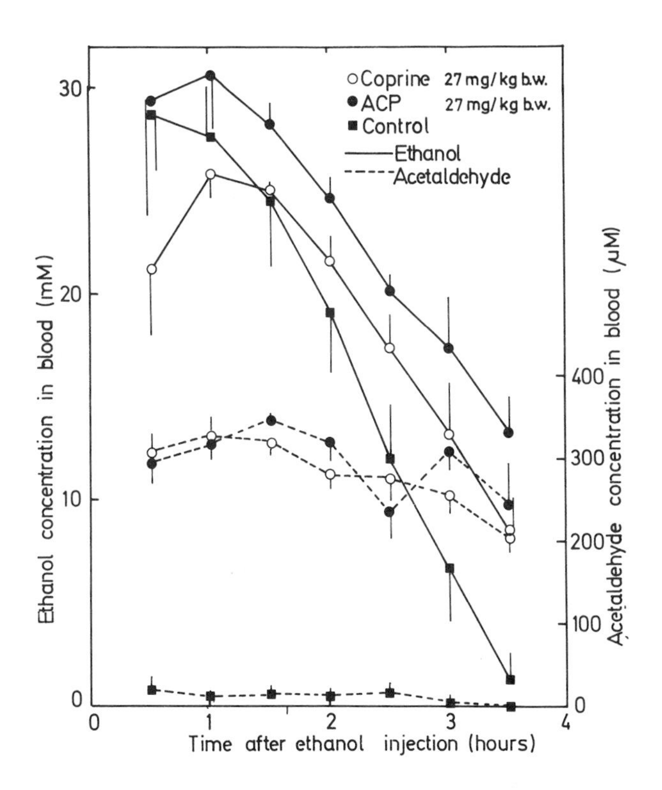

Fig. 3. Effects of coprine and 1-aminocyclopropanol on the acetaldehyde level and the rate of ethanol elimination in vivo. Rats were given ethanol (1.50 g/kg) intraperitoneally 24 h after the administration of the inhibitors (27 mg/kg). The values are the means ± S.D. from experiments on 5 rats.

D. Effects of Administration of ALDH-inhibitors to Intoxicated Rats

The low-K_m ALDH was much less susceptible to inhibition by the inhibitors in intoxicated rats than in control rats (Table 1). The acetaldehyde level in rats given coprine or calcium cyanamide 15 min after administration of a single injection of ethanol did not increase during the period of ethanol elimination (3-4 h). In agreement with the results obtained *in vitro* (see above), this suggests that acetaldehyde formed during ethanol metabolism protects the low-K_m ALDH from inhibition by these agents.

TABLE 1
Effects of administration of ALDH-inhibitors to intoxicated rats.

	Activity of low-K_m ALDH (n mole NADH/min/mg of prot.)		
	Ethanol	*Ethanol + Inhibitor*	*0.9% NaCl + Inhibitor*
Control	7.1 ± 0.7	-	-
Disulfiram (300 mg/kg)	-	6.8 ± 1.2	4.6 ± 0.7
Calcium cyanamide (30 mg/kg)	-	4.6 ± 2.4	0.8 ± 0.3
Coprine (30 mg/kg)	-	5.1 ± 0.5	2.8 ± 0.3

The inhibitors were given by stomach tube to rats 15 min after an intraperitoneal injection of ethanol (1.50 g/kg) or 0.9% NaCl-solution. The ALDH-activity was measured in liver homogenates 3 h later. Mean ± S.D. (n = 6).

These observations may be of clinical interest, since it is known that the hypersensitivity to alcohol produced by calcium cyanamide can easily be circumvented by intake of alcohol prior to the drug.

E. Effects of Feeding Rats on Diets Contaminated with Calcium Cyanamide on the Experimental Results in Alcohol Studies

In rats given a 15% solution of ethanol for 3 months and fed on a diet containing calcinated bonemeal, the activity of the low-K_m ALDH in the liver was 2 times higher, the rate of ethanol elimination was 26% higher, and the acetaldehyde level in blood during ethanol metabolism was 2-3 times lower, as compared to pair-fed control rats (Table 2). The results obtained for the alcohol-treated rats were similar to these obtained in experiments on control rats fed on a diet not containing calcinated bonemeal (Table 2).

These effects of chronic alcohol intake on the oxidation of ethanol and acetaldehyde, as observed in this study, were probably caused mainly by the presence of calcium cyanamide in the diet. A plausible explanation, as judged from the results previously discussed in this paper (see under studies *in vitro* and Table 1), is that acetaldehyde in the alcohol-treated rats protected the low-K_m from inhibition by cyanamide.

TABLE 2
Effects of Long-term Ethanol Feeding on the Metabolism of Acetaldehyde and Ethanol in Rats Fed on a Diet Containing Calcinated Bonemeal

Experimental group		*Activity of low K_m ALDH (nmole NADH/min/ mg of protein)*	*Acetaldehyde level in blood (μM) Time after ethanol administration*		*Rate of ethanol elimination (mg/h/kg of b.w.)*
			30 min	*90 min*	
Diet containing calcinated bonemeal					
Controls	(4)	2.6 ± 0.8	170 ± 54	114 ± 32	335 ± 14
Ethanol-treated	(4)	6.2 ± 0.3	74 ± 40	40 ± 19	453 ± 23
Diet containing no calcinated bonemeal					
Controls	(7)	6.8 ± 0.7	52 ± 16	45 ± 13	478 ± 41

Details about animal treatment and diets have been given in the Materials and Methods section. Mean ± S.D. (n = 4-7)

Disulfiram (100-300 mg/kg of b.w.) caused very little inhibition of ALDH in rats fed diets containing calcinated bonemeal.

Calcinated bonemeal is widely used in animal diets. As is evident from the present study, feeding animals diets containing calcinated bonemeal may cause serious misinterpretations of the results obtained in studies on the metabolism and pharmacology of ethanol and acetaldehyde, and also, in studies involving other drugs and biogenic amines and aldehydes.

IV. ACKNOWLEDGEMENTS

This investigation was supported by grants from the Swedish Medical Research Council (grant # 04743) and from the Swedish Board for Technical Development. The authors wish to thank Miss Kerstin Holmberg for valuable technical assistance.

V. REFERENCES

(1). Truitt, E.B., Jr., and Walsh, M.J., in "The Biology of Alcoholism" (B. Kissin and H. Begleiter, Eds.), Vol. 1, pp. 161-195. Plenum Press, New York - London, 1971.

(2). Wallgren, H., and Barry, H., III, "Actions of Alcohol", Vol. 2, pp. 674-679. Elsevier, Amsterdam, London, New York, 1970.

(3). Lindberg, P., Bergman, R., and Wickberg, B., J. Chem. Soc. Chem. Commun. 946 947 (1975).

(4). Hatfield, C.M., and Schaumberg, J.P., Lloydia 38, 489-496 (1975).

(5). Tottmar, O., and Lindberg, P., Acta Pharmacol. et Toxicol. (1976), in press.

(6). Moench, G.L., New York J. Med. 50, 308 (1950).

(7). Clark, W.C., and Hulpieu, H.R., J. Pharmacol. Exp. Therap. 123, 74-80 (1958).

(8). Tottmar, O., and Marchner, H., in "The Role of Acetaldehyde in the Actions of Ethanol" (K.O. Lindros and C.J.P. Eriksson, Eds.), Vol. 23, pp. 47-66. The Finnish Foundation for Alcohol Studies, 1975.

(9). Marchner, H., and Tottmar, O., Acta Pharmacol. et Toxicol. 38, 59-71 (1976).

(10). Marchner, H., and Tottmar, O., Acta Pharmacol. et Toxicol. 39, 331-343 (1976).

(11). Tottmar, O., Pettersson, H., and Kiessling, K.-H., Biochem. J. 135, 577-586 (1973).

(12). Tottmar, O., and Marchner, H., Acta Pharmacol. et Toxicol. 38, 366-375 (1976).

(13). Deitrich, R.A., and Hellerman, L., J. Biol. Chem., 238, 1683-1689 (1963).

(14). Kitson, T.M., Biochem. J., 151, 407-412 (1975).

(15). Deitrich, R.A., and Erwin, V.G., Mol. Pharmacol., 7, 301-307 (1971).

(16). Deitrich, R.A., and Erwin, V.G., Fed. Proc., 34, 1962-1968 (1975).

(17). Ferguson, J.K.W., Can. Med. Ass. J., 74, 793-795 (1956).

(18). Deitrich, R.A., and Worth, W.S., Fed Proc., 27, 237 (1968).

CORRELATION BETWEEN THE ACTIVITY OF ACETALDEHYDE DEHYDROGENASE AND THE STRUCTURE OF MITOCHONDRIA

In-Young Lee and Britton Chance

University of Pennsylvania

The turnover number of cytochrome c, k_3, which is limited by the activity of acetaldehyde dehydrogenase, was measured in rat liver mitochondria in the normal condensed state and in the swollen state. In normal mitochondria, the value of k_3 reached a maximum below 0.033 mM acetaldehyde and decreased rapidly with an increase of the concentration of acetaldehyde which is indicative of substrate inhibition. On the other hand, in hypotonically swollen mitochondria, the value of k_3 was very low and scarcely dependent upon the concentration of acetaldehyde. When mitochondria were induced to be swollen either in 10 mM phosphate or in 100 nmoles Ca^{++} ions/mg mitochondrial protein, the value of k_3 was lower than that of the control at low concentrations of acetaldehyde. The value was also less dependent upon the concentration of acetaldehyde under those conditions. These findings strongly suggest that the characteristics of mitochondrial acetaldehyde dehydrogenase vary with the structural state of mitochondria. In the condensed state, the dehydrogenase has a higher turnover number and shows higher sensitivity toward substrate inhibition. In the swollen state, it has a lower turnover number and exhibits lower sensitivity toward substrate inhibition.

I. INTRODUCTION

Hepatic oxidation of ethanol to acetate involves two oxidation steps, each catalyzed by an NAD-dependent enzyme. Most of the NADH generated during the metabolism of ethanol has to be oxidized by the mitochondrial electron transport chain, since cytoplasmic reactions are inadequate for reoxidation of NADH. Thus, the capacity of mitochondria to oxidize NADH becomes rate-limiting in the over-all metabolism of ethanol. In view of this importance of mitochondria in the oxidation reaction of NADH, the present studies were undertaken to measure the rate of electron flux from acetaldehyde dehydrogenase located in the mitochondrial matrix to the cytochrome chain bound to the inner membrane, under various conditions. By employing very sensitive fluorometric and spectrophotometric techniques, redox changes of NADH and cytochrome c upon addition of acetaldehyde were monitored directly from intact mitochondria, without treatment with detergent or addition of an excess amount of exogenous NAD+. Thus, the characteristics of mitochondrial acetaldehyde dehydrogenase were explored under the condition which was more close to a native environment. All studies described in the present paper were performed on liver mitochondria isolated from male Sprague-Dawley rats starved for approximately 18 hours.

II. RESULTS AND DISCUSSION

A. Typical Redox Changes of NADH and Cytochrome c after Additions of Various Amounts of Acetaldehyde

The steady state reduction levels of NADH and cytochrome c in rat liver mitochondria were found to be very low even in the presence of acetaldehyde. However, treatment with a small amount of azide (3.3 mM) of mitochondria significantly increased the steady state reduction levels of both NADH (approximately 75% of the total) and cytochrome c (approximately 84% of the total) and allowed an accurate measurement of redox changes of the two

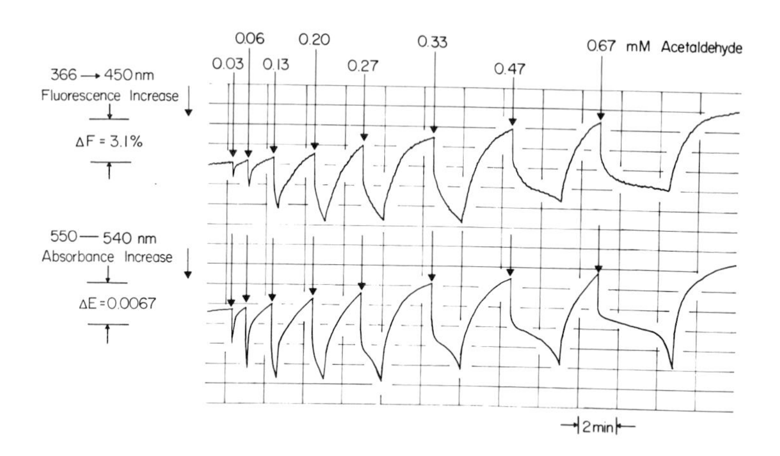

Fig.1. Redox changes of NADH and cytochrome c in rat liver mitochondria after additions of various amounts of acetaldehyde. Mitochondria were suspended in a reaction mixture containing 225 mM mannitol, 75 mM sucrose, 50 mM Na-morpholinopropane sulfonate (MOPS, pH 7.4), 2 mM K-phosphate (pH 7.4) and 3.3 mM Na-azide at a final concentration of 3.05 mg protein/ml.

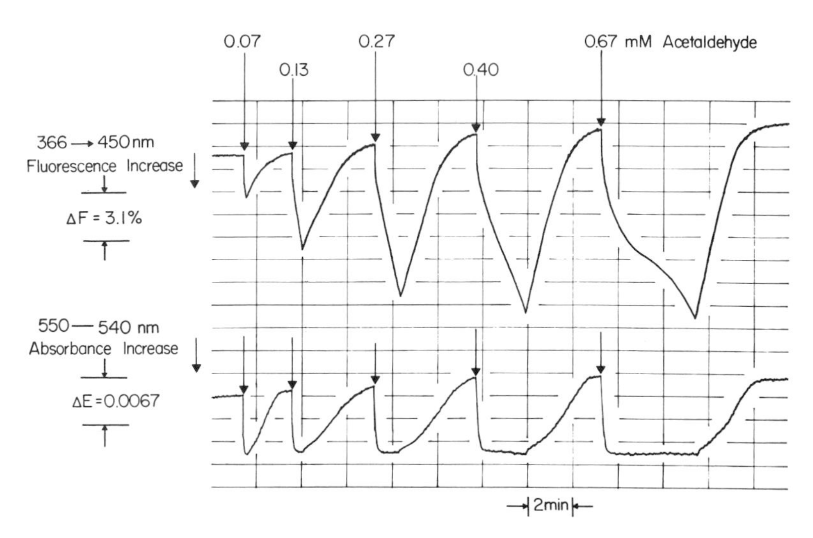

Fig.2. Redox changes of NADH and cytochrome c in hypotonically swollen mitochondria after additions of various amounts of acetaldehyde. Rat liver mitochondria were incubated in a reaction mixture containing 15 mM mannitol. 5 mM sucrose and 1 mM K-phosphate (pH 7.4) for 30 min at 0°C. Just before the measurement the temperature of the mixture was brought to 25°C and 3.3 mM Na-azide was added. The final concentration of mitochondrial protein was 3.13 mg/ml.

even when a very small amount of acetaldehyde (0.033 mM) was added to mitochondria. Fig. 1 shows recordings of reduction-oxidation changes of NADH and cytochrome c after additions of various amounts of acetaldehyde. The interesting feature of this titration experiment is that the characteristics of reduction kinetics varied with the concentration of acetaldehyde. Below 0.06 mM, the reduction consisted of a fast phase, the magnitude of which increased with acetaldehyde concentration. As the concentration was increased above 0.06 mM, a slow phase appeared in the course of the reduction, giving a characteristic biphasicity. At the same time, the area under the curve which is the measure of the decrease in k_3 (see above) increased disproportionately with the concentration. Above 0.20 mM acetaldehyde, the magnitudes of the fast and the slow phases remained more or less the same. Only the duration of the slow phase increased with the concentration.

B. Effects of Swelling of Mitochondria on the Reduction Kinetics of NAD+ and Cytochrome c by Acetaldehyde

1. Hypotonic Swelling

As shown in Fig. 2, when mitochondria were induced to be swollen by suspending in a hypotonic reaction mixture, the magnitude of NAD+ reduction by acetaldehyde increased markedly. The increase was entirely due to the increase of the slow phase. At the same time, the reduction of cytochrome c proceeded in monophasic kinetics: at all concentrations of acetaldehyde the reduction consisted entirely of a fast phase. This change in the kinetic behavior from biphasicity to monophasicity could be reversed when the tonicity of the reaction mixture was brought to isotonicity by addition of an appropriate amount of saturated sucrose (unpublished observation).

2. Phosphate-Induced Swelling

Incubation with phosphate of mitochondria is known to induce the osmotic swelling of mitochondria (1). As shown in Fig. 3, addition of acetaldehyde to mitochondria swollen in 10 mM K-phosphate caused reductions of both NAD+ and cytochrome c, the kinetics of which were quite different from those of normal mitochondria. At all concentrations of acetaldehyde, the reduction kinetics were devoid of the slow phase.

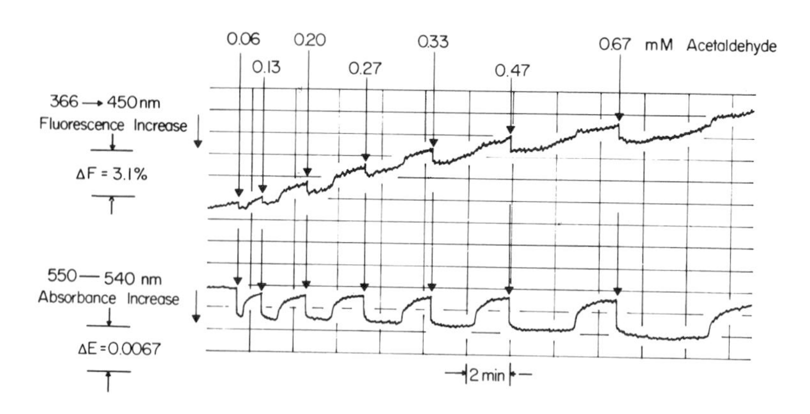

Fig. 3. Redox changes of NADH and cytochrome c in rat liver mitochondria swollen in 10 mM K-phosphate. Mitochondria were incubated in a reaction mixture containing 225 mM mannitol, 75 mM sucrose, 50 mM Na-morpholinopropane sulfonate (pH 7.4) and 10 mM K-phosphate (pH 7.4) for 30 min at room temperature. Just before the measurement, 3.3 mM Na-azide was added to the mixture. The final concentration of mitochondrial protein was 3.05 mg/ml.

3. Calcium-Induced Swelling

A large amount of Ca++ ions can be accumulated within the mitochondrial matrix in an energy-dependent manner, which results in the osmotic swelling of mitochondria (2). In the experiment of Fig. 4, mitochondria were induced to be swollen by incubating with Ca++ ions. Under this condition, addition of acetaldehyde caused reductions of both NAD+ and cytochrome c, the kinetics of which were very similar to those of mitochondria swollen in the presence of phosphate. A small fraction of the slow phase in the course of cytochrome c reduction is likely due to incomplete swelling of the calcium-treated mitochondria.

The extent of NAD+ reduction by acetaldehyde was much less in the calcium-induced or in the phosphate-induced swollen mitochondria than in the hypotonically swollen mitochondria. The cause for this difference remains obscure and has to be investigated in the future studies.

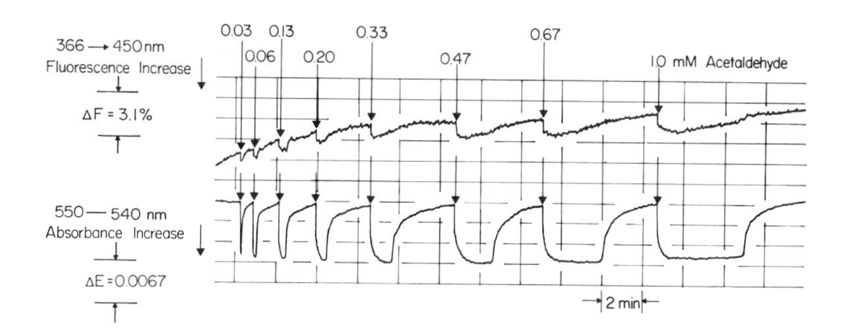

Fig.4. Redox changes of NADH and cytochrome c in rat liver mitochondria swollen in the presence of calcium ions. Mitochondria were incubated in a reaction mixture containing 225 mM mannitol, 75 mM sucrose, 50 mM Na-morpholineopropane sulfonate (pH 7.4) and 100 nmoles Ca++ ions/mg mitochondrial protein for 3 hours at 0°C. Just before the measurement the temperature of the mixture was brought to 25°C and 3.3 mM Na-azide was added. The final concentration of mitochondrial protein was 5.04 mg/ml.

4. Comparison of the Rate of Electron Flux at the Level of Acetaldehyde Dehydrogenase in the Normal Condensed and in the Swollen Mitochondria

In the mitochondrial matrix, acetaldehyde dehydrogenase and other NAD-dependent dehydrogenases are in an equilibrium with each other through the common NAD pool. Thus, in intact mitochondria, the rate of electron flux at the level of acetaldehyde dehydrogenase is difficult to measure directly from NADH changes caused by acetaldehyde. However, the rate can be measured indirectly at the level of cytochrome c, since the electron transfer from NADH to cytochrome c can occur within 90 msec (3-7) and since cytochrome c is in the equilibrium with only cytochromes c_1 and a. Under the experimental condition of Fig. 1, the oxidation rate of acetaldehyde in rat liver mitochondria of State 3 (active state) was estimated to be 11 nmoles O_2/min/mg protein. This oxidation rate is approximately one-fourth the oxidation rate of malate plus glutamate and is about one-sixth that of succinate under the same experimental condition. The highest turnover number of cytochrome c in the presence of acetaldehyde is on the average 14 sec^{-1} (cf. Fig. 5), which is approximately one-fifth of the maximal value for cytochrome c in mitochondria (8). Thus, during the oxidation of acetaldehyde, the turnover number of cytochrome c is limited by the rate of electron flux at the level of acetaldehyde dehydrogenase. Under this condition any change in the turnover number of cytochrome c will be directly related with the change in the activity of acetaldehyde dehydrogenase. The turnover number of cytochrome c, k_3, which is limited by the activity of acetaldehyde dehydrogenase, is then calculated according to the equation (9):

$$k_3 = \frac{[\text{Acetaldehyde}]}{P_{max} \cdot T_{1/2}}$$

where P_{max} represents the maximal reduction of cytochrome c (in moles/liter) and $T_{1/2}$ refers to the time for a half maximal reduction (in seconds).

The plots of k_3 against the concentration of acetaldehyde in the normal condensed state and in the swollen state are compared in Fig.5. In all three control experiments in which mitochondria were in the normal condensed state, the value of k_3 reached a maximum below 0.033 mM and decreased very rapidly as the concentration of acetaldehyde was increased from 0.03 to 0.3 mM. Above 0.3 mM the value of k_3 remained low and did not change any further with an increase of the concentration of acetaldehyde. This finding is highly indicative that the activity of acetaldehyde dehydrogenase in intact mitochondria is subjected to substrate inhibition well below 0.033 mM acetaldehyde. It is understandable that the appearance of the slow phase at high concentration of acetaldehyde is attributable to the inhibition of the dehydrogenase by substrate. As shown in Fig. 5A, the value of k_3 in hypotonically swollen mitochondria was approximately one-third that of the control, at 0.03 mM acetaldehyde. Moreover,

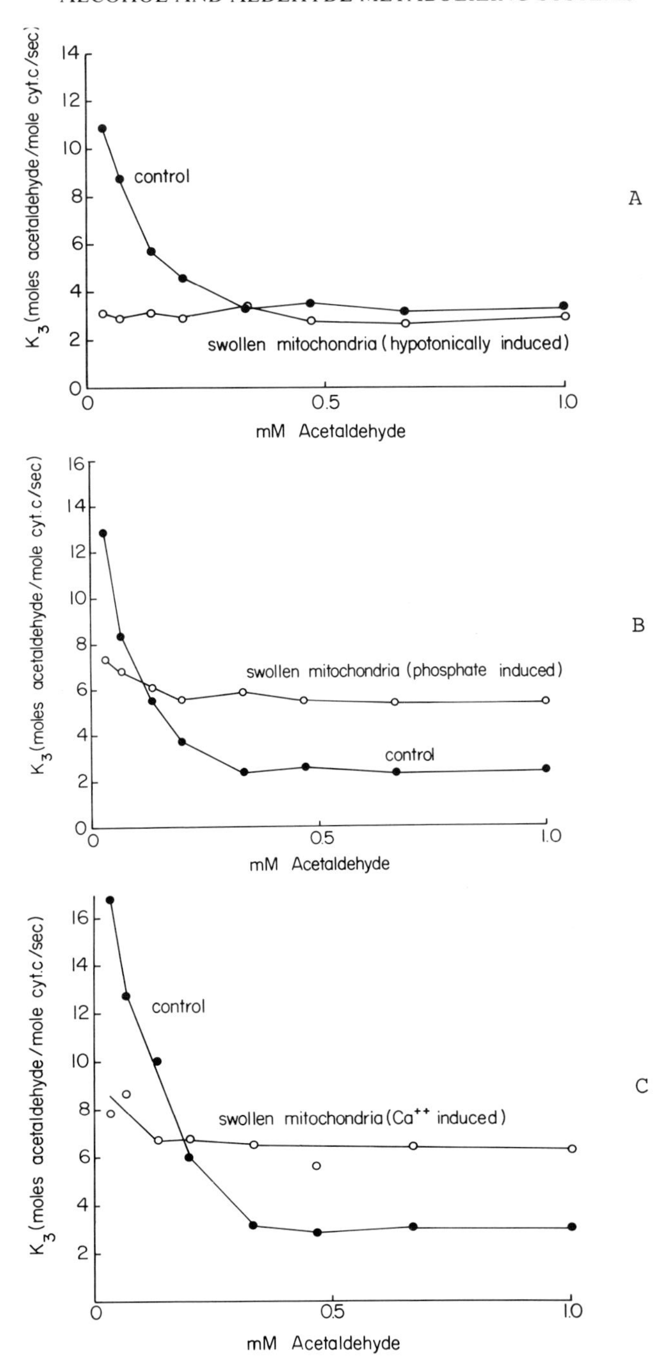

Fig. 5. Plots of k_3 against the concentration of acetaldehyde in the normal condensed and in the swollen mitochondria.

the value was scarcely dependent upon the concentration of acetaldehyde. In mitochondria swollen in the presence of either phosphate (Fig. 5B) or Ca++ ions (Fig. 5C), the value of k_3 was approximately half that of the control, at 0.03 mM acetaldehyde. The value was also much less dependent upon the concentration of acetaldehyde. Thus, the value became higher than the control at the concentration where substrate inhibition should occur. The possibility that substrate inhibition of mitochondrial acetaldehyde dehydrogenase is reversed by phosphate or by Ca++ ions remains to be clarified in the future studies.

Addition of exogenous NAD+ (2.0 mM) and/or cytochrome c (0.67 μM) to swollen mitochondria slightly increased the steady state reduction levels of both NADH and cytochrome c. This suggests that during the process of mitochondrial swelling a part of NAD+ and cytochrome c leaked out from mitochondria. However, addition of the two in an excess amount to swollen mitochondria did not restore the characteristics of acetaldehyde dehydrogenase in the normal condensed state of mitochondria.

The loss of sensitivity toward substrate inhibition and the marked decrease in k_3 upon swelling of mitochondria clearly indicate that characteristics of mitochondrial acetaldehyde dehydrogenase vary with the structural state of mitochondria. The simplest explanation for this finding may be that mitochondrial acetaldehyde dehydrogenase probably exists in two forms: one with a higher turnover number and higher sensitivity toward substrate inhibition, and the other with a lower turnover number and lower sensitivity toward substrate inhibition. In the swollen state of mitochondria, the dehydrogenase predominantly exists in the latter form.

III. ACKNOWLEDGEMENT

This work was supported by the National Institute on Alcohol Abuse and Alcoholism under the Frant AA00292.

IV. REFERENCES

(1). Hunter, F.E. Jr., Biochim. Biophys. Acta, 20, 237 (1956).
(2). Lehninger, A.L., Physiol. Revs. 42, 467 (1962).
(3). Chance, B., Azzi, A., Lee, I.Y., Lee, C.P., and Mela, L., in:"Mitochondria:Structure and Function" (L. Ernster and Z. Drahota, Eds.), Vol. 17, p. 233, Academic Press, New York 1969.
(4). Beinert, H., and Palmer, G., J. Biol. Chem. 240, 475 (1965).
(5). Kröger, A., and Klingenberg, M., Eur. J. Biochem. 34, 358 (1973).
(6). Boveris, A., Erecinska, M., and Wagoner, M., Biochim.

Biophys. Acta 256, 223 (1972).
(7). Chance, E.M., Erecinska, M., and Chance, B., in:"Analysis and Simulation" p. 303 (1972).
(8). Chance, B., and Williamson, J.R., J. Biol. Chem. 217, 395 (1955).
(9). Chance, B., J. Biol. Chem. 151, 553 (1943).

INDUCTION[1] OF CYTOSOLIC ALDEHYDE DEHYDROGENASE ISOZYMES: INFLUENCE ON ACETALDEHYDE METABOLISM

D.R. Petersen, A.C. Collins
and R.A. Deitrich

University of Colorado

There are several cytosolic aldehyde dehydrogenase (ALDH) isozymes in rat liver. One of these is inducible by phenobarbital treatment in genetically selected (RR) rats, while it is not inducible in other (rr) rats. Administration of 2,3,7,8-tetrachlorodibenzo-p-dioxin (TCDD) causes induction of another isozyme in the liver cytosol in both genotypes The isozymes induced by phenobarbital and TCDD have been purified and are known to be different proteins. The influence of ALDH induction on ethanol disappearance rates and blood acetaldehyde levels was measured. Phenobarbital administration led to an increased rate of ethanol disappearance and decreased blood acetaldehyde levels in RR rats only. TCDD treatment was without effect. The finding of an increase in blood acetaldehyde levels following pargyline treatment was attributed to inhibition of the mitochondrial matrix ALDH. Phenobarbital or TCDD treatment resulted in reversal of the pargyline effect in that normal or near normal blood acetaldehyde levels were seen after ethanol treatment. These results are rationalized by taking into account the kinetics of the mitochondrial matrix enzyme and the induced cytosolic ALDH isozymes. It is proposed that the mitochondrial matrix ALDH is primarily responsible for acetaldehyde metabolism, while the fine control of circulating acetaldehyde levels may rest with other ALDH isozymes such as those in the cytosol.

I. INTRODUCTION

Our understanding of the subcellular distribution of various isozymes of aldehyde dehydrogenase and their roles in acetaldehyde metabolism has been advanced considerably in the last few

[1]We have referred to "induction" of aldehyde dehydrogenase throughout this paper to signify an increase in enzyme activity. Although we have some evidence from immunological studies that there is a corresponding increase in enzyme protein, these studies are not yet complete.

years. ALDH activity has been detected in mitochondria (1), cytoplasm (2,3) and microsomes (3,4,5). The relative importance of the enzymes in these fractions to aldehyde metabolism varies with the substrate, its concentration and the reaction conditions. Recent evidence suggests that more than one isozyme of ALDH exists in both the cytoplasm and the mitochondria. In the case of the cytosolic ALDH isozymes, this evidence has been supported by the observation that one or more of the isozymes can be induced by foreign compounds. Redmond and Cohen (6) detected a 2-fold induction of ALDH activity by phenobarbital in the 700 x *g* supernatant fraction of mouse liver homogenate. Deitrich (7) studied the same phenomenon in rat liver and determined that the induction is seen only in the cytosol and is controlled by a single gene in this species. A selective breeding program was initiated and resulted in the production of some animals that show induction (RR) and some that do not (rr).

Other evidence supporting the notion that more than one ALDH isozyme exists in the cytosol was also obtained (7). The phenobarbital-induced isozyme, referred to as the phi (ϕ) enzyme, differs in substrate specificity and heat stability from another cytosolic ALDH which utilizes p-carboxybenzaldehyde as a substrate. More recently, evidence for a third cytosolic ALDH isozyme has been obtained (8). This enzyme is induced nearly 100-fold by TCDD and is referred to as the tau (τ) enzyme. Other investigators (9,10) have detected additional cytosolic enzymes, at least one of which may originate in the mitochondria.

Kinetic analyses suggested the presence of at least two ALDH isozymes in mitochondria (5,11), and these enzymes subsequently were separated and purified (12). One of these isozymes, designated Enzyme I, has a Km for acetaldehyde of 10 μM and utilizes formaldehyde as a substrate; the other, Enzyme II, has a Km nearly 100 times higher (0.9-1.7 mM). Enzyme II uses NADP as a cofactor, is similar to the microsomal enzyme (5,12) and does not use formaldehyde as a substrate (9,12,13,14). Enzyme I appears to be confined in the mitochondrial matrix, while Enzyme II is associated with the intermembrane space (14)or with the outer mitochondrial membrane (5,11,15). Several investigations have suggested that the major site of acetaldehyde oxidation is in the mitochondria (16,17,18), with the greater portion being restricted to the matrix (16) where Enzyme I is found.

The aims of the current study were to elucidate further the nature of phenobarbital- and TCDD-induced cytosolic ALDH isozymes and to ascertain the effects of this induction on *in vivo* acetaldehyde metabolism.

II. METHODS

Rats of known genotype, RR (reactor) and rr (nonreactor) with respect to the phenobarbital induction of cytosolic ALDH, were used in these experiments. All animals were maintained on a standard laboratory diet and water *ad libitum*.

Both RR and rr rats, weighing 130 to 320 g, were used in the first experiment to study the effect of induction of the ϕ and τ ALDH isozymes on the circulating levels of acetaldehyde which arise during ethanol metabolism. Phenobarbital was administered in the drinking water (1.0 mg/ml) to rats of both genotypes for a minimum of 3 days. This method and duration of phenobarbital treatment has been shown to result in a maximal induction of cytosolic ALDH in liver of RR animals in our laboratory (unpublished data). TCDD was dissolved in dioxane (100 µg/ml) and injected intraperitoneally (75 µg/kg). Within 24 hours after the last exposure to phenobarbital and 12 days following TCDD injection, the rats were given an intraperitoneal injection of 2.5 g/kg ethanol. Blood samples were obtained 30, 90 and 150 min after ethanol injection.

A 40-µl sample was obtained from each animal by puncturing the retroorbital sinus with a capillary pipet. The blood sample was immediately placed in a 16 x 100 mm tube containing 0.96 ml of a .01 mg% isopropyl alcohol solution, which served as an internal standard. To avoid the possibility of spontaneous acetaldehyde formation in blood hemolysates, the isopropyl alcohol solution contained 25 mM thiourea (19). Immediately after the blood sample was placed in the tube, the tube was sealed by means of a rubber stopper and placed on ice until determination of ethanol and acetaldehyde by head-space gas chromatography.

The tubes containing blood samples were incubated at 65 $^{\circ}$C for 15 min. An aliquot (0.7-1.0 ml) of head-space gas was injected into a Beckman GC-45 gas chromatograph equipped with a Porapak column and flame ionization detector. Peak areas were computed with a Hewlett-Packard 3373B integrator and compared with acetaldehyde and ethanol standards which were prepared and run daily. Acetaldehyde concentrations of standard solutions were determined spectrophotometrically each day.

Immediately after the last blood sample was taken from each animal, it was sacrificed by a blow on the head. The liver was quickly excised and placed in 9 volumes of .25 M sucrose. Cytosolic ALDH activity was measured in the 48,000 x *g* supernatant (20).

As described in detail in the following section, results of this study suggested that induction of the ϕ enzyme serves to decrease blood acetaldehyde levels, but induction of the τ enzyme was apparently without effect. A second experiment utilized pargyline to inhibit mitochondrial ALDH in order to ascertain whether induction of the τ enzyme alters acetaldehyde metabolism when higher levels of acetaldehyde are attained. RR rats, weighing 100 to 160 g, were treated with phenobarbital or TCDD as previously described. Within 24 hours after phenobarbital treatment and 12 to 14 days following TCDD injection, the rats were injected intraperitoneally with saline or pargyline HCl (100 mg/kg). Ethanol (2.5 g/kg) was injected 90 min after pargyline administration. Blood samples were collected for determinations of ethanol and acetaldehyde content. After collection of the last blood sample, the animals was sacrificed, the liver was removed, and mitochondria were prepared as described by Siew *et.al.* (14).

Washed mitochondria were suspended in a volume of .25 M sucrose, containing 2 mM mercaptoethanol and 10 mM sodium phosphate buffer, pH 7.4. The final volume was equal to the original wet liver weight. The mitochondrial fraction was stored overnight at $-20^{\circ}C$. After the mitochondrial preparations were thawed, a 100 mg/ml sodium deoxycholate solution (20 µl per 1.0 ml of preparation) was added. Each mitochondrial suspension was rehomogenized and centrifuged for 1 hour at 48,000 x *g*.The resulting supernatant was used for mitochondrial ALDH assays.

The ALDH activities of cytosolic and mitochondrial preparations were determined by measuring changes in absorption at 340 nm with a Gilford 240 recording spectrophotometer. The typical reaction mixture contained 50 mM sodium pyrophosphate buffer, pH 8.8; 1 mM pyrazole, 1 mM NAD; and .05 ml of supernatant or mitochondrial fraction in a final volume of 1.0 ml. Cytosolic ALDH activity was determined in the presence of 5.0 mM propionaldehyde. Each mitochondrial preparation was assayed in the presence of two different propionaldehyde concentrations (5.0 mM and 50 µM) and one formaldehyde concentration (1.0 mM). Reactions were started by the addition of substrate. Blanks which contained no aldehyde were run with each sample.

III. RESULTS AND DISCUSSION

Poland *et.al.* (21) have found that TCDD is a potent inducer of microsomal arylhydrocarbon hydroxylase (AHH) even in mouse strains that are resistant to 3-methylcholanthrene induction. This suggests that these strains have a deficient receptor mechanism for inducing agents. Since we have identified lines of rats that do (RR) and do not (rr) react to phenobarbital, it was of interest to test TCDD as an inducing agent in these animals. We have found that the ED50 for induction of a cytosolic ALDH by TCDD is about 30 μg/kg and that a maximal induction is obtained at 75 μg/kg in both RR and rr rats. We have purified both phenobarbital- and TCDD-induced enzymes from rat liver cytosol by ammonium sulfate precipitation and by DEAE cellulose, CM cellulose and Biogel 200 column chromatography. The enzymes resulting from this purification have been used to produce antibodies in rabbits (unpublished data). All of these studies have shown that the two enzymes are different proteins. A striking aspect of these enzyme inductions is the large increase in enzyme activity afforded. This is shown in Table 1, along with data on the relative Km values for enzymes purified from animals treated with either TCDD or phenobarbital.

TABLE 1
Increase in Cytosolic Aldehyde Dehydrogenase Activity in RR and rr Rats after Phenobarbital or TCDD Treatment[a]

Genotype	Treatment	Fold Induction		Km Values for Acetaldehyde (nmolar)
RR	Phenobarbital	23 ± 4[b]	(6)	2.7
RR	TCDD	115 ± 16[b]	(5)	22.0
rr	Phenobarbital	1.2 ± 0.1	(4)	
rr	TCDD	148 ± 19[b]	(6)	

a. Measures were taken within 24 hours of phenobarbital administration (1.0 mg/ml in drinking water for 3 days) and within 12 days after one intraperitoneal injection (75 μg/kg) of TCDD. Control level of cytosol enzyme activity was 4.0 ± 0.3 nmoles NADH/min/mg protein for RR animals; for rr animals, it was 2.8 ± 0.4 nmoles NADH/min/mg protein. Data are expressed as mean $\pm$ S.E.M. Number of animals in each treatment group is given in parentheses.

b. Significantly different from controls ($p < 0.001$).

To ascertain the importance of these enzymes in aldehyde and alcohol metabolism *in vivo*, we treated RR and rr rats with phenobarbital or TCDD and determined the rates of ethanol and acetaldehyde elimination. Following a dose of 2.5 g/kg, ethanol disappearance was essentially linear over the 150-min time period examined in both RR and rr rats. The control disappearance rate for RR animals was 9.6 ± 1.8 nmoles/kg/hr, while ethanol disappeared from the blood of control rr animals at the rate of 8.3 ± 1.2 mmoles/kg/hr. TCDD did not significantly alter the rate of ethanol disappearance in either line. Phenobarbital, on the other hand, caused a 59% increase ($p < 0.05$) in ethanol elimination rate only in the RR animals.

Figure 1 illustrates blood acetaldehyde levels at 30, 90 and 150 min after ethanol injection for control and for phenobarbital- and TCDD-treated animals. Neither drug significantly altered acetaldehyde levels in rr rats. In RR animals, however,

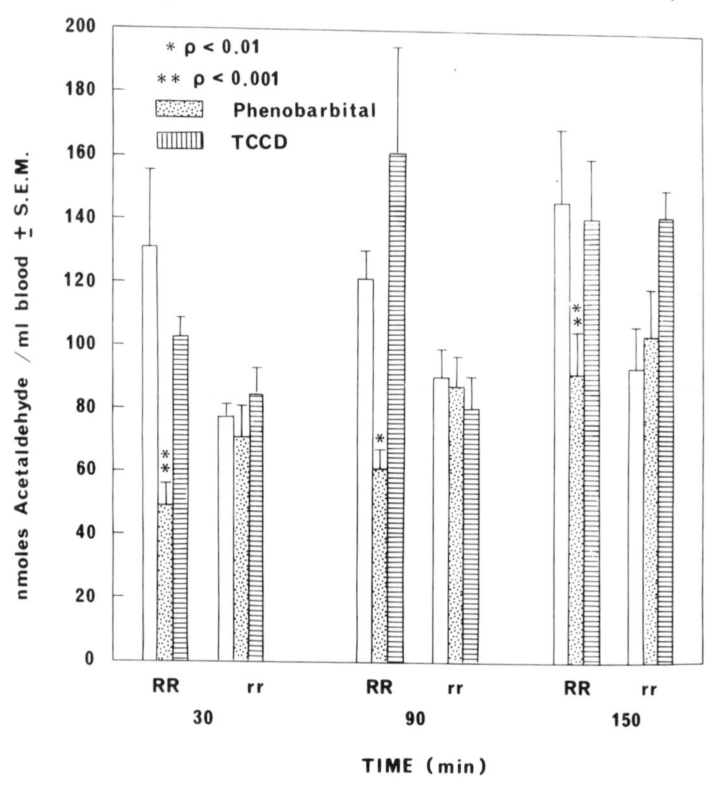

Fig.1. Blood acetaldehyde content of control and of phenobarbital- and TCDD-treated RR and rr rats 30,90, and 150 min after injection of 2.5 g/kg ethanol. Each group consisted of six animals.

phenobarbital treatment resulted in significant decreases in blood acetaldehyde at all time points when compared with RR controls. TCDD was without effect in RR animals. Thus, phenobarbital treatment increased the rate of acetaldehyde production (ethanol disappearance) at the same time that circulating acetaldehyde levels were decreased in RR rats, whereas no effect on either measure was observed in rr animals.

A comparison of the effects of phenobarbital and TCDD on cytosolic ALDH activity with the effects on circulating acetaldehyde levels revealed that induction of the ϕ enzyme changes blood acetaldehyde content but that a significantly greater induction of the τ enzyme is without effect. This may be explained by reference to basic principles of enzymology. If we assume a liver acetaldehyde level of 0.5 mM (22) and use Km values of 2.7 mM for the ϕ enzyme and 22 mM for the τ enzyme, it can be calculated from the Michaelis-Menten equation ($v/Vmax = S/Km + S$) that the 23-fold increase in cytosolic activity as measured *in vitro* which arises due to induction of the ϕ enzyme will result in a 3.7-fold increase in the contribution of the cytosol to acetaldehyde metabolism. Similar calculations using the mean induction of the τ enzyme (a 131-fold increase) result in a 2.9-fold increase in the cytosolic contribution to acetaldehyde metabolism. If the cytosol contributes 15% to the total acetaldehyde metabolism under normal circumstances, a 23-fold increase in activity of the ϕ enzyme should result in a 41% increase in total capacity of the liver to metabolize acetaldehyde, whereas a 131-fold increase in activity of the τ enzyme should result in a 28% increase in acetaldehyde metabolism.

If the assumptions made in these calculations are valid, it would seem reasonable to assume that a further elevation in hepatic acetaldehyde level should result in an increased contribution of the cytosolic enzymes to acetaldehyde metabolism. This hypothesis was tested by examining the effects of induction of the ϕ and τ enzymes on circulating acetaldehyde levels in RR rats which had been treated to elevate blood acetaldehyde content. This was accomplished by administration of pargyline. A previous study in our laboratory (unpublished data) had corroborated the observation of Dembiec *et.al.* (23), who noted that pargyline inhibits hepatic ALDH activity and results in elevated circulating acetaldehyde levels. Our study determined that the pargyline inhibition was restricted to an effect on mitochondrial ALDH activity and most probably to Enzyme I.

Table 2 presents the effects of pargyline treatment on mitochondrial ALDH activity as measured at two propionaldehyde concentrations (5 mM or 50 µM) and in the presence of 1 mM formaldehyde. When either formaldehyde or the lower concentration of propionaldehyde was used as substrate, greater than 90% inhibition of mitochondrial ALDH activity was evident. At the higher propionaldehyde concentration, the greatest inhibition was 60% for the phenobarbital + pargyline group.

The mitochondrial Enzyme I has a low Km (31 µM) for aldehydes and utilizes formaldehyde as substrate, whereas Enzyme II has a high Km (450 µM) for aldehydes and does not use formaldehyde as substrate (14). The data presented in Table 2 suggest that Enzyme I is completely inhibited by pargyline treatment, while Enzyme II is essentially unaffected. From the Michaelis-Menten

Table 2
Mitochondrial Aldehyde Dehydrogenase Activities in Control and Pargyline-Treated RR Rats[a]

Genotype	Treatment	Activity (nmoles NADH/min/mg protein)[b] Propionaldehyde 5 mM	Propionaldehyde 50 μM	Formaldehyde 1 mM
RR	Control	60±2*	24.0±0.8*	41.0±0.2*
RR	Pargyline	26±2^{+}	2.2±0.4^{+}	1.2±0.2^{+}
RR	Phenobarbital +pargyline	24±4^{+}	2.3±0.4^{+}	1.5±0.2^{+}
RR	TCDD + pargyline	43±4*	1.5±0.2^{+}	1.9±0.1^{+}

a. Animals were treated with 100 mg/kg pargyline, livers were excised, mitochondria were isolated, and ALDH measurements were carried out as described in the text. Data are expressed as mean ± S.E.M.. Each group consisted of six animals.

b. Means in the same column with different superscripts (or $^{+}$) are significantly different ($p < 0.05$).*

equation, one may calculate that Enzyme I is operating at 62% of Vmax and Enzyme II is at 10% of Vmax when 50 μM propionaldehyde is the substrate. When 5 mM propionaldehyde is the substrate, Enzyme I is at 99.4% of Vmax and Enzyme II is at 92% of Vmax. These values plus the the velocities in Table 2 for control rats (60 and 24 nmoles NADH/min/mg protein) allow one to solve two simultaneous equations ($0.994\ Vmax_{I} + 0.92\ Vmax_{II} = 60$; $0.62\ Vmax_{I} + 0.10\ Vmax_{II} = 24$) to arrive at Vmax values for Enzyme I of 34.1 nmoles/NADH/min/mg protein and for Enzyme II of 28.3 nmoles/NADH/min/mg protein. The rates obtained when 50 μM propionaldehyde was substrate for pargyline-treated animals are very close to 10% of 28.3, or the amount of activity that Enzyme II contributes to the velocity of this low substrate concentration. Similarly, if one assumes complete inhibition of Enzyme I and subtracts its Vmax (34) from the calculated Vmax of 65, one obtains a value of 31. This value is very close to the values of 26 and 24 obtained for pargyline and pargyline + phenobarbital treatments. The value for TCDD may reflect a small contamination by an active cytoplasmic enzyme.

When cytosolic ALDH activity was measured in these animals, pargyline alone did not alter activity in this fraction nor did it seem to influence the induction of the ϕ and τ enzymes by phenobarbital and TCDD. Control RR animals demonstrated a cytosolic activity of 9.2 ± 1.2 nmoles/NADH/min/mg protein. Pargyline treatment did not alter this value. Phenobarbital-treated animals displayed an activity of 34.4 ± 7.6 nmoles NADH/min/mg protein, while the activity of TCDD-treated animals was 467.9 ± 28.1 nmoles NADH/min/mg protein. Although the increase in cytosolic ALDH activity seen after phenobarbital treatment in this experiment was not as great as that reported for RR rats in Table 1, it was nevertheless sufficient to reverse the effects of pargyline on circulating acetaldehyde levels.

These results are presented in Fig. 2, which illustrates both the effects of a 100 mg/kg dose of pargyline on blood acetaldehyde

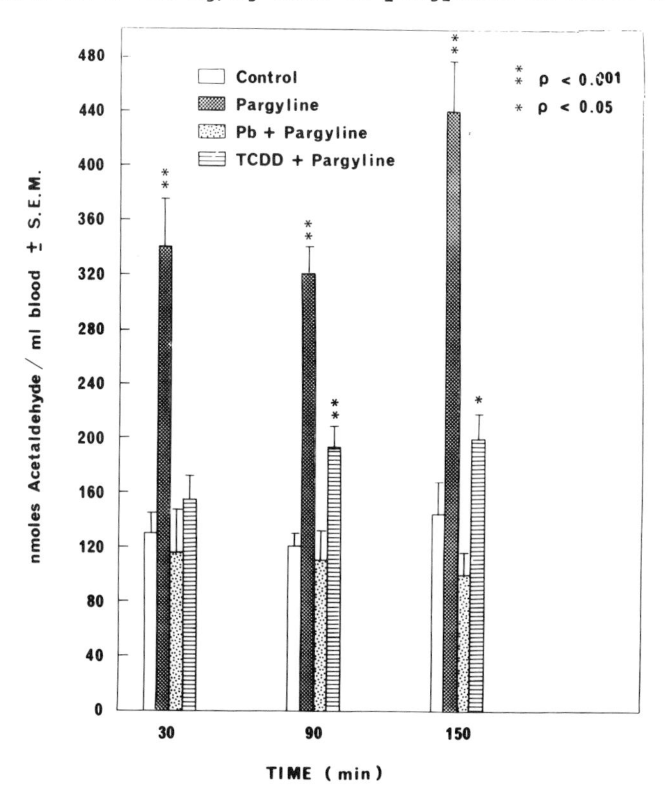

Fig.2. Blood acetaldehyde content of control RR animals and of animals treated with pargyline, phenobarbital (Pb) + pargyline, and TCDD +pargyline. Animals were injected with 100 mg/kg pargyline 90 min prior to injection of 2.5 g/kg ethanol, and blood samples were taken at 30,90, and 150 min after ethanol administration. Each group consisted of six animals.

levels in RR rats and the consequences of induction of the ϕ and τ enzymes with respect ot pargyline's actions. It can be seen that pargyline caused a significant increase in blood acetaldehyde content at all time points after ethanol injection. Treatment with phenobarbital completely reversed this effect, while TCDD treatment was only partially successful in reversing pargyline's effects on blood acetaldehyde levels.

IV. CONCLUSIONS

The results of the first experiment provide several interesting observations. Phenobarbital treatment altered the rate of ethanol disappearance and blood acetaldehyde levels following administration of 2.5 g/kg ethanol only in the RR animals. The increase in rate of ethanol elimination is not likely to be due to an increase in hepatic alcohol dehydrogenase (ADH) activity, since phenobarbital treatment has been shown previously not to affect hepatic ADH activity in RR or rr animals (20). Two explanations for these data may be proposed: (1) If ALDH activity becomes rate limiting in ethanol metabolism at high levels of ethanol, induction of an ALDH isozyme, even one with a relatively high Km for acetaldehyde, could increase the rate of ethanol metabolism. (2) A second explanation involves the liver microsomal oxidizing system (MEOS). According to Lieber and DeCarli (24), the MEOS is inducible by phenobarbital and has a relatively high Km for ethanol. Phenobarbital treatment did not change the rate of ethanol metabolism or blood acetaldehyde content following administration of 2.5 g/kg ethanol in the rr animals. Therefore, it may be that RR and rr rats differ not only in inducibility of the ϕ enzyme but also in inducibility of the MEOS.

It is also interesting to note that both genotypes demonstrated a greater than 100-fold increase in cytosolic ALDH activity following TCDD treatment, indicating that genetic control of induction of the τ enzyme is not present in these animals. The τ enzyme does not appear to be involved extensively in acetaldehyde metabolism, at least under normal circumstances. Thus, induction of the ϕ enzyme by phenobarbital appears to decrease blood acetaldehyde levels, whereas a significantly greater induction of the τ enzyme by TCDD does not alter circulating acetaldehyde levels.

The second experiment was designed to determine whether the ϕ and τ enzymes become more important in acetaldehyde metabolism when the activity of mitochondrial Enzyme I is inhibited. A 100 mg/kg dose of pargyline resulted in a substantial increase in blood acetaldehyde content. This effect was completely reversed by induction of the ϕ enzyme and partially reversed by induction of the τ enzyme. These results are consistent, since induction of the enzyme with the lower Km for acetaldehyde (the

ϕ enzyme) should be more effective in reversing the pargyline effect even though it is not induced to nearly as great an extent. In view of the fact that induction of the τ enzyme partially reversed the effect of pargyline, it can be concluded that the elevation of hepatic aldehyde concentration following pargyline treatment is sufficient to provide a role for the τ enzyme in acetaldehyde metabolism under these circumstances.

In summary, data obtained in the current study support the notion that acetaldehyde oxidation occurs primarily in the mitochondria and most likely involves the matrix enzyme. However, the fact that altering the activities of the ϕ enzyme, and , to a lesser extent, the τ enzyme does have an effect upon blood acetaldehyde content leads to the conclusion that fine control of circulating acetaldehyde levels may be influenced by cytosolic ALDH isozymes.

V. ACKNOWLEDGEMENT

This research was supported by grants from the National Institute of Alcohol Abuse and Alcoholism to D.R.P. (AA-05011) and R.A.D. (AA-00263).

VI. REFERENCES

(1). Walkenstein, S.S., and Weinhouse,S., J. Biol. Chem. 200, 515 (1953).
(2). Büttner, H., Biochem Z. 341, 300 (1965).
(3). Deitrich, R.A., Biochem. Pharmacol. 15, 1911 (1966).
(4). Tietz, A., Lindberg, M., and Kennedy, E.P., J. Biol. Chem. 239, 4081 (1964).
(5). Tottmar, S.O.C., Petersson, H., and Kiessling K.H., J. Biochem. 135, 577 (1973).
(6). Redmond, G., and Cohen, G., Science 171, 387 (1971).
(7). Deitrich, R.A., Science 173, 334 (1971).
(8). Roper, M., Stock, T., and Deitrich, R.A., Fed. Proc. 35, 282 (1976).
(9). Koivula, T., and Koivusalo, M., Biochim. Biophys. Acta 397, 9 (1975).
(10). Weiner, H., King, P., Hu, J.H.J., and Bensch, W.R., in: "Alcohol and Aldehyde Metabolizing Systems" (R.G. Thurman, T. Yonetani, J.R. Williamson, and B. Chance, Eds.) p. 125, Academic Press, New York (1974).
(11). Smith, L., and Packer, L., Arch. Biochem Biophys. 148, 270 (1972).
(12). Siew, C., Deitrich, R.A., and Erwin, V.G., Fed. Proc. 33, 538 (1974).
(13). Cinti, D.L., Keyes, S.R., Lemelin, M.A., Denk, H., and

Schenkman, J.B., J. Biol. Chem., 251, 1571 (1976).
(14). Siew, C., Deitrich, R.A., and Erwin, V.G., Arch. Biochem. Biophys., in press (1976).
(15). Horton, A.A., and Barrett, M.C., Arch. Biochem. Biophys. 167, 426 (1975).
(16). Grunnet, N., Eur. J. Biochem. 35, 236 (1973).
(17). Parrilla, R., Ohkawa, K., Lindros, K.O., Zimmerman, J.P., Kopayashe, K., and Williamson, J.R., J. Biol. Chem. 249, 4926 (1974).
(18). Corrall, R.J.M., Havre, P., Margolis, J., Kong, M., and Landau, R., Biochem. Pharmacol. 25, 17 (1976).
(19). Eriksson, C.J.P., Sippel, H.W., and Forsander, O.A., in "The Role of Acetaldehyde in the Actions of Ethanol" (K. O. Lindros and C.J.P. Eriksson, Eds.), p. 9. The Finnish Foundation for Alcohol Studies, Helsinki, (1975).
(20). Deitrich, R.A., Collins, A.C., and Erwin, V.G., J. Biol. Chem.247, 7232 (1972).
(21). Poland, A., Glover, E., and Kende, A. S., J. Biol. Chem. 251, 4926 (1976).
(22). Eriksson, C.J.P., Biochem. Pharmacol. 22, 2283 (1973).
(23). Dembiec, D., MacNamee, D., and Cohen, G., J. Pharmacol. Exp. Ther. 197, 332 (1976).
(24). Lieber, C.S., and DeCarli, L.M., J. Pharmacol. Exp. Ther. 181, 279 (1972).

THE PLASTIC AND FLUID NATURE OF THE MITOCHONDRIAL ENERGY TRANSDUCING MEMBRANE

C.R. Hackenbrock and M. Höchli

University of North Carolina School of Medicine

Differential scanning calorimetry combined with freeze-fracture electron microscopy reveals a high potential for free lateral translational diffusion of integral proteins in the energy transducing membrane of the mitochondrion. The observations suggest that the proteins of the energy transducing membrane are not immobilized, anchored, or organized in a fixed, rigid protein-protein lattice throughout the membrane. Some integral proteins such as cytochrome c oxidase co-diffuse laterally with other integral proteins in the lateral plane of the membrane lipid bilayer. Other integral proteins diffuse laterally in the lipid bilayer, totally independent of the oxidase. The rate of lateral translational motion by metabolically active integral proteins in the mitochondrial energy transducing membrane may account for the time structure of electron transfer and ATP synthesis.

I. INTRODUCTION

The mitochondrial energy transducing membrane is remarkable in its structural and functional complexity relative to most other membranes of eukaryotic cells. A unique function of this membrane, which is of paramount importance to the structural integrity and metabolic effectiveness of the cell, is the conservation of metabolic energy mediated by coupling the free energy of respiratory chain electron transfer to the production of ATP. Although intensively studied for more than twenty years, the definitive mechanisms of electron transfer and oxidative phosphorylation remain elusive. However, as in all biological membranes, these metabolic events will relate ultimately to the specific molecular composition and structure of the membrane. While the molecular composition of the energy transducing membrane is reasonably well known, a clear understanding of its molecular structure is obscured by its high degree of complexity.

The energy transducing membrane contains up to thirty different integral and peripheral enzymes and ion-substrate transport and electron transfer proteins that collectively account for approximately 75% of the membrane's dry mass. Many of these proteins have not yet been isolated. Of major consideration regarding the structural details of the membrane is the geometric relationship between the electron transfer proteins of the respiratory chain during the rapid and sequential events of electron transfer, and the relationship of the electron transfer proteins to ATPase during the transduction of redox energy into the chemical bond energy of ATP.

The majority of the proteins of the electron transfer and phosphorylating systems occur in direct association with both hydrophilic and hydrophobic portions of phospholipid amphiphiles in the lipid bilayer of the membrane. The definitive details of this association are essentially unknown. The lipid bilayer is unusual compared to most other cell membranes in that it possesses a high mole fraction of cardiolipin which, along with several other phospholipids, contains a large amount of unsaturated hydrocarbon chains. Further, the membrane is virtually free of cholesterol. This composition of bilayer lipid results in an intrinsically low viscosity for a biological membrane and may account for the extreme configurational plasticity demonstrated in the energy transducing membrane (Hackenbrock, 1966, 1968).

In this paper, we shall consider further the results of the unique molecular composition of the mitochondrial energy transducing membrane and present new data which reveal that the energy transducing membrane is not only highly plastic, but indeed more fluid than heretofore recognized. The data will reveal further that the integral proteins of this complex membrane are neither crowded laterally nor rigidly latticed, but on the contrary possess considerable potential for free lateral translational diffusion. In this regard, special consideration will be given to the lateral translational diffusion of cytochrome *c* oxidase. Finally, we shall discuss our data in terms of the free lateral translational motion of metabolically active integral proteins in the energy transducing membrane related to the time structure of electron transfer and ATP synthesis.

II. DISCUSSION

A. Configurational Plasticity

Energy-linked ultrastructural transformations were first observed to occur in mitochondria isolated from mouse and rat liver (Hackenbrock, 1966, 1968). Of particular interest is the observation that metabolically and structurally intact mitochondria undergo reversible ultrastructural transformations in spatial folding of the energy transducing membrane that parallels the onset of oxidative phosphorylation (Figs. 1-4). Such dramatic energy-linked configurational changes are completely blocked by inhibitors of electron transfer such as antimycin and cyanide, as well as by inhibitors and uncouplers of oxidative phosphorylation such as oligomycin and dinitrophenol respectively (Hackenbrock, 1968). Considerable evidence now exists that energy-linked, inhibitor-sensitive configurational changes occur in the energy transducing membrane in a wide variety of mitochondria, whether the mitochondria are isolated or occur within the intact cell (Hackenbrock *et al.*, 1971; Hackenbrock, 1972a). Similar configurational changes that occur in isolated mitochondria from cardiac muscle appear under special experimental conditions to be insensitive to metabolic inhibitors (Weber, 1972; Stoner and Sirak, 1973; Scherer and Klingenberg, 1974).

Perhaps the most dramatic example of an energy-linked ultrastructural transformation in the energy transducing membrane is the topological change that occurs at the membrane surface during oxidative phosphorylation and can be observed by scanning electron microscopy (Andrews and Hackenbrock, 1975). In these studies, the surface of the energy transducing membrane is examined in different energy states after removal of the outer mitochondrial membrane (Figs. 5-8).

The ultrastructural transformations that occur in the energy transducing membrane, especially during uncoupler-sensitive oxidative phosphorylation, clearly reveals the degree to which this membrane can undergo rapid topological reorganization and presumably molecular reorganization. Such observation suggest that the energy transducing membrane is highly fluid as well as plastic in nature. In the sections to follow we shall examine the structural nature of the energy transducing membrane in terms of the potential for lateral translational diffusion of bilayer lipids and integral proteins.

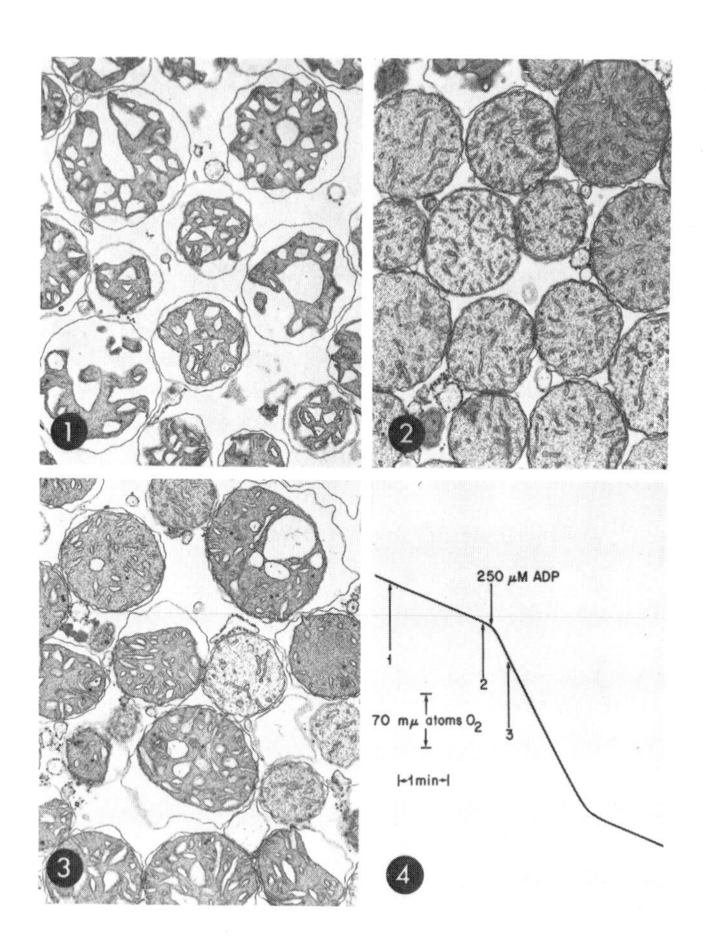

Figs. 1-4. Metabolically-linked structural transformations in isolated rat liver mitochondria: Fig. 1. Freshly isolated mitochondria; Fig. 2. Mitochondria during succinate-supported electron transfer; Fig. 3. Mitochondria during oxidative phosphorylation, X 27,000; Fig. 4. Polarographic trace showing ADP-induced oxidative phosphorylation and consecutive times (arrows) at which mitochondria in Figs. 1-3 were chemically fixed for electron microscopy.

B. Bilayer Lipid Phase Transitions

We have determined the relative degree of fluidity in the bilayer lipid and the potential for lateral translational diffusion of integral proteins in the energy transducing membrane by combining differential scanning calorimetry (DSC) with

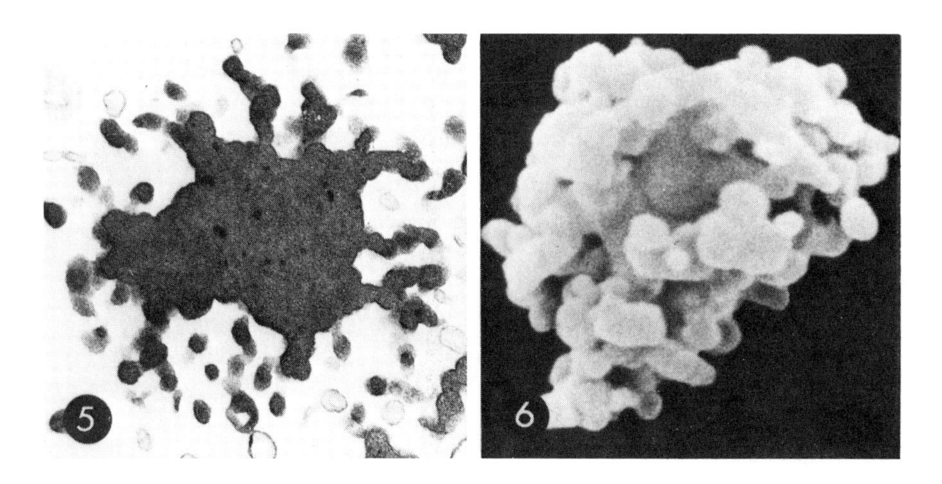

Figs. 5,6. Surface configuration of the energy transducing membrane of an inner membrane-matrix particle before oxidative phosphorylation; Fig. 5. Transmission image of thin section, X 48,000; Fig. 6. Scanning image of whole mount, X 42,000.

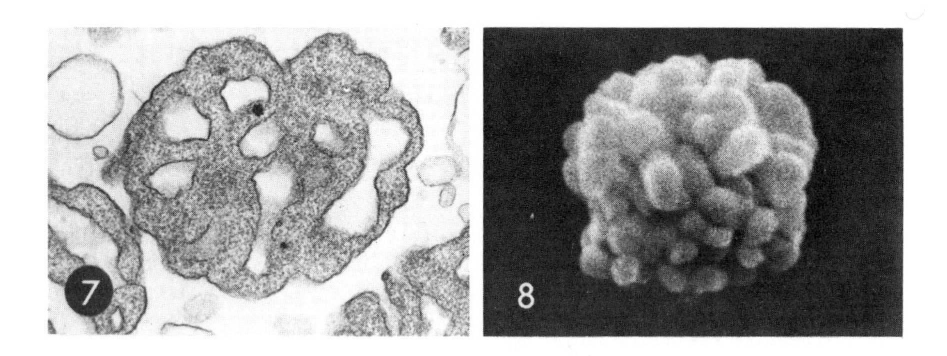

Figs. 7,8. Surface transformations in the energy transducing membrane of an inner membrane-matrix particle during oxidative phosphorylation; Fig. 7. Transmission image of thin section, X 48,000; Fig. 8. Scanning image of whole mount, X 42,000.

freeze fracture electron microscopy (Hackenbrock, 1976; Hackenbrock *et al.*, 1976). In these studies isolated whole mitochondria from rat liver (Fig. 1), an inner membrane-matrix fraction (Fig. 5), and an osmotically unfolded, spherical inner membrane-matrix fraction (Fig. 9) (Hackenbrock, 1972b were used.

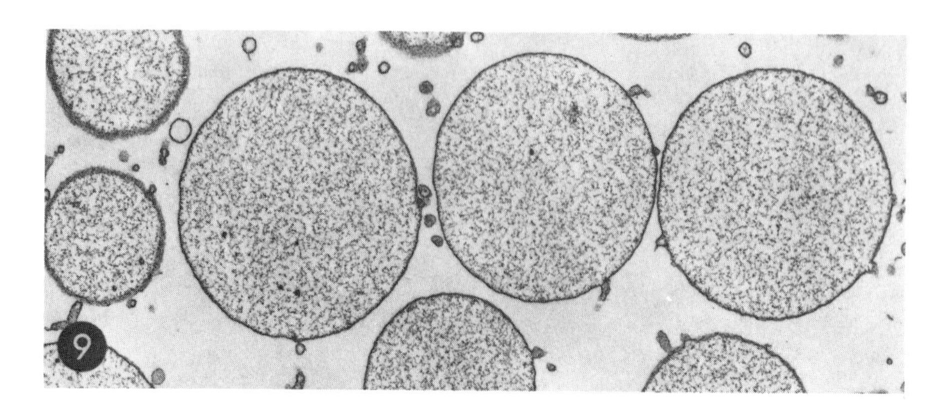

Fig. 9. Purified spherical inner membrane-matrix fraction. X 20,000.

TABLE 1
RELATIVE PURITY OF FRACTIONATED MITOCHONDRIAL MEMBRANES

	Specific Activity*	
	Monoamine Oxidase	Cytochrome c Oxidase
Whole Mitochondria	12.8	720
Condensed Inner Membrane-Matrix Fraction	0.257	1060
Spherical Inner Membrane-Matrix Fraction	0.014	1200
Outer Membrane Fraction	352.0	150

*Specific Activity: for monoamine oxidase in nmoles benzylamine oxidized · $min^{-1} \cdot mg^{-1}$ protein; for cytochrome c oxidase in natoms oxygen reduced · $min^{-1} \cdot mg^{-1}$ protein.

The purified spherical inner membrane-matrix fraction is essentially free of outer membrane (Table 1) and lends itself well to freeze fracture studies in which large area fracture faces are required for adequate observation by electron microscopy. Since the two mitochondrial membranes, i.e. the outer and the energy transducing membranes, are distinct in their lipid and protein composition, a highly purified energy transducing membrane is required if the lipid phase transition determinations of this membrane are to be accurately assessed by differential scanning calorimetry. The purified spherical energy transducing membrane is active in electron transport and oxidative phosphorylation (Lemasters and Hackenbrock, 1973) and differs from the energy transducing membrane of the intact whole mitochondrion primarily in its configuration.

In practice, the membranes are cooled slowly in a differential scanning calorimeter (Perkin-Elmer DSC-2) to various temperatures while thermograms are obtained over a wide temperature range of between 30°C and -30°C (Hackenbrock *et al.*, 1976). Similarly, for comparison with the calorimetric data, freeze fracture electron microscopy is carried out by slowly cooling the membranes to various temperatures prior to rapid freezing.

As the temperature is lowered slowly, DSC reveals an onset temperature of a lipid phase transition exothern to occur at -4°C in both the purified, condensed inner membrane-matrix preparation and the spherical energy transducing membrane (Fig. 10). Of interest is that whole mitochondria show a biphasic onset temperature in the transition exothern at 9°C and at 1°C with a greater transition enthalpy than that determined for the purified energy transducing membrane (Fig. 10). This difference is due to the summation of lipid phase transitions occurring in both the outer and energy transducing membrane of the intact, whole mitochondrion (Fig. 16). Blazyk and Steim (1972) previously reported a broad transition temperature centering at 0°C for whole mitochondria.

Table 2 summarizes the transition onset temperatures as well as the extents of the transition regions for the various mitochondrial membrane preparations that we have studied. That the full extent of the transition exotherm for the energy transducing membrane occurs at subzero temperatures suggests considerable molecular motion in the bilayer lipid at physiological temperatures. This finding is in agreement with Carbon-13 NMR spectra that indicate a high degree of mobility of phospholipid hydrocarbon chains in the total lipid component of mitochondria (Keough *et al.*, 1973).

Our DSC data also reveal that the bilayer lipid of the energy transducing membrane is more fluid than that of the outer mitochondrial membrane (Figs. 10, 16). This finding is consistent with the virtual absence of cholesterol and the low saturated to unsaturated phospholipid ratio in the energy transducing membrane (Colbeau *et al.*, 1971). It is generally well

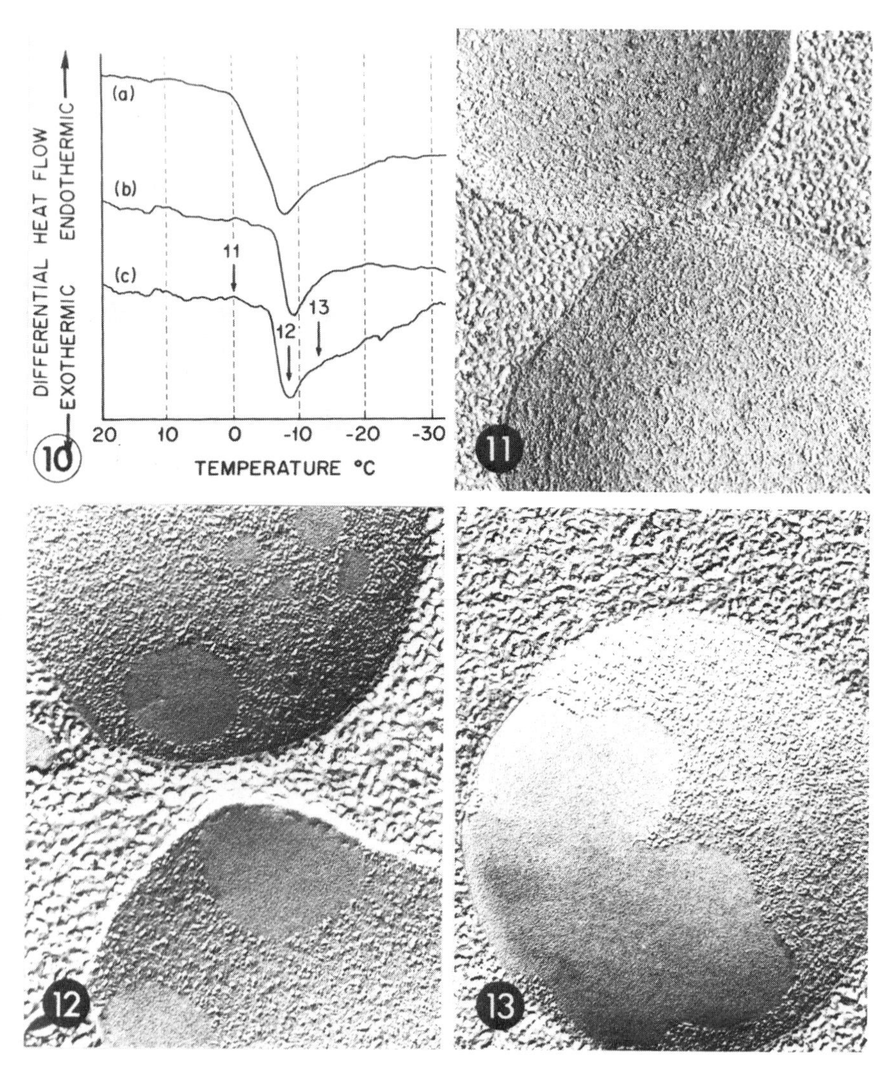

Figs. 10-13. Lipid phase transition and lateral mobility of integral proteins; Fig. 10. DSC cooling runs of whole mitochondria (a), condensed inner membrane-matrix fraction (b), and spherical inner membrane-matrix fraction (c). Arrows identify temperatures on the curve for the spherical inner membrane-matrix fraction which correspond to Figs. 11-13; Fig. 11. Fracture faces of both halves of the spherical energy transducing membrane cooled to 0°C, then frozen; Fig. 12. Cooled to -8°C, then frozen; Fig. 13. Cooled to -13°C, then frozen. Note smooth areas of gel state lipid and lateral displacement of integral proteins. X 67,500.

TABLE 2
RELATIVE LIPID PHASE TRANSITION TEMPERATURES OF WHOLE MITOCHONDRIA AND FRACTIONATED MEMBRANES[a]

	Onset Temperature in 50% Ethylene Glycol (%)		Onset Temperature Corrected (°C)[b]		Extent of Transition Region (°C)
	Exothermal	Endothermal	Exothermal	Endothermal	
Whole Mitochondria	+3, −5	−21, −14	+9, +1	−15, −8	24
Condensed Inner Membrane-Matrix Fraction	−10	−21, −14	−4	−15, −8	11
Spherical Inner Membrane-Matrix Fraction	−10	−21, −14	−4	−15, −8	11
Outer Membrane Fraction	+3, −5	−21	+9, +1	−15	24

[a]All temperatures given are averages of 2 or 3 heating or cooling runs and are ± 1°C.
[b]Corrected for 6°C depressions induced by 50% ethylene glycol.

known that cholesterol decreases the motional freedom of phospholipid hydrocarbon chains above the transition temperature (Phillips, 1972) and that unsaturated phospholipids have characteristically lower transition temperatures than do saturated phospholipids (Ladbrook & Chapman, 1969). Such a lipid composition explains the unusually low viscosity (0.1 P at 30°C) determined for the lipid component of mitochondria (Keith *et al.*, 1970). The lack of cholesterol and high content of unsaturated phospholipids, the subzero lipid phase transition region, the high molecular motion of the phospholipid hydrocarbon chains, and the low viscosity are collectively consistent with our previous finding that the energy transducing membrane is highly plastic and further suggests that the bilayer lipid of this membrane is highly fluid.

C. Lateral Translational Diffusion of Integral Proteins

Our DSC data reveal that an exothermal liquid crystalline to gel state phase transition is initiated at −4°C in the energy transducing membrane and progresses, quantitatively peaking at approximately −9°C and reaching completion at approximately −22°C (Fig. 10). The exothermal phase transition of bilayer phospholipids consists primarily of the conversion of *gauche* to *trans* conformations in the phospholipid hydrocarbon chains. In the all-*trans* conformation, chain motion becomes restricted and anisotropic as the phospholipids become packed and highly ordered in hexagonal arrays (Salsbury and Chapman, 1968; Levine, 1973). In a complex natural biomembrane containing a wide variety of phospholipids and little or no cholesterol, such as occurs in the energy transducing membrane, the long chain saturated phospholipids of similar structure will preferentially and cooperatively adopt the gel state at the onset temperature of the transition exotherm. As cooling continues below the onset temperature, fractional mixtures of miscible phospholipids of similar composition will progressively form segregated pools of gel state lipid in the bilayer and laterally exclude dissimilar, shorter chain and more unsaturated phospholipids which have characteristically lower transition temperatures. Since in the gel state the packing density of bilayer phospholipids increases and the volume occupied by each phospholipid decreases (Chapman *et al.*, 1967; Trauble and Haynes, 1971; Levine, 1973), it can be anticipated that should integral proteins occur in the bilayer, they too will be laterally excluded, provided the proteins are not overly crowded and/or are not anchored or immobilized through a rigid protein-protein lattice in the lateral plane of bilayer. At some low temperature, i.e., at the end of the exotherm transition as identified by DSC, the liquid crystalline to gel state transition in the bilayer lipid will be completed and thermotropic lipid-lipid and lipid-protein separations will have occurred in the lateral plane of the membrane. Thus, the potential for lateral translational diffusion of integral proteins in the lipid bilayer may be ascertained.

Utilizing our DSC data combined with these physicochemical considerations as a basis for experimentation, we investigated the potential for freedom of lateral translational diffusion of integral proteins in the purified energy transducing membrane by inducing liquid crystalline $\rightleftarrows$ gel state lipid phase transitions while following the lateral distribution of integral proteins (intramembrane particles) in the hydrophobic interior of the lipid bilayer with freeze fracture electron microscopy (Höchli and Hackenbrock, 1976a; Hackenbrock *et al.*, 1976; Höchli and Hackenbrock, 1976b; Hackenbrock, 1977; Hackenbrock and Höchli, 1977). When the spherical energy transducing membrane is rapidly frozen from 30, 10, or 0°C, i.e. from above the onset temperature of the transition exotherm, freeze fracture electron microscopy reveals a completely random distribution of integral proteins in the hydrophobic interior of the membrane (Fig. 11). However, when frozen from below -4°C, i.e. from below the onset temperature of the transition exotherm, progressive lateral separations are observed to occur between smooth regions of ordered gel state lipid and integral protein-rich regions in the hydrophobic interior of the membrane (Figs. 12,13). At -13°C, which is well into the transition exotherm, the area of the gel state lipid is quite extensive while the laterally excluded integral proteins become highly aggregated, often occupying only one third to one half of the total surface area of the lipid bilayer (Fig. 13). Such observations are consistent with calculations which predict that the total integral protein fraction of the energy transducing membrane occupies only one third of the total membrane surface area (Vanderkooi, 1974). In addition, although the composition of the energy transducing membrane is approximately 75% protein (Lévy *et al.*, 1969; Colbeau *et al.*, 1971), only 50% of this is integral to the membrane (Capaldi and Tan, 1974; Harmon *et al.*, 1974).

Clearly, up to two thirds of the hydrophobic surface area of the lipid bilayer of the native energy transducing membrane can be unoccupied by integral proteins. Thus, considerably more area exists in the lipid bilayer than previously recognized and represents a more than adequate space requirement for lateral translational diffusion of the integral proteins to occur in this membrane.

The thermotropic lateral separations between gel state lipid and integral proteins that occur in the energy transducing membrane during the transition exotherm are rapidly and completely reversible. After decreasing the temperature to -25°C, DSC reveals an onset temperature in the transition endotherm at about -15°C with completion of the transition at about 10°C for both configurations of the energy transducing membrane as well as for whole mitochondria (Fig. 14). When returned to above the transition temperature for a few seconds, the smooth areas of ordered gel state lipid that occur in the transition region as revealed by freeze fracture (Figs. 12,13) are not observed, thus the integral proteins are found to be completely dispersed (Fig. 15). These observations clearly reveal that the integral proteins mix rapidly with the bilayer lipid during the melt above the transition temperature and that rapid lateral translational diffusion of the proteins can occur in the purified energy transducing membrane.

As would be predicted from the DSC traces presented in Figures 10 and 14, reversible thermotropic lateral separation between gel state lipid and integral proteins occurs not only in the purified energy transducing membrane but in the energy transducing membrane of the intact, whole mitochondrion as well (Figs. 16-19). Such lateral separations in membrane lipids and proteins are observed first at approximately 7°C in the outer membrane and at approximately -4°C in the energy

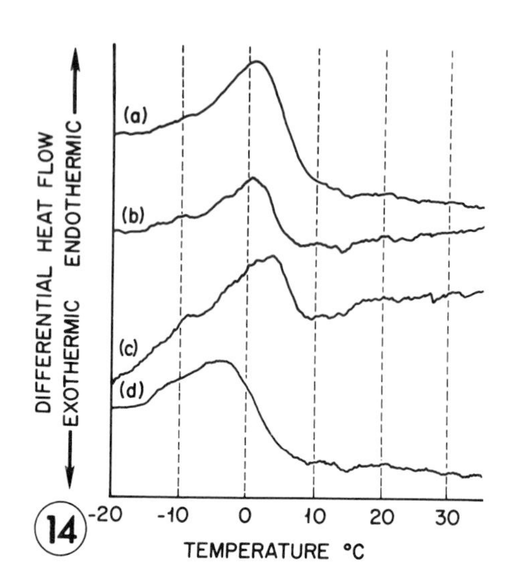

Fig. 14. DSC heating runs of whole mitochondria (a), condensed inner membrane-matrix fraction (b), spherical inner membrane-matrix fraction (c), and outer mitochondrial membrane fraction (d).

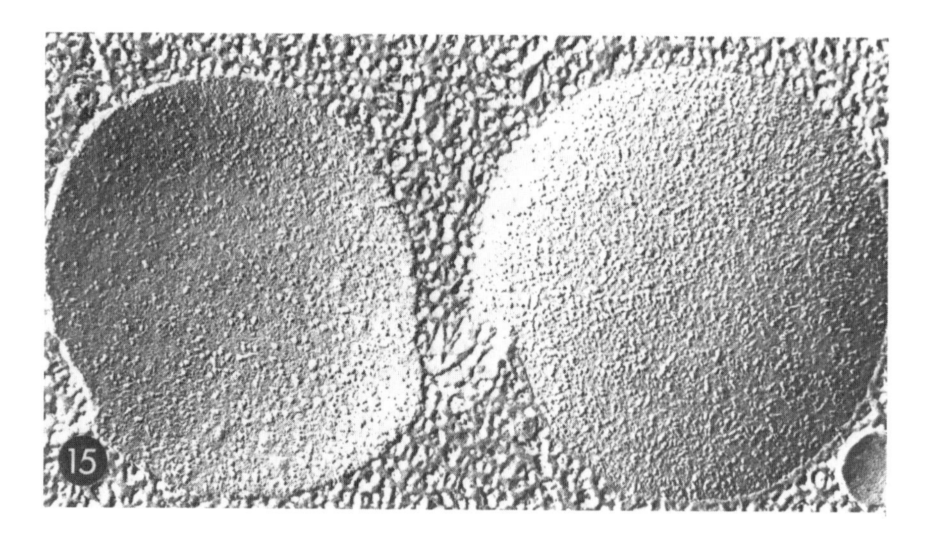

Fig. 15. Free lateral diffusion of integral proteins in the energy transducing membrane. Membranes cooled to -10°C, returned to 25°C and then frozen show completely random integral proteins in both fracture faces (cf. membranes at -8 to -13°C, Figs. 12,13). X 67,500.

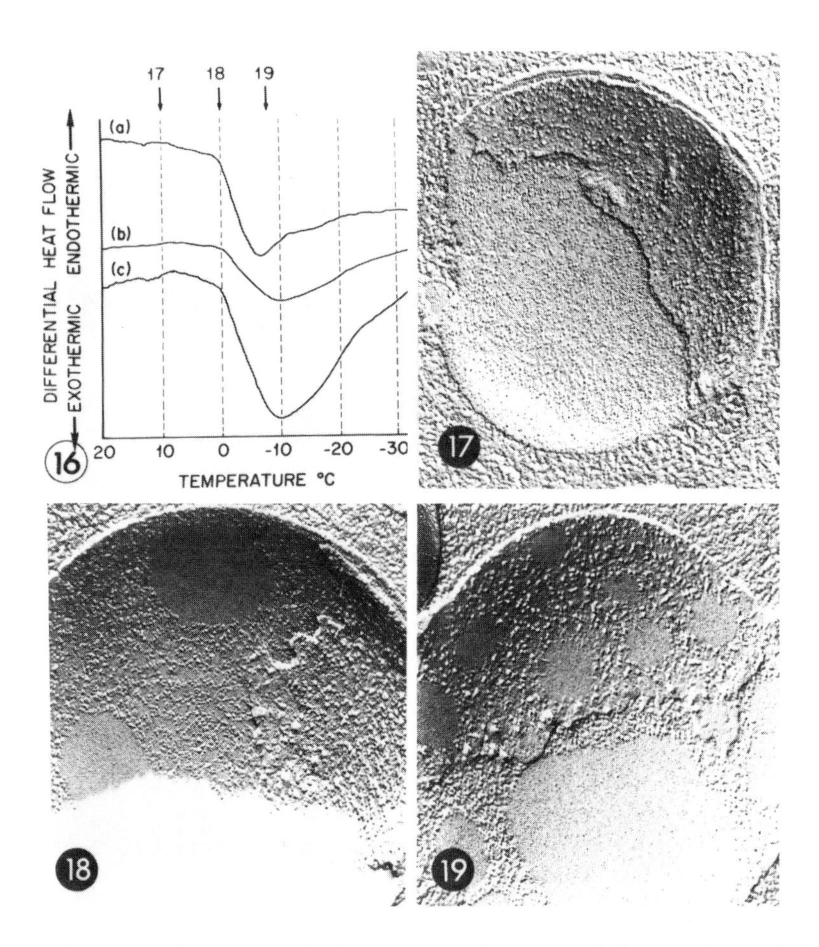

Figs. 16-19. Lipid phase transition and lateral mobility of integral proteins; Fig. 16. DSC cooling runs of whole mitochondria (a) and purified outer membrane (b,c). Recording sensitivity 0.5 mCal·sec^{-1} (a,b), 0.2 mCal·sec^{-1} (c). Arrows identify temperatures on the curve for whole mitochondria which correspond to Figs. 17-19; Fig. 17. Concave fracture faces of both membranes of a mitochondrion frozen from 30°C; Fig. 18. Cooled to 0°C, then frozen; Fig. 19. Cooled to -8°C, then frozen. Note smooth areas of gel state lipid and lateral displacement of integral proteins. X 67,500.

transducing membrane. This difference fits closely with the difference in the onset temperatures in the transition exotherms of the two membranes as observed by DSC and can be accounted for by the cholesterol content and the higher saturated to unsaturated phospholipid ratio in the outer membrane (Hackenbrock *et al.*, 1976).

Of particular significance is the determination that the reversible thermotropic lateral separations between gel state lipid and integral proteins in the energy transducing membrane are not destructive to electron transfer or oxidative phosphorylation (Höchli and Hackenbrock, 1976a). Mitochondria equilibrated at temperatures well within the transition exotherm for as long as 15 min. are as efficient in oxidative phosphorylation when returned to room temperature as mitochondria in which lateral separations between membrane components were not induced (Table 3). These observations indicate that no rigid protein-protein lattice is necessary to sustain molecular organization or functional integrity in the energy transducing membrane of the mitochondrion.

TABLE III

ADP:O AND ACCEPTOR CONTROL RATIOS IN MITOCHONDRIA BEFORE AND AFTER LOW TEMPERATURE-INDUCED LATERAL SEPARATIONS IN THE INNER MEMBRANE

	Pretreatment with Glycerol Medium[1]			
Exp.	Temp. in °C	Time in Min.	ADP:O	A.C.
1	No Pretreatment		1.80	5.50
2	25	2	1.48	3.38
3	25	2	1.45	3.33
4	0	8	1.50	3.19
5	0	8	1.59	3.05
6	−8	8	1.42	3.43
7	−8	8	1.48	3.00
8	−8	15	1.56	3.05

[1]*Mitochondria were pretreated in glycerol medium (30% glycerol; 250 mM sucrose; 10 mM Tris, pH 7.4) at temperatures and times indicated prior to polarographic analysis. Succinate was the electron donor.*

D. Lateral Motional Freedom of Cytochrome *c* Oxidase

We have prepared an affinity purified rabbit I_gG monospecific for cytochrome *c* oxidase and a ferritin conjugate of the IgG for the purpose of identifying the site-by-site distribution of the oxidase on both surfaces of the energy transducing membrane (Hackenbrock and Miller Hammon, 1975). Binding of the oxidase IgG-ferritin probe to the outer surface of the energy transducing membrane completely inhibits cytochrome *c* oxidase activity and reveals a random distribution of the oxidase in the lateral plane of the membrane (Figs. 20,21).

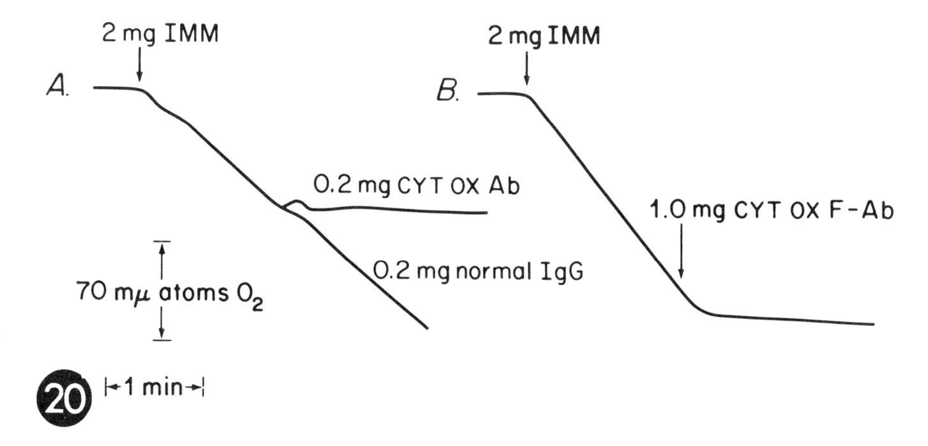

Fig. 20. Polarographic traces of the inhibition of cytochrome c oxidase activity in the inner membrane-matrix fraction (IMM) by affinity pure monospecific IgG (cyt ox Ab) and the ferritin conjugate of the affinity pure IgG (cyt ox F-Ab). Succinate was the electron donor in all curves.

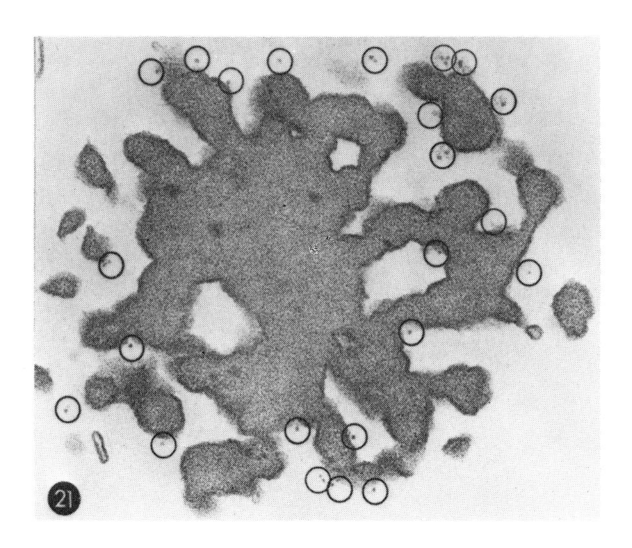

Fig. 21. Site-by-site distribution of cytochrome c oxidase on the outer surface of the energy transducing membrane of an inner membrane-matrix particle. Fixed for electron microscopy after complete oxidase inhibition by the ferritin conjugate of cyt ox Ab. X 113,000.

The oxidase typically shows a non-equidistant intermolecular spacing over the surface of the membrane and thus a random distribution typical of an integral membrane protein which is free to translate laterally. The oxidase-oxidase nearest neighbor distance appears often to be 100 nm (Fig. 21).

The affinity purified rabbit IgG monospecific for cytochrome c oxidase, which we abbreviate cyt ox Ab, can be used in various ways to attempt to follow lateral translational diffusion of the oxidase on the surface of the membrane during thermotropic lipid-protein separations. Furthermore, since cytochrome c oxidase is a completely transmembranous integral protein having a molecular weight of 200,000 and occurs in the membrane most likely in a polymeric cluster (Hackenbrock and Miller Hammon, 1975), the oxidase represents a large intramembrane particle, and its lateral diffusion in the hydrophobic interior of the energy transducing membrane can be followed by freeze fracture electron microscopy (Hackenbrock, 1977).

Rabbit cyt ox Ab was bound to the outer surface of the spherical energy transducing membrane at 25°C at concentrations which completely inhibit oxidase activity. This was followed by addition of a goat anti-rabbit IgG in order to at least partially crosslink the cyt ox Ab bound to the oxidase exposed at the membrane surface. In this way the immunoglobulins represent membrane surface artificial peripheral proteins which specifically lattice the cytochrome c oxidase. Since we determined previously the oxidase-oxidase nearest neighbor distance to be approximately 100 nm, such latticing of course could not be complete.

Under the conditions of such partial latticing, a lowering of the temperature to -13°C results in the appearance of unusually small regions of gel state lipid while a significant inhibition of thermotropic lateral translational diffusion of integral proteins is obvious (Fig. 22). This finding is in dramatic contrast to the appearance of the large pools of gel state lipid and the long-range lateral translational diffusion of integral proteins that take place at -13°C in the membrane in which cytochrome c oxidase is not latticed (Fig. 13). After such limited separation between gel state lipid and integral proteins the immunoglobulin lattice can be resolved on the membrane surface directly over areas rich in gel state lipid (Fig. 23). Thus, the IgG lattice, monospecific for cytochrome c oxidase tends to inhibit lateral motional freedom of the oxidase. In addition, the limited lateral diffusion of integral proteins under these conditions shows a general co-migration with the IgG specific for the oxidase.

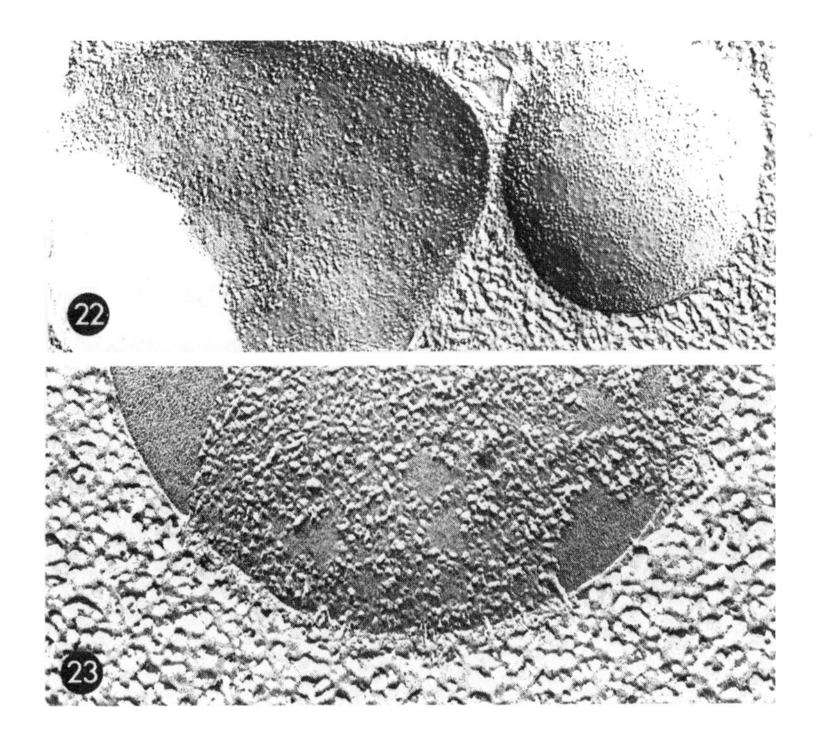

Figs. 22,23. Lateral mobility of cytochrome c oxidase in the energy transducing membrane and its hindrance by immunoglobulin latticing. Rabbit cyt ox Ab was permitted to bind to the membrane surface followed by addition of goat anti-rabbit IgG. After latticing, membranes were cooled to -13°C and then frozen; Fig. 22. Both fracture faces reveal only slight lateral displacement of integral proteins, X 67,500 (cf. non-latticed membranes, Fig. 13); Fig. 23. High magnification reveals immunoglobulin lattice in the etched ice surface in register with the integral proteins and no immunoglobulin in regions of smooth gel state lipid. X 132,000.

In another experimental approach, lateral separation of integral proteins and gel state lipid was induced in the spherical energy transducing membrane at -8 to -10°C (Fig. 12). Maintaining the temperature at -10°C, rabbit cyt ox Ab was bound to the membrane surface followed by goat anti-rabbit IgG-ferritin conjugate. The membranes were then fixed with glutaraldehyde and deep etched to expose large areas of the true membane surface. Since -10°C is well within the region of the transition exotherm the intermolecular distance between individual oxidases is considerably decreased owing to lateral aggregation of all integral proteins. Therefore, addition of the goat anti-rabbit IgG-ferritin probe maximally crosslinks the oxidases through the cyt ox Ab and in addition identifies directly and specifically the distribution of cytochrome c oxidase on the membrane surface. Freeze etching reveals the distribution of the immunoglobulin-ferritin probe for cytochrome c oxidase on the membrane surface to be crowded at -10°C and to be lacking in various regions on the membrane surface which are equated with regions of gel state lipid (Fig. 24). This experiment demonstrates unequivocally the thermotropic lateral translational diffusion of cytochrome c oxidase on the membrane surface.

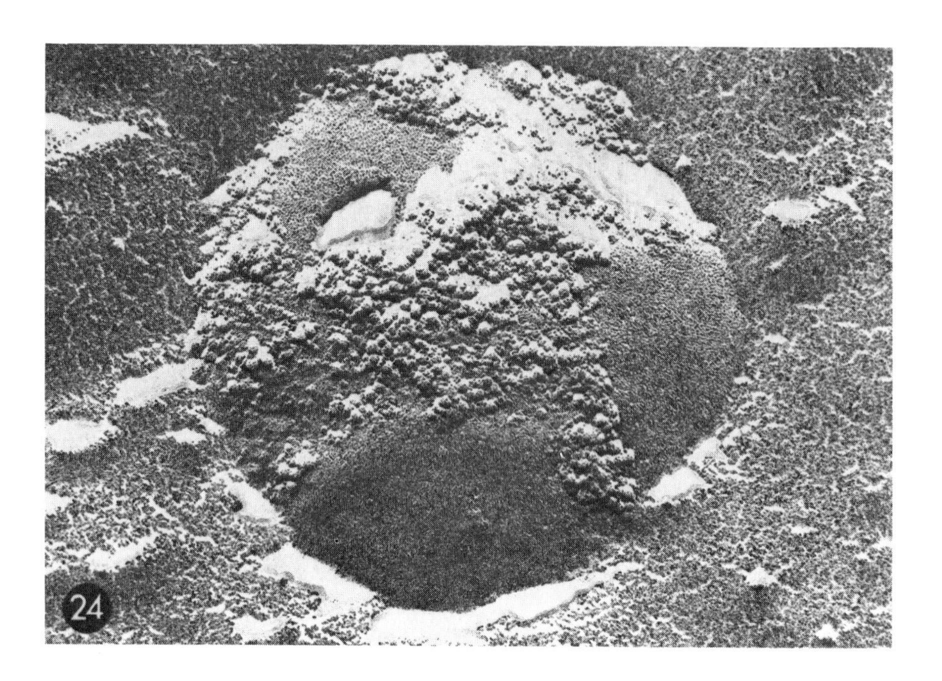

Fig. 24. Lateral mobility of cytochrome c oxidase on the surface of the energy transducing membrane. Membranes were cooled to -10°C. Rabbit cyt ox Ab was then permitted to bind to the membrane surface followed by addition of goat anti-rabbit IgG-ferritin conjugate. Membranes were then fixed with glutaraldehyde and deep etched. Note the smooth gel state lipid regions and lateral displacement of cytochrome c oxidase identified by the non-random distribution of immunoglobulin-ferritin probe on the surface of the membrane. X 132,000.

As pointed out earlier, the aggregation of integral proteins that occurs through lateral exclusion from the gel state lipid formed in the transition exotherm is rapidly and totally reversible. Complete randomization of the integral proteins occurs rapidly as the temperature is raised above 0°C and the gel state to liquid crystalline phase transition develops in the bilayer lipid (Figs. 13,15). However, the complete cross-linking of cytochrome c oxidase in the transition exotherm at -10°C with rabbit cyt ox Ab followed by goat anti-rabbit IgG prevents the oxidase from undergoing free lateral translational diffusion and thus disaggregation and randomization when the membrane is raised to well above the transition temperature (Figs. 25,26).

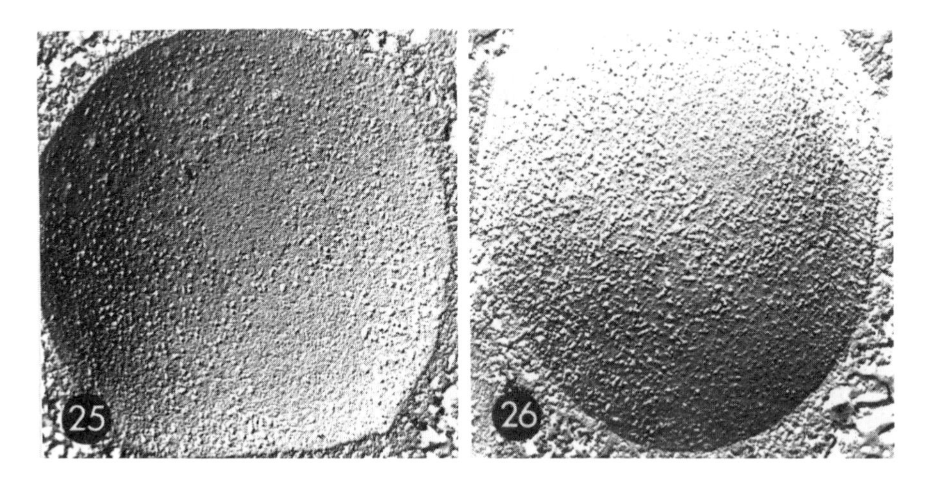

Figs. 25,26. Free lateral diffusion of integral proteins in the energy transducing membrane independent of cytochrome c oxidase mobility. Membranes were cooled to -10°C, then latticed with rabbit cyt ox Ab and goat anti-rabbit IgG, then returned to 25°C and finally frozen; Fig. 25. Concave fracture face shows smooth, integral protein-free regions at 25°C; Fig. 26. Convex fracture face shows large areas containing small but not large integral proteins at 25°C. (Compare with unlatticed control membranes, Fig. 15). X 67,500.

After return of the temperature to 25°C the oxidase, which appears predictably as large intramembrane particles up to 18 nm in diameter, is revealed to be in a non-random lateral distribution in the hydrophobic interior of the membrane. Virtually no integral proteins appear to diffuse into large pools of liquid crystalline lipid in the outer half of the membrane (Fig. 25) while a large number of small integral proteins (4.4 to 7.4 nm in diameter) diffuse laterally into these pools of liquid crystalline lipid in the inner half of the membrane (Fig. 26). Thus a number of small integral proteins can diffuse laterally in the hydrophobic interior of the energy transducing membrane independent of cytochrome c oxidase.

We conclude from these observations that cytochrome c oxidase can diffuse laterally in the bilayer lipid of the energy transducing membrane. Other integral proteins, as yet unidentified, can diffuse laterally in the membrane independent of the oxidase. Lateral translational diffusion of cytochrome c oxidase can be inhibited by latticing through monospecific IgG which suggests that no native latticing of the oxidase of comparable effectiveness occurs in the energy transducing membrane. The observations are clear, however, that the total number of large integral proteins inhibited in free lateral translational diffusion by monospecific IgG is greater than can be accounted for by the estimated number of oxidase molecules or molecule polymers located in the membrane (Figs. 25,26). Thus other unidentified integral proteins may codiffuse laterally in physical union with the oxidase. Of considerable importance will be to determine whether or not the oligomycin-sensitive ATPase complex diffuses laterally in the energy transducing membrane independent of cytochrome c oxidase and other respiratory integral proteins. The geometric relationship between ATPase and the redox components, whether in physical union or spatially separated, will of course ultimately relate to the mechanism of energy conservation in the mitochondrion.

E. Fluidity in the Lipid Bilayer, Lateral Diffusion of Integral Proteins and Metabolic Function

We have observed that the lipid bilayer of the structurally and functionally intact energy transducing membrane of the mitochondrion is considerably plastic and fluid. This finding is consistent with the unique lipid composition of the bilayer. The phospholipids of the membrane are collectively 53% unsaturated with cardiolipin alone being 90% unsaturated (Colbeau *et al.*, 1971). In addition, the membrane lacks cholesterol. Such a composition accounts for the subzero liquid crystalline to gel state phase transition determined in our studies and presumably for the low viscosity (0.1 P at 30°C) and high mobility in the phospholipid hydrocarbon chains reported for mitochondrial lipids (Keith *et al.*, 1970; Keough *et al.*, 1973). It is well known that unsaturated phospholipids have characteristically lower transition temperatures, which appears to be related to their less cooperative interaction than do saturated phospholipids of similar hydrocarbon chain length (Chapman and Wallach, 1968). The occurrence of cholesterol tends to decrease the fluidity of phospholipids in the liquid crystalline state (Ladbrook *et al.*, 1968). Thus, the result of the unique molecular composition of the lipid bilayer of the energy transducing membrane is to partition a highly effective concentration of metabolically active proteins in a phospholipid-pure bilayer of low viscosity and high plasticity and fluidity.

By utilizing covalent, electrostatis, and immunoligand latticing, we have recently demonstrated that no natural, rigid protein-protein latticing exists in the energy transducing membrane (Höchli and Hackenbrock, 1976b; Hackenbrock, 1976; Hackenbrock and Höchli, 1976). These observations, together with the finding that cytochrome c exists in a random distribution in the energy transducing membrane after chemical fixation (Hackenbrock and Miller Hammon, 1975), do not support inferences that the metabolically active proteins are stabilized in the membrane through a continuous, rigid protein-protein lattice (Fleisher *et al.*, 1967; Sjöstrand and Barajas, 1970; Capaldi and Green, 1972) or that the proteins exist in an ordered array and with recurring intermolecular spacing in the plane of the membrane (Klingenberg, 1968).

Our observations, combined with earlier calculations by Vanderkooi (1974), reveal that the integral proteins of the energy transducing membrane occupy only one third of the membrane surface area. Since only 50% of the total protein of the membrane is integral to the membrane (Capaldi and Tan, 1974; Harmon *et al.*, 1974), the quantity of membrane phospholipid immobilized through hydrophobic interactions between the apolar surfaces of the globular proteins and hydrocarbon chains of the phospholipid is considerably less than heretofore assumed. The fraction of protein-immobilized lipid which does not participate in thermotropic liquid crystalline to gel state transitions in the bilayer usually accounts for only 15 to 20% of the total lipid component of membranes (Steim *et al.*, 1969; McConnell *et al.*, 1972; Trauble and Overath, 1973). Consistent with these estimates is our direct observation by freeze fracture electron microscopy that as much as two thirds of the surface area of the lipid bilayer of the energy transducing membrane is converted to gel state lipid at -13°C. This indicates that only a negligible fraction of the total lipid of the energy transducing membrane is immobilized as boundary lipid by the integral proteins.

In addition to considerable intrinsic motion in the phospholipid hydrocarbon chains of mitochondrial membranes (Keough *et al.*, 1973), studies of lecithin bilayers in the liquid crystalline state indicate the existence of axial rotation of the entire phospholipid molecule with a rotational diffusion coefficient of 7 x 10^9 sec^{-1} at 52°C (Levine *et al.*, 1972) and lateral translational diffusion with a lateral diffusion coefficient of 1 x 10^{-7} cm^2 sec^{-1} at 40°C (Scandella *et al.*, 1972). In natural membranes, e.g. sarcoplasmic reticulum and rat liver microsomes, the lateral diffusion coefficient of lipids is 6 x 10^{-8} cm^2 sec^{-1} to 1.4 x 10^{-7} cm^2 sec^{-1} at 40°C (Scandella *et al.*, 1972; Stier and Sackmann, 1973). Transverse diffusion of phospholipids, i.e. flip-flop across the bilayer, is negligible in the mitochondrial energy transducing membrane (Rousselet *et al.*, 1976).

With such rotational and lateral motional freedom in lipid bilayers and provided the integral membrane proteins are not immobilized through anchoring or latticing as in the energy transducing membrane, lateral translational diffusion of the proteins may be considerable in both rate and distance. For those integral proteins of natural membranes studied so far, lateral diffusion occurs at approximately an order of magnitude less than for phospholipids, with diffusion coefficients centering at approximately 5 x 10^{-9} cm^2 sec^{-1} at 20°C (Edidin and Fambrough, 1973; Poo and Cone, 1974; Liebman and Entine, 1974). This motion is equal to an approximate molecular linear displacement of 100 nm in one second. The potential for free lateral translational diffusion of integral proteins in the energy transducing membrane may therefore be greater than 1 nm 10 $msec^{-1}$ at physiological temperature which is well within the time structure of electron transfer between some redox components of the respiratory chain and the transduction of the free energy of electron transfer into the chemical bond energy of ATP.

The relationship between the half time of oxidation of some electron transfer proteins and the half time of ATP synthesis in the energy transducing membrane is difficult to reconcile in terms of a permanent physical association of electron transfer proteins to ATPase. For example, although the half time oxidation of cytochrome c oxidase is approximately 0.5 msec (Chance, 1967), the half time of ATP synthesis coupled to this oxidation is approximately 100 msec (Lemasters and Hackenbrock, 1976). This considerable delay between electron transfer and ATP synthesis can represent the time required for a chemical reaction or conformational motion to occur in redox and/or ATPase components of the membrane which may be a prerequisite for energy transduction. Alternatively, such delay may be rationalized in terms of the time required to generate a protonmotive gradient across the energy transducing membrane adequate to produce ATP.

Just as likely, the highly fluid nature of the membrane may support a time requiring lateral translational diffusion of electron transfer and/or ATPase proteins prior to their collision at which time the free energy realized from electron transfer is transduced. We would point out that the time delay between the oxidation of cytochrome c oxidase and ATP synthesis is approximately 100 msec during which, as our calculations show, an integral membrane protein can diffuse laterally at least 10 nm at physiological temperature.

Lateral translational diffusion of electron transfer proteins in the energy transducing membrane may also account for various delays in electron transfer times, e.g. in the extended oxidation half times of cytochrome b and of flavoproteins in oxygen pulse experiments (Chance, 1967). Time delays in the transfer of reducing equivalents to the cytochrome chain from succinate and NADH dehydrogenase is consistent with the ratio of the dehydrogenase to cytochrome b ratio of 1:10. Most likely in this case ubiquinone functions as a redox component diffusing laterally in the membrane between the dehydrogenases and b cytochromes thereby delivering reducing equivalents to the cytochrome chain (Kröger *et al.*, 1973a, 1973b). In the ADP-induced State 4 to 3 transition, the half time of cytochrome b oxidation at 26°C may be as long as 70 msec (Chance, 1965), with the half time transition of NADH at 18°C as long as 1.7 sec (Klingenberg, 1963). Electron transfer chain cross-reactivity is also highly suggestive of free lateral translational diffusion by electron transfer proteins in the energy transducing membrane. It has been determined, e.g. that with the reduction of each cytochrome b by succinate dehydrogenase, three molecules of cytochrome c and nine molecules of cytochrome c oxidase are reduced (Lee *et al.*, 1965). It remains to be determined whether or not all of these temporal phenomena of electron transfer and oxidative phosphorylation can be accounted for by lateral translational motion of the metabolically active proteins in the energy transducing membrane.

III. CONCLUSIONS

The low viscosity and considerable degree of phospholipid fluidity provide the basis for the high potential for lateral translational diffusion of integral proteins in the hydrophobic interior and on the surface of the mitochondrial energy transducing membrane determined in our studies. The extreme configurational plasticity and fluidity of the energy transducing membrane is consistent with the existence of the vast number of diversified metabolically active proteins which catalyze the various ion and substrate transport, electron transfer, and energy transducing activities in this membrane. Most likely all of these metabolic activities require proteins with an intrinsic freedom for conformational motion and an extrinsic freedom for lateral diffusional motion as well. We would suggest that both conformational and lateral diffusional motion of metabolic proteins occur normally during the complex and diversified activities of the energy transducing membrane and that a highly plastic, fluid and laterally compressable lipid bilayer is required for such motional freedom. The rate of lateral translational motion by the metabolically active integral proteins in the energy transducing membrane can account for the time structure of electron transfer and ATP synthesis in this membrane.

IV. ACKNOWLEDGEMENTS

We wish to thank our associates who have contributed to the research present here: Drs. Luci Hochli, Peter Andrews, Katy Hammon and Mary Tobleman. Dr. Matthias Hochli was a Fellow of the Swiss National Foundation and the Muscular Dystrophy Association of America. Our studies were supported by research grants from the National Science Foundation and the National Institutes of Health.

V. REFERENCES

Andrews, P. and Hackenbrock, C.R., Exp. Cell Res. 90, 127 (1975).

Blazyk, J.F. and Steim, J.M., Biochim. Biophys. Acta 266, 737 (1972).

Capaldi, R.A., and Green, D.E., FEBS Lett. 25, 205 (1972).

Capaldi, R.A., and Tan, P.F., Fed. Proc. 33, 1515 (1974).

Chance, B., J. Biol. Chem. 240, 2729 (1965).

Chance, B., DeVault, D., Legallais, V., Mela, L. and Yonetani, T. in "Nobel Symposium 5, Fast Reactions and Primary Processes in Chemical Kinetics", (S. Claesson, ed), p. 437. Interscience Pub., New York, 1967.

Chapman, D., Williams, R.M., and Ladbrook B.D., Chem. Phys. Lipids 1, 445 (1967).

Chapman, D. and Wallach, D.F.H., in "Biological Membranes, Physical Fact and Function", (D. Chapman, ed.). Academic Press, New York, 1968.

Colbeau, A., Nachbaur, J., and Vignais, P.M., Biochim. Biophys. Acta. 249,462 (1971).

Edidin, M. and Fambrough, D., J. Cell Biol. 57,27 (1973).

Fleisher, S., Fleisher, B. and Stoeckenius, W., J. Cell Biol. 32, 193 (1967).

Hackenbrock, C.R., J. Cell Biol. 30, 269 (1966).
Hackenbrock C.R., J. Cell Biol. 37, 345 (1968).
Hackenbrock, C.R., Rehn, R.G., Weinbach, E.C., and Lemasters, J.J., J. Cell Biol. 51, 123 (1971).
Hackenbrock, C.R., Ann. N.Y. Acad. Sci. 195, 492 (1972a).
Hackenbrock, C.R., J. Cell Biol. 53, 450 (1972b).
Hackenbrock, C.R. and K. Miller Hammon, J. Biol. Chem. 250, 9185 (1975).
Hackenbrock, C.R. in "The 34th Nobel Symposium: The Structure of Biological Membranes," (S. Abrahamsson, ed). Plenum Publishers, London, 1977, in press.
Hackenbrock, C.R. and Höchli, M., in "Biochemistry of Membrane Transport", (G. Semenza and E. Carafoli, eds.). Springer-Verlag, Heidelberg, 1977, in press.
Hackenbrock, C.R., Höchli, M. and Chau, R.M., Biochim. Biophys. Acta 455, 466 (1976).
Harmon, H.J., Hall, J.D. and Crane, F.L., Biochim. Biophys. Acta 344, 119 (1974).
Höchli, M. and Hackenbrock, C.R., Proc. Natl. Acad. Sci. USA 73, 1636 (1976a).
Höchli, M. and Hackenbrock, C.R., J. Cell Biol. (1977b, in press).
Keith, A., Bulfield, G., and Snipes, W., Biophysic. J. 10, 618 (1970).
Keough, K.M., Oldfield, E., Chapman, D., and Beynon, P., Chem. Phys. Lipids 10, 37 (1973).
Klingenberg, M., in "Energy-Linked Functions of Mitochondria", (B. Chance, ed.). Academic Press, New York, p. 270, 1963.
Klingenberg, M., in "Biological Oxidations", (T.P. Singer, ed.), p. 3. Wiley, New York, 1968.
Kröger, A., Klingenberg, M., and Schweidler, S., Eur. J. Biochem. 34, 358 (1973a).
Kröger, A., Klingenberg, M., and Schweidler, S., Eur. J. Biochem. 39, 313 (1973b).
Ladbrook, B.D., Williams, R.M., and Chapman, D., Biochim. Biophys. Acta. 150, 333 (1968).
Ladbrook, B.D., and Chapman, D., Chem. Phys. Lipids 3, 304 (1969).
Lee, C.P., Estabrook R.W., and Chance, B., Biochim. Biophys. Acta 99, 32 (1965).
Leibman, P.A. and Entine, G., Science 185, 457 (1974).
Lemasters, J.J. and Hackenbrock, C.R., Fed. Proc. 32, 516 (1973).
Lemasters, J.J. and Hackenbrock, C.R., Eur. J. Biochem. 67, 1 (1976).
Levine, Y.K., Partington, P., Roberts, G.C.K., Birdsall, N.J. M., Lee, A.G. and Metcalfe, J.C., FEBS Lett. 23, 203 (1972).

Levine, Y.K., in "Progress in Surface Science 3", (S.G. Davison, ed.). Pergamon Press, Oxford, 1973.

Lévy, M., Toury, R., Sauner, M-T., and André, J., in "Mitochondria, Structure and Function", (L. Ernster and Z. Drahota, eds.), p. 33. Academic Press, New York, 1969.

McConnell, H.M., Wright, K.L. and McFarland, B.G., Biochem. Biophys. Res. Commun. 47, 273 (1972).

Phillips, M.C., in "Progress in Surface and Membrane Science 5", (J.F. Dannielli, M.D. Rosenberg, and D.A. Cadenhead, eds.). Academic Press, New York, 1972.

Poo, M-M., and Cone, R.A., Nature 247, 438 (1974).

Rousselet, A., Colbeau, A., Vignais, P.M., and Devaux, P.F., Biochim. Biophys. Acta 426, 372 (1976).

Salsbury, N.J., and Chapman, D., Biochim. Biophys. Acta 163, 314 (1968).

Scandella, C.J., Devaux, P. and McConnell, H.M., Proc. Natl. Acad. Sci. USA 69, 2056 (1972).

Scherer, B. and Klingenberg, M., Biochem. 13, 161 (1974).

Sjöstrand, F.S. and Barajas, L., J. Ultrastruct. Res. 32, 293 (1970).

Steim, J.M., Tourtellotte, M.E., Reinert, J.C., NcElhaney, R.N. and Rader, R.L., Biochem. 63, 104 (1969).

Stier, A., and Sackmann, E., Biochim. Biophys. Acta 311, 400 (1973).

Stoner, C.D., and Sirak, H.D., J. Cell Biol. 56, 51 (1973).

Trauble, H., and Haynes, D.H., Chem. Phys. Lipids 7, 324 (1971).

Trauble, H., and Overath, P., Biochim. Biophys. Acta 307, 491 (1973).

Vanderkooi, G., Biochim. Biophys. Acta 344, 307 (1974).

Weber, N.E., J. Cell Biol. 55, 457 (1972).

PEROXIDE GENERATION IN MITOCHONDRIA AND UTILIZATION BY CATALASE

B. Chance*, A. Boveris+ and N. Oshino0

*University of Pennsylvania School of Medicine
+University of Buenos Aires
0Nihon Schering K.K., Osaka, Japan

Peroxide metabolism of mitochondria was an unknown and unsuspected phenomenon a few years ago; few clues were available and accurate measurements were not possible. Sensitive optical methods for H_2O_2 quantitation show that the generation of H_2O_2 is intimately linked to cell respiration in an inverse fashion. H_2O_2 production is maximal when cytochrome oxidase activity is minimal as in the resting state 4. This paper is presented as a review of the mechanisms of H_2O_2 production and utilization and their relationship to ethanol metabolism under normoxic and hyperoxic conditions.

I. DISCUSSION

A. Regulation of O_2^- and H_2O_2

The goal of cell respiration is to produce ATP and water from oxygen and NADH via cytochrome oxidase, which is specially constructed to sequester dangerous intermediates of oxygen reduction. However, it is thermodynamically feasible to accumulate such toxic intermediates as O_2^- or H_2O_2 in appreciable quantities, either via specific enzyme activities or by unexpected "leaks" in the mitochondrial respiratory chain. H_2O_2 can be generated in other organelles as well. The schematic diagram of Figure 1 shows the pathways of oxygen metabolism in the cell, with O_2^- and H_2O_2 generation in the mitochondrial matrix and the cytosol, and H_2O_2 generation in the peroxisome, with superoxide dismutase rapidly converting O_2^- to H_2O_2 in both the mitochondria and the cytosol. Nevertheless, a very low O_2^- concentration (about 10^{-11} M) and a somewhat higher level

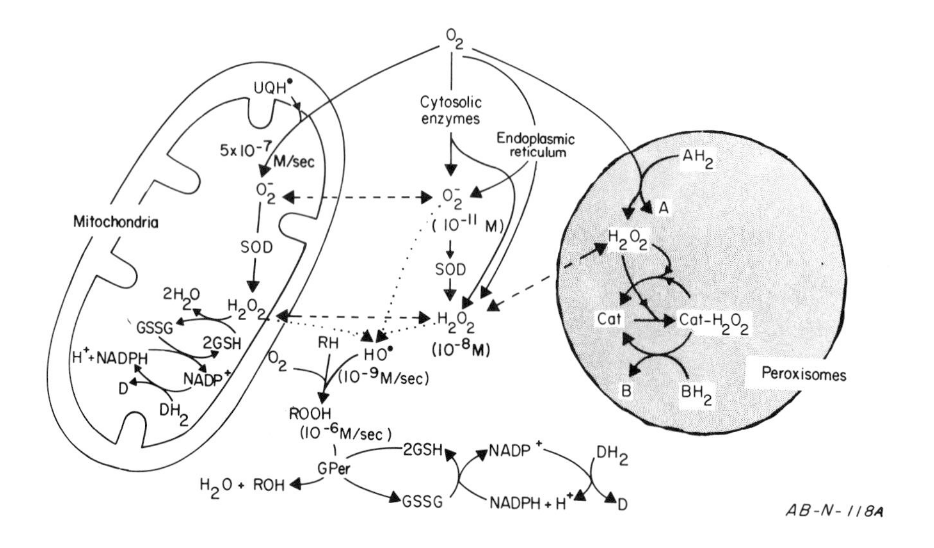

Fig. 1. A general diagram indicating the sources and sinks for oxygen reduction products in the mitochondrial, cytosolic, and peroxisomal spaces. Those abbreviations not given in the text are UQH·, ubiquione radical; GSSG, oxidized glutathione; GSH, reduced glutathione; DH_2 and D, non-specific donors to the respiratory chain; SOD, superoxide dismutase; NADPH and NADP, nicotine adenine dinucleotide; O_2^- , superoxide anion; HO·, a free radical intermediate of oxygen reduction; ROOH, an alkyl hydroperoxide ; GPer, glutathione peroxidase; Cat, catalase; B and BH_2, hydrogen donors of a specificity appropriate to catalase, such as ethanol.

of H_2O_2 (10^{-8}M) accumulate; the latter can be utilized in alcohol (BH_2) oxidation in the peroxisomes. The subsequent reactions of H_2O_2 may activate glutathione (GSH) peroxidase directly, or indirectly through lipid peroxidation (ROOH) caused by radical intermediates (HO·). The possible functions of endoplasmic reticulum are discussed elsewhere in this volume (1).

Catalase affords an unique regulatory control of cellular H_2O_2 levels (2) and at the same time is capable of converting alcohol to aldehyde without expending any of the NADH store as in the case of alcohol dehydrogenase, or of NADPH as in the case of any microsomal alcohol-oxidizing system dependent upon this coenzyme. Thus, we shall first take up the properties of catalase and then discuss the nature of O_2 and H_2O_2 generation in mitochondria.

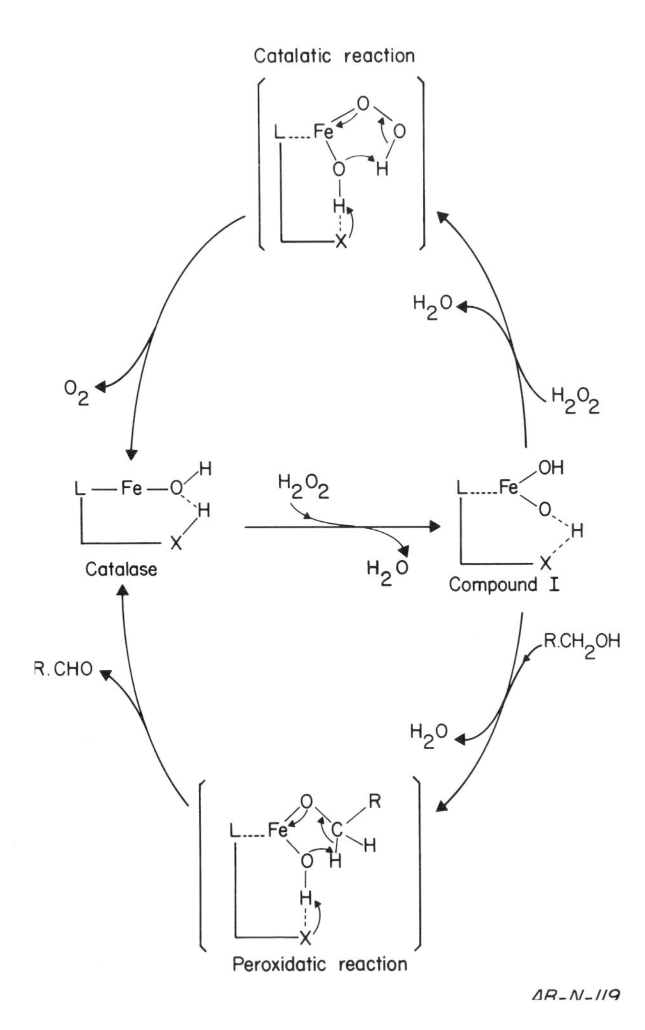

Fig. 2. A schematic diagram of the reactions of catalase, illustrating various forms of the several intermediates. The cycle starts on the left-hand side with the enzyme combining with H_2O_2 to form compound I (right-hand side), which is identified as Compound I. This compound may be, but is not necessarily, the first intermediate in the reaction of iron and peroxide. The reduction of Compound I by a second molecule of H_2O_2 yields the catalatic reaction (top) in which two molecules of H_2O_2 are decomposed with the formation of a molecule of oxygen. At the bottom of the figure is represented the peroxidatic activity, in which a molecule of alcohol is oxidized to aldehyde. The detailed structures of the intermediates remain unknown.

B. Catalase

Catalase is a unique dual-purpose enzyme, whose reaction mechanism involves a consecutive bimolecular reaction of the enzyme with two substrate molecules, a phenomenon unknown in enzymology except for catalase (2) and possibly superoxide dismutase (Figure 2). The catalatic reaction (top of Fig. 2) affords a safety valve for high levels of H_2O_2, and as the blood alcohol levels rise, the slower peroxidatic reaction is activated (bottom of Figure 2). The key linkage between these two pathways is the common intermediate, Compound I (Figure 2), discovered about thirty years ago while working in Hugo Theorell's laboratory in Stockholm (3). This is a very reactive peroxide compound whose exact structure is not yet known in detail, but is pictured here according to recent proposals (4). This compound is capable of rapid destruction of H_2O_2 or oxidation of short-chain alcohols. Such activities are demonstrated most precisely by the direct reaction of the catalase·H_2O_2 compound with a variety of compounds in addition to alcohols such as nitrate and formate. They show rapid bimolecular kinetics with the peroxide intermediate (4,5). Figure 3 shows such kinetics in terms of the 45° slope of the log-log plot of the pseudo first order velocity constant (k_4a_o) against the donor concentration.

Similarly, continuous generation of H_2O_2 by glucose oxidase forms aldehyde from alcohol in generous amounts via the catalase pathway, as shown by Keilin and Hartree for ethanol and methanol (6), and at lower rates and higher apparent K_M's for propanol and butanol, in agreement with direct kinetic studies (4,5).

As shown in Figure 8 below one can readily calculate activities from these apparent K_M values by the simple equation

$$k_4 = \frac{0.8}{a_{\frac{1}{2}}} \qquad (8)$$

from which similar values are obtained for methanol, ethanol, propanol and butanol as with the assay of aldehyde formation.

It is useful to inquire what factors govern the activity of the catalase ·H_2O_2 complex towards alcohols. The first is obvious: there is a chain-length phenomenon suggesting that either the active site itself, or the channel to the active site, of catalase fits - or, as Dr. Koshland might say, can be induced to fit - only the lower alcohols. A summary of values for 2 , 3 , and 4 carbon alcohols appear in Table 1 as the second order velocity constants (k_4 M^{-1} sec^{-1}) (4). It seems

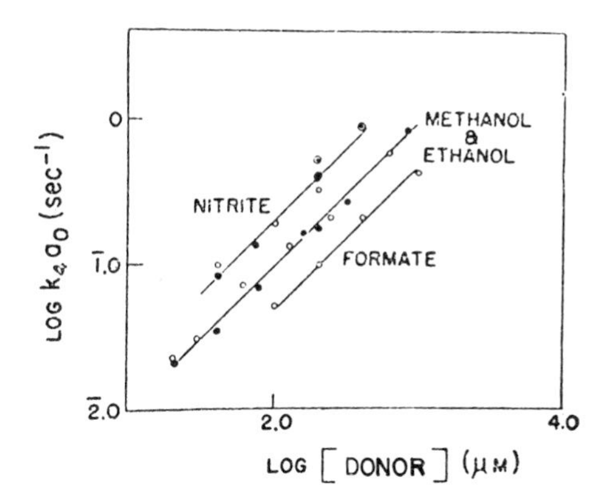

Fig. 3. An illustration of the specificity of catalase for four hydrogen donors: nitrite, methanol, ethanol and formate. The positions of the lines on the traces represent the relative activities, nitrate being largest, ethanol and methanol identical, and formate the least (as calculated from the total formate added). The 45° slope of the line indicates a bimolecular reaction of the donor and the catalase intermediate. The abscissa in this case is labelled "Acceptor" because of the concept that the oxygen atom is transferred to the reducing substrate to form water as a product.

TABLE 1
Reduction of Compound I[a] by Two, Three, and Four Carbon Alcohols

	Alcohol				
	Ethyl	Propyl	Butyl	Allyl	Propargyl
k_4 app ($M^{-1}sec^{-1}$)	1020	6.5	0.4	330	2500

[a]Horse erythrocyte catalase; 5 mM phosphate, pH 7, 25° (See ref. 4).

that chain length is a dominant factor in the reaction of ethyl, propyl and butyl alcohol but other factors enter with some larger alcohols such as allyl and propargyl (acetylenyl carbonol) alcohols that react very rapidly (4). A mechanism which takes into account steric effects, and a nonrandom and specific substrate-binding site is included in Fig. 4.

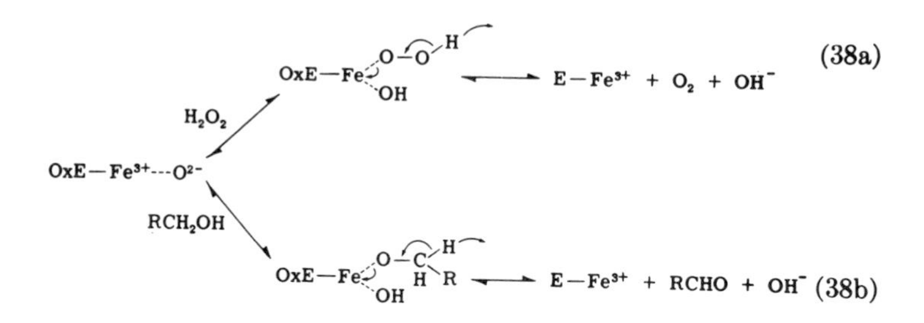

Fig. 4. A schematic diagram of the reduction of Compound I by H_2O_2 (top line) or by alcohols (bottom line). The nature of the intermediates is, as yet, speculative but representative of the general type of reaction that is likely to occur.

Consideration of the substrate-binding site is facilitated by the observation of Theorell (9) and ourselves (10) that acetate and formate bind heme proteins as shown in Figure 5, and

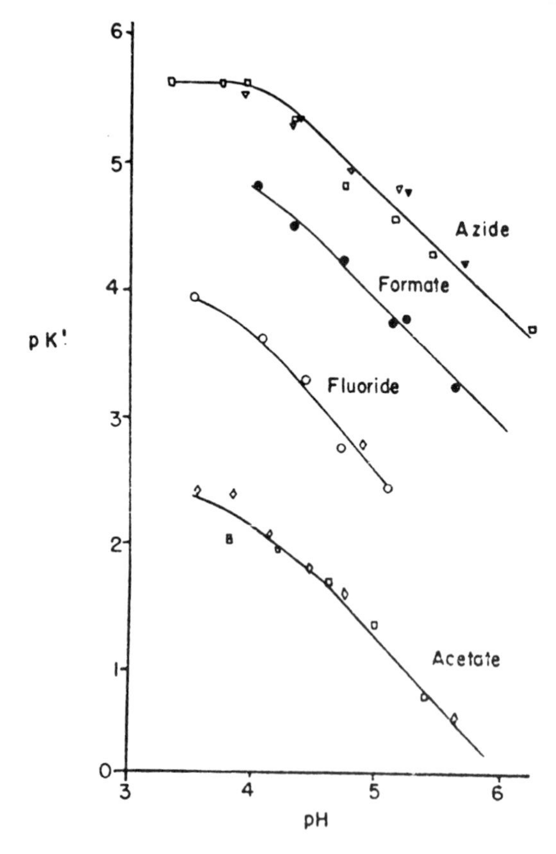

Fig. 5. The binding of anions to catalase, including formate and acetate. Titrations of the enzyme with various concentrations of anions such as formate acetate azide and fluoride as a function of pH. The curves suggest that the protonated species is the active form.

[Courtesy *J. Biol. Chem.*, Ref. (10)].

by our observation that formate but not acetate is a substrate for catalase (10). Making the assumption usually made by enzymologists that a substrate-binding site is also a catalytic binding site, proton NMR as described here by Dr. Cohn (11) can be used to measure the effect of the paramagnetic iron of catalase upon the relaxation of the formate protons, and thus calculate the distance between the two. As shown by the diagram of Figure 6 the distance between the heme iron and a proton of a ligand such as water is 2.8 Å (12), and that to a formate molecule about 4 Å; i.e., formate is too far away to be a direct iron ligand. Thus the formate binding site is likely to be separate from the peroxide binding site which appears to bind the iron atom directly. Presumably, catalysis occurs with the bound and appropriately oriented molecules such as H_2O_2, ethanol, or formate.

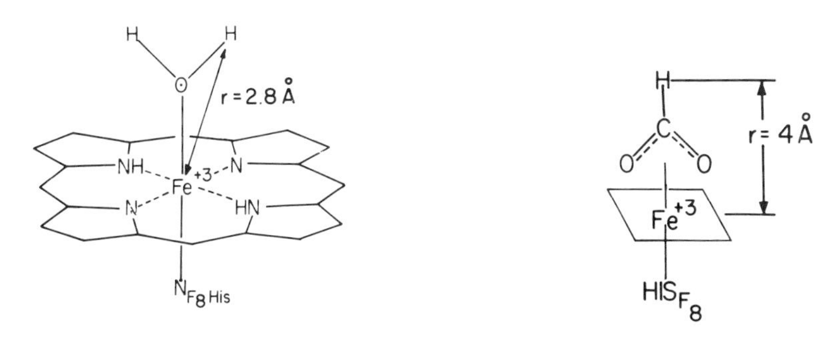

Fig. 6. Schematic diagrams indicating the distance of the water and formate molecules from the iron atom based upon studies of myoglobin. These models afford the interpretation of similar data on catalase, which indicate that formate does not bind directly to the iron atom (12).

The performance of catalase in coupled oxidations can be analyzed mathematically on the basis of the mechanism which was founded upon spectroscopic observations made in Theorell's laboratory (3). One of the simple formulae tells us that the ethanol concentrations ($a_{\frac{1}{2}}$) (see Figure 7) that causes half-maximal saturation of the catalase·H_2O_2 intermediate is proportional to the turnover number of the catalase; i.e., the rate of H_2O_2 utilization, $\frac{dx_n}{dt}$, divided by the molar concentration of catalase. This relationship is illustrated by Figure 7 left, in which titration of Compound I with methanol is illustrated and the value of $a_{\frac{1}{2}}$ is almost 0.02 mM for the particular concentration of catalase and the rate of generation of H_2O_2. In the right-hand portion of the figure are plots of the equation showing how $a_{\frac{1}{2}}$ varies with the two variables,

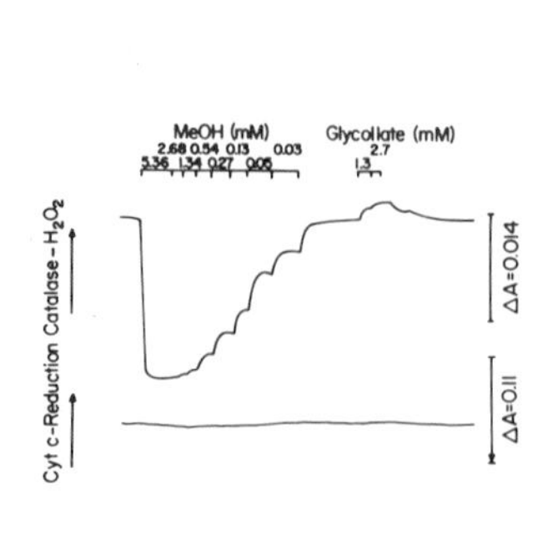

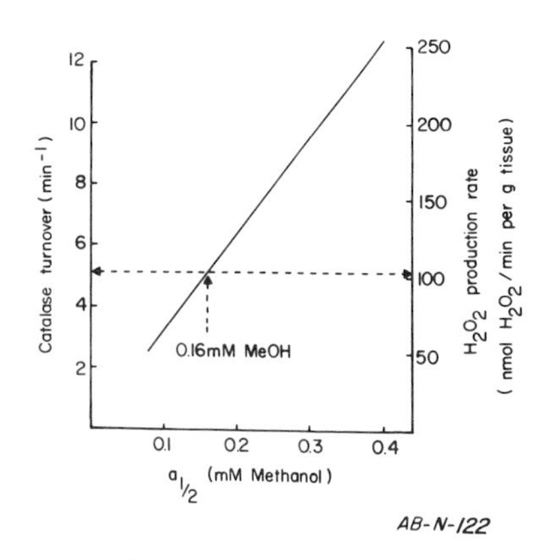

Fig. 7. Illustrations of the existence of catalase Compound I in the perfused liver, its reactivity towards methanol, the very near saturation of catalase with H_2O_2 in the normally metabolizing perfused liver, and finally, the calculation of the catalase turnover number and the H_2O_2 production rate from titrations of the perfused organ with methanol. The figure illustrates, by the abrupt downward deflection of the trace, the reaction of Compound I with methanol in the intact, perfused liver. As the methanol concentration decreases, the intensity of the reaction decreases until the concentration of the intermediate returns to the initial level. Justification that the enzyme is nearly saturated with H_2O_2 under the endogenous conditions is afforded by the very small increment in its concentration caused by perfusion with glycollate.

catalase turnover and rate of production of H_2O_2. Alcohol oxidation by the peroxidatic reaction is favored by a high catalase concentration as is characteristic of the peroxisomes; in fact, the catalase concentration in the peroxisomes is so high that it resistant to inhibitors of the enzyme (8) - a fact that has led to much more controversy than it should have.

C. H_2O_2 Generation

Figure 7 also demonstrates the reactivity of catalase, Compound I in perfused and intact livers. The generation of H_2O_2 can be followed by the spectroscopic measurement of the direct reaction of the catalase·H_2O_2 intermediate with alcohol in the organ, thus providing the ultimate proof of the function of this catalase intermediate in alcohol oxidation in liver tissue. The intermediate reacts with alcohol; it disappears as the alcohol concentration is increased and is restored to full concentration as the alcohol concentration is decreased in stepwise fashion (Figure 7, left). Saturation of catalase with H_2O_2 is ensured by adding glycollate, which further increases H_2O_2 generation in the liver. The equation above is plotted for various conditions, affording an indication of the wide range over which the rate of alcohol oxidation by the catalase pathway of the liver may be controlled by appropriate diets of H_2O_2-generating substrates.

TABLE 2

H_2O_2 (or O_2^-) Producing Enzymes of Liver and Their Subcellular Localization

EC Number	*Enzyme (Trivial Name)*	*Localization*
1.1.3.1	Glycollate oxidase	Peroxisome
1.1.3α	L-α-hydroxy acid oxidase	Peroxisome
1.1.3.8	L-gulonolactone oxidase	
1.2.3.1	Aldehyde oxidase	
1.2.3.2	Xanthine oxidase	Cytosol
1.4.3.3	D-amino acid oxidase	Peroxisome
1.4.3.4	Monoamine oxidase	Mitochondrial outer membrane
1.4.3.5	Pyridoxamine oxidase	
1.4.3.6	Diamine oxidase	Endoplasmic reticulum
1.6.99.1	NADPH cytochrome c reductase	Endoplasmic reticulum
1.6.99.3	NADH cytochrome c reductase	
1.7.3.3	Urate oxidase	Peroxisome "core"
1.15.1.1	Superoxide dismutase	Cytosol, mitochondrial matrix

Table 2 shows the plethora of direct H_2O_2 generators, and indirect H_2O_2 generators via O_2^- and superoxide sidmutase. It is apparent that we have at our disposal a great variety of enzymes and enzyme systems for designing suitable H_2O_2 generation rates for optimal alcohol oxidation.

One of the hitherto unsuspected H_2O_2 generators is the respiratory chain itself - unsustected for two historic reasons. First and most important is Otto Warburg's opposition to the idea of alternate pathways of cell respiration; in spite of Wieland's protests (13) he convinced most contemporary workers that all cell respiration passes through "atmungsferment" (14). For a summary of these views see references 15 and 16. Secondly, no one could have imagined in 1925 that Keilin had really opened the lid of Pandora'a box when he identified the cytochromes as components of the respiratory chain (17). Theorell suggested flavins on the basis of studies of the "old yellow enzyme" (18). Now, more than sixteen components of the respiratory chain are recognized, including Fe-S centers and quinone components as well as several novel flavins and cytochromes (Figure 8) (19).

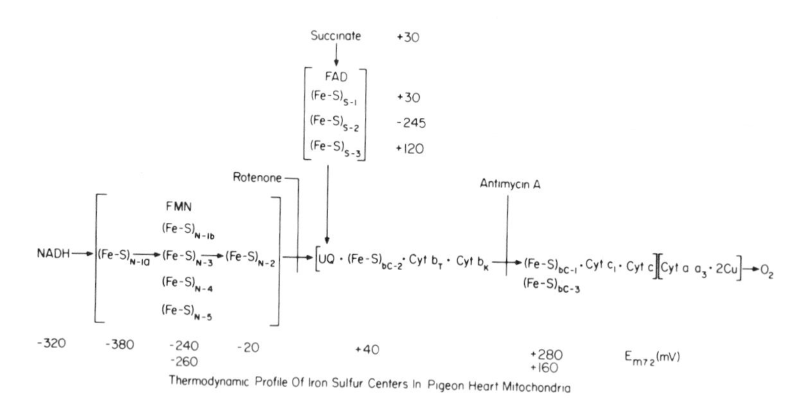

Fig. 8. The components of respiratory chain, including iron sulfur compounds, ubiquinone, and cytochromes, together with the values of their mid-potentials according to Ohnishi. The diagram indicates the main pathways of electron transfer from the citric acid cycle substrates and from succinate. In addition, the sites of action of inhibitors, rotenone and antimycin A, are shown.

What is remarkable is that so many components channel their electrons through to cyrochrome oxidase, and so few react directly with oxygen. Proof that they can react with oxygen, "short-circuiting" cytochrome oxidase and generating H_2O_2 instead of H_2O, awaited the use of Yonetani's pure yeast cytochrome *c* peroxidase (20) as an enzyme reagent for detecting free H_2O_2 (21). This peroxidase is highly sensitive and specific, and has been used to identify possible sites at which H_2O_2 is generated in the mitochondria (Figure 9A). The slow rate of H_2O_2 generation with succinate as substrate falls nearly to zero

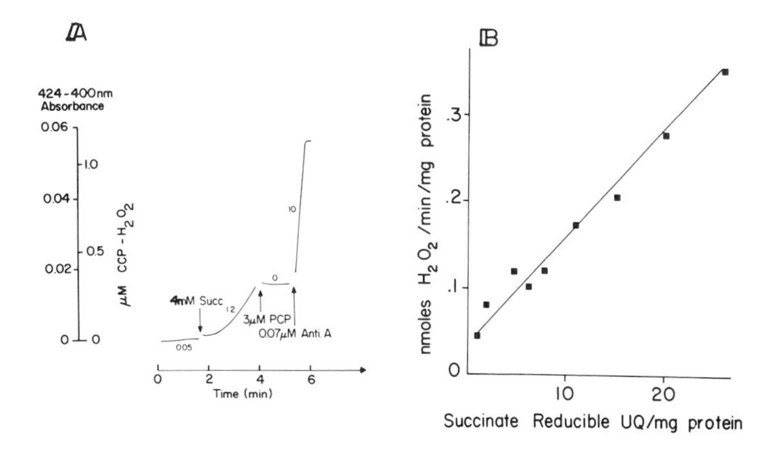

Fig. 9. The properties of mitochondrial H_2O_2 generation. A, the assay of mitochondrial H_2O_2 generation, using yeast peroxidase as an H_2O_2 indicator, following activation of H_2O_2 generation by succinate addition, inhibition by pentachlorophenol (PCP), and further inhibition of electron transport to cytochrome oxidase by antimycin A addition. B, the reconstitution of H_2O_2 generation in ubiquinone-depleted mitochondria by ubiquinone repletion.

on uncoupling and accelerates to a maximum in the presence of antimycin A, the inhibitor of cytochrome *b* and ubiquinone oxidation. Thus the H_2O_2 generator is localized in the portion of the chain on the substrate side of the antimycin A sensitive site.

Just which of the many components is involved in H_2O_2 generation cannot be surely determined without complete purification and reconstitution (Fig. 9B). However, selective extraction of the quinone component decreases the activity, and reconstitution of the quinone linearly increases the H_2O_2 generation rate (22) so quinone may be at least one of the mitochondrial H_2O_2 generators (Fig. 9B). The mechanism most likely involves the formation of ubiquinone radicals which react in a one-step reaction with oxygen to give O_2^-, which is then converted to H_2O_2 by superoxide dimutase. Boveris finds that over half the H_2O_2 formed in the mitochondria is generated by O_2^- and superoxide dismutase (22).

Such data are plotted in Figure 10A which illustrates how the physiological control of H_2O_2 generation shows a considerable sensitivity to the metabolic state of the mitochondria and to the pH of the reaction medium. The resting or near-resting State 4 which seems to characterize the normal liver *in situ*

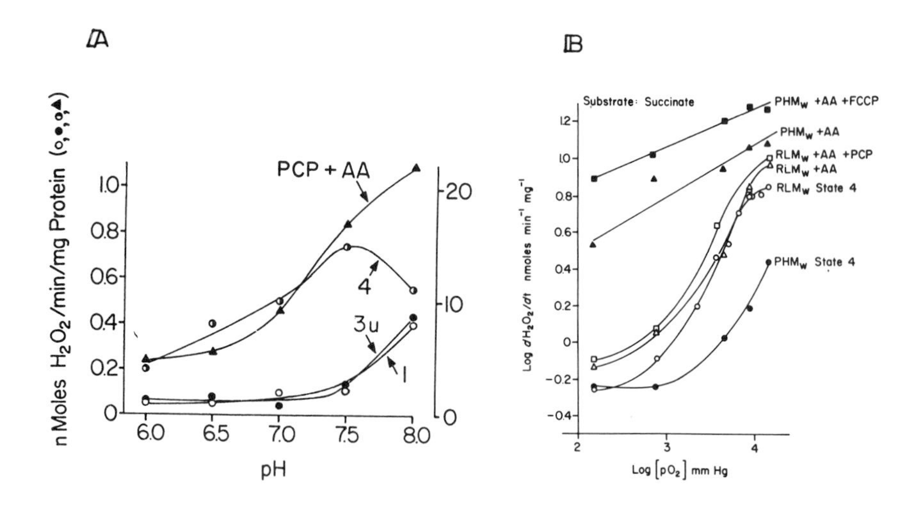

Fig. 10. Further properties of H_2O_2 generation as a function of pH (10A) and as a function of PO_2 (10B). The assay is essentially as in Figure 9A. In 10A is illustrated the effect of metabolic states 1, 3U, and 4, together with metabolic inhibitors and uncouplers, PCP + AA (pentachlorophenol plus antimycin A). 10B indicates the effect of oxygen pressure for state 4 and for the antimycin A treated and uncoupled states (AA + PCP, or AA + FCCP) for rat liver mitochondria (RlMw) and pigeon heart mitochondria (PHMw), succinate being the substrate.

is appropriate to maximal H_2O_2 generation and ethanol oxidation. The antimycin A inhibited, uncoupled state also shows high rates, while the uncoupled state 3a and the starved state 1 show low rates (21).

A most interesting parameter is oxygen tension, which exerts a dramatic effect on mitochondrial H_2O_2 generation in various metabolic states. Figure 10B illustrates the effect of oxygen pressure upon H_2O_2 generation in two types of isolated mitochondria, cardiac and (rat) liver (21). It is apparent that the near resting State 4, shown for these two types of mitochondria, shows a marked increase of H_2O_2 generation which is larger for the rat liver mitochondria than for the pigeon heart mitochondria. In the former case, a roughly tenfold increase of H_2O_2 production is obtained under hyperbaric conditions. This signifies that most of the electron transport rate of State 4 mitochondria can be converted to H_2O_2 production under hyperbaric conditions. The effects are smaller *in vivo*, presumably due to vasomotor effects which regulate the tissue oxygen tension at lower values than those obtained *in vitro* (23). In any case, the direct reaction with oxygen of components of the respiratory chain other than cytochrome oxidase is assured, since cytochrome oxidase is already saturated at the lowest oxygen concentrations.

When all these parameters are brought together a truly remarkable range of H_2O_2 generation rates can be obtained, from the very low rates with endogenous substrate to the high rates found with hyperbaric oxygenation and a "cocktail" of substrates appropriate to the cellular H_2O_2 generators, providing a multifaceted control system for H_2O_2 generation and for alcohol or formate oxidation via the catalase pathway (24).

II. SUMMARY

In conclusion, H_2O_2 is here to stay as a cell metabolite in the range from 10^{-8} to 10^{-6} M, where we find a remarkable multiplicity of cellular sources of H_2O_2 and a remarkable range of controls, both physiological as in the transition from the active State 3 to the resting State 4, and biochemical as in the use of appropriate substrates and inhibitors. These afford a rational control of the cellular redox state in the course of alcohol metabolism, and as well an opportunity to study the events on the pathway to liver damage that occurs in some alcoholics.

III. REFERENCES

(1). References to section in Microsomal Alcohol Oxidase, this volume.

(2). Chance, B. and Higgins, J.J., Arch. Biochem. Biophys., 41, 432-441(1952).

(3). Chance, B., Acta Chem. Scand. 1, 236-269 (1949). Reprinted in Alcohol and Aldehyde Metabolizing Systems, pp. 569-599. Academic Press, New York, 1974.

(4). Schonbaum, G.R. and Chance, B., The Enzymes 13, 363-408 (1976). Academic Press, NY.

(5). Chance, B., Journal Biol. Chem. 179, 1341-1369 (1949).

(6). Keilin, D., and Hartree, E.F., Biochem. J., 39, 148-157 (1945).

(7). Vatsis, G., this volume.

(8). Chance, B. and Oshino, N., Biochem. J. 131, 564-567 (1973).

(9). Agner, K., and Theorell, H., Arch., Biochem. 10, 321-338 (1946).

(10). Chance, B., J. Biol. Chem. 194, 483-490 (1952).

(11). Cohn, M., personal communication.

(12). Hershberg, R.D. and Chance, B., Biochemistry 14, 3885-3891 (1975).

(13). Wieland, H., Zeits. Ang. Chem. 44, 579-590(1931).
(14). Warburg, O., Biochem. Zeits, 189, 354-380(1927).
(15). Oppenheimer, C., Die Fermente, Leipzig, 1925-1926.
(16). Oppenheimer, C. and Stern, K.G., "Biological Oxidation", Nordemann, New York, 1939.
(17). Keilin, D., Proc. Roy. Soc. B., 98, 312 (1925).
(18). Theorell, H., Biochem. Zeitz, 288, 317 (1936).
(19). Ohnishi, T. Europ. J. Biochem., 64, 91-103(1976)
(20). Yonetani, T. and Ray, G., J. Biol. Chem. 240, 4503-4508 (1965).
(21). Boveris, A. and Chance, B., Biochem. J., 134, 707-716 (1973).
(22). Boveris, A., Cadenas, E., and Stoppani, A.M.O. Biochem. J. 444 (1976).
(23). Oshino, N., Jameison, D., Chance, B., Biochem. J., 146, 53-65 (1975).

METABOLISM OF H_2O_2 AND FATTY ACIDS BY RAT LIVER PEROXISOMES: IMPLICATIONS FOR ETHANOL DETOXIFICATION

P.B. Lazarow

The Rockefeller University

Peroxisomes are centers of H_2O_2 metabolism. Their oxidases and catalase directly oxidize a few substrates; through hypothetical coupled reactions NADH may also be oxidized. Rat liver peroxisomes oxidize palmitoyl-CoA, reducing O_2 to H_2O_2 and NAD to NADH. The activity of this system is increased approximately one order of magnitude by treatment with clofibrate and other hypoliqidemic drugs. Peroxisomes purified from the livers of clofibrate-treated rats contain crotonase, β-hydroxybutyryl-CoA dehydrogenase and thiolase. Peroxisomes oxidize ethanol by means of the peroxidatic reaction of catalase. They may also contribute to ethanol detoxification by their hypothetical oxidation of NADH. Clofibrate could plausibly increase both mechanisms.

I. H_2O_2 METABOLISM

Peroxisomes, by definition (1), are centers for the metabolism of H_2O_2. The association of H_2O_2-producing oxidases with catalase in a single cell organelle was discovered by de Duve and his collaborators in Belgium (2,3), and has since been found in many cell types (4). In addition, numerous other catalase-positive structures have been recognized morphologically (5,6), but have yet to be assayed for oxidases.

A. Oxidases

Rat liver contains three oxidases acting on urate, D-amino acids and L-α-hydroxyacids, including lactate and glycolate.

As pointed out by de Duve and Baudhuin in their review

(7), this latter oxidase activity, when coupled to lactate dehydrogenase, may serve for the oxidation of cell sap NADH (Fib. 1). As originally envisaged, this pathway involves

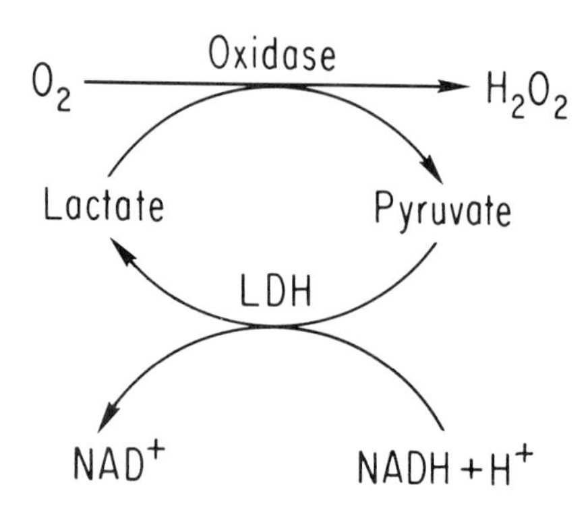

Fig. 1. Hypothetical coupled oxidation of NADH. Peroxisomal lactate oxidase has an activity of 0.6 μmol/min/g (8). LDH = lactate dehydrogenase.

the peroxisomes and the cell sap, which contains LDH. Since the peroxisomal membrane is permeable to lactate and pyruvate, it poses no impediment to such a coupled system. McGroarty *et al.* (9) have subsequently reported the presence of small amounts of LDH in rat liver peroxiomes, and recent experiments (P.B. Lazarow, unpublished) tend to support this finding. About 5 μmol/min/g liver or 2% of the cell's LDH appear to be associated with the peroxisomes, which is quite sufficient to support the cycle of Fig. 1. The possibility that this LDH could be artifactually adsorbed cannot be excluded at present, but LDH is not found adsorbed to mitochondria.

With or without LDH in peroxisomes, the cycle of Fig. 1 appears to be physiologically quite inactive because, according to Poole (10), the K_m of the oxidase for lactate is approximately 9 mM, while tissue lactate concentrations are generally much smaller, although they may approach this value during exercise. This peroxisomal pathway could have played a much more important role at an early point in evolution, as pointed out by de Duve and Baudhuin (7):

"the acquisition of the ability to transfer electrons to an exogenous acceptor represents a considerable gain to an anaerobe, even when the transfer is not coupled to any phosphorylation. While it leaves the burden of energy conservation to substrate-level phosphorylations, as in fermentation, it frees the organism from the necessity of eliminating energy-rich metabolites and makes these available for biosynthesis and for further phosphorylative degradation. It also allows the utilization of new substrates, such as glycerol, that are unfermentable owing to an excess of electrons. Oxidative phosphorylation brings an additional and immense advantage, but answers a less immediate need; it may have appeared later in evolution."

B. Catalase

The H_2O_2 produced in the peroxisomes is decomposed by catalase, either catalatically or by the peroxidatic mechanism discovered by Keilin and Hartree (11).

According to Chance and Oshino (12), the latter reaction may account for as much as 70% of the H_2O_2 flux in normal liver. The known peroxidatic substrates include ethanol and formate; the high percentage of peroxidation suggests the possible existence of other unknown hydrogen donors.

Asnoted by de Duve and Baudhuin (7), the peroxidatic reaction of catalase may also serve for the oxidation of cell sap NADH in normal liver by means of coupled reactions with ethanol and alcohol dehydrogenase (Fig. 2). Ethanol is

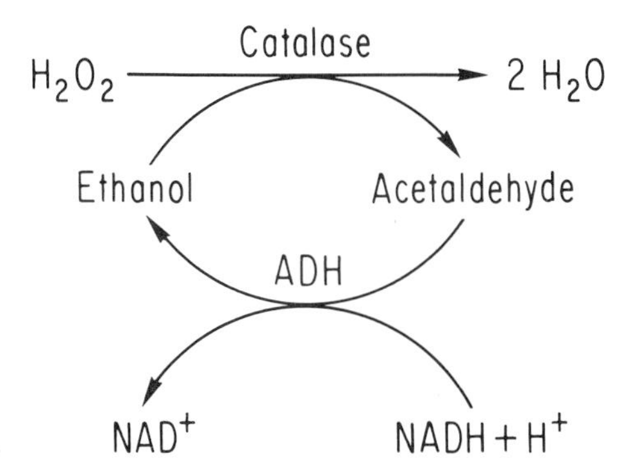

Fig. 2. Hypothetical coupled peroxidation of NADH. ADH = alcohol dehydrogenase.

supplied by the portal circulation (13) and also may form via pyruvate dehydrogenase (14). de Duve (15) has estimated that this pathway may account for 0.1 μmol of NADH oxidized/min/g normal liver.

The coupled reactions of Fig. 2 have also been employed *in vitro* to assay H_2O_2 production (P.B. Lazarow and C. de Duve, unpublished results). In the presence of excess catalase, alcohol dehydrogenase and ethanol, the rate of NADH oxidation becomes a quantitative measure of the rate of H_2O_2 production. This assay has the advantage of employing catalase, in contrast to the peroxidase-based H_2O_2 assays, with which catalase may interfere (see Herzog and Fahimi (16) for example).

C. Physiological Role

In summary, peroxisomes, through their metabolism of H_2O_2, function as a primitive oxidizing organelle. A few substrates appear to be oxidized directly; through coupled reactions NADH may hypothetically also be oxidized, and thus indirectly, many other metabolites. The peroxisomes liberate the energy of their oxidations as heat, unlike the mitochondria which conserve much of this energy in the form of ATP. Therefore, the peroxisomes could play a role in thermoregulation.

The dissipative peroxisomal oxidation of NADH, free of the throttle of respiratory control, could also be useful to cells when they are overloaded with reducing equivalents, as in ethanol intoxication. Under other conditions, however, it could be harmful.

Determining the physiological importance of the peroxisomal oxidation system, vis a vis the mitochondrial electron transport chain, as well as the mechanisms that may control the relative fluxes through them, is a problem for future study.

II. FATTY ACID OXIDATION

Recent experiments of Lazarow and de Duve (17) have demonstrated that rat liver peroxisomes are also capable of the oxidation of palmitoyl-CoA.

A. Pathway

Fig. 3 illustrates a pathway of fatty acyl-CoA oxidation by rat liver peroxisomes compatible with all the data obtained thus far. In the first dehydrogenation the electrons are transferred to O_2, producing H_2O_2. This step is similar to that described for plant glyoxysomes by Cooper and Beevers (18), but differs from the corresponding mitochondrial reaction

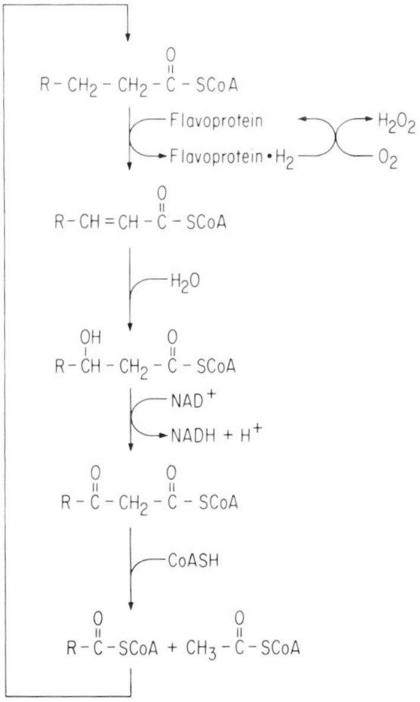

$R-CH_2-CH_2-\overset{O}{\overset{\|}{C}}-SCoA$

Flavoprotein → Flavoprotein•H_2 ; O_2 → H_2O_2

$R-CH=CH-\overset{O}{\overset{\|}{C}}-SCoA$

H_2O

$R-\overset{OH}{\overset{|}{C}H}-CH_2-\overset{O}{\overset{\|}{C}}-SCoA$

NAD^+ → $NADH + H^+$

$R-\overset{O}{\overset{\|}{C}}-CH_2-\overset{O}{\overset{\|}{C}}-SCoA$

CoASH

$R-\overset{O}{\overset{\|}{C}}-SCoA + CH_3-\overset{O}{\overset{\|}{C}}-SCoA$

Fig. 3. Peroxisomal pathway of fatty acid oxidation (from 17).

where the electrons are believed to be transferred to the electron transport chain. After addition of water across the double bond, a second dehydrogenation occurs using NAD as the electron acceptor. The terminal 2-carbon fragment is cleaved off by thiolysis with Coenzyme A, and the cyclic β-oxidation process continues.

Experimentally, when palmitoyl-CoA is added to purified peroxisomes, we detect the consumption of O_2, the formation of H_2O_2, and the reduction of NAD (17). In the absence of added free CoA as acceptor, the reduction of NAD is roughly stoichiometric with added palmitoyl-CoA, and additional substrate causes further NAD reduction. When CoA is present, however, approximately 3-5 moles of NAD are reduced per mole of palmitoyl-CoA, indicating that β-oxidation has proceeded through 3-5 cycles. The theoretical maximum of 7 has not been observed. Experiments with various acyl-CoA substrates demonstrate a decreasing activity of the peroxisomal system with decreasing chain length (P.B. Lazarow, unpublished results). The system appears inactive with butyryl-CoA, explaining the incomplete oxidation observed with longer substrates.

If peroxisomes are added anaerobically from the side-arm of a Thunberg cuvette, to a reaction mixture containing palmitoyl-CoA, no reaction occurs (Fig. 4). When O_2 is

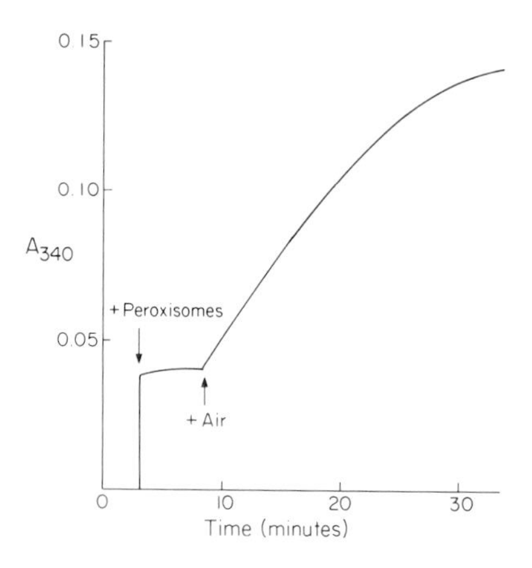

Fig. 4. Oxygen-dependent reduction of NAD by peroxisomes and palmitoyl-CoA (from 17).

admitted, NAD reduction proceeds (17). This dependence upon O_2 for NAD reduction demonstrates that O_2 serves as electron acceptor for the first dehydrogenation, and NAD serves for the second. These reactions are insensitive to 1 mM KCN.

B. Effect of Clofibrate

The results described thus far demonstrate the oxidation of palmitoyl-CoA by normal rat liver peroxisomes. In addition, we find that the peroxisome is also involved in the drug-induced lowering of plasma lipids. Ten years ago, Hess, Stäubli and Reiss (19) and Svoboda and Azarnoff (20) demonstrated that the hypolipidemic drug clofibrate, used clinically in the treatment of hyperlipemias, caused a striking increase in the number of liver peroxisomes in the rat, and a doubling of the liver catalase activity. The biochemical mechanism of action of the drug has remained unknown, however,

We treated rats with clofibrate for one week. Fig. 5 illustrates the biochemical effects on the liver (17). The rate of palmitoyl-CoA oxidation by the liver is strikingly increased, whether assayed by NAD reduction or by H_2O_2 production. In order to determine whether this increase was due to a change in the peroxisomal system of fatty acid oxidation, a similar experiment was performed in which rats were treated with clofibrate for 2 weeks and peroxisomes were purified by

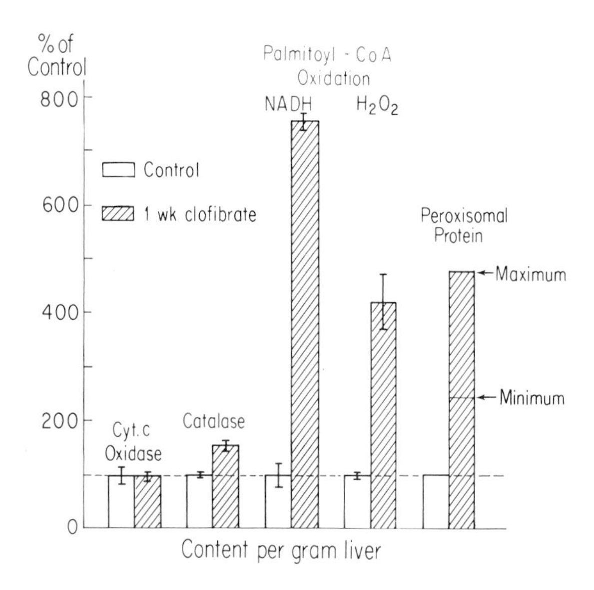

Fig. 5. Effects of clofibrate on the liver (data from 17).

differential and equilibrium density centrifugation according to Leighton et al. (21) (Fig. 6). The peroxisomes from the drug-treated animals showed the same order of magnitude increase over the controls in their ability to oxidize palmitoyl-CoA as had been seen previously in homogenates.

In addition, the amount of peroxisomal protein in the liver was increased, at least 2.5 times and perhaps as much as 5 times (Fig. 5). These results provide a plausible biochemical explanation for the action of clofibrate in reducing serum lipids.

Furthermore, two other hypolipidemic drugs (tibric acid and Wy-14,643) also cause similar substantial increases in the rate of fatty acid oxidation by hepatic peroxisomes (P.B. Lazarow, unpublished results), suggesting that peroxisomes may play a central role in lowering plasma lipid levels during therapy with hypolipidemic drugs.

C. Individual Enzymes of β-oxidation

The high rates of acyl-CoA oxidation by peroxisomes isolated from the livers of clofibrate-treated rats has permitted the direct assay of individual enzymes of β-oxidation

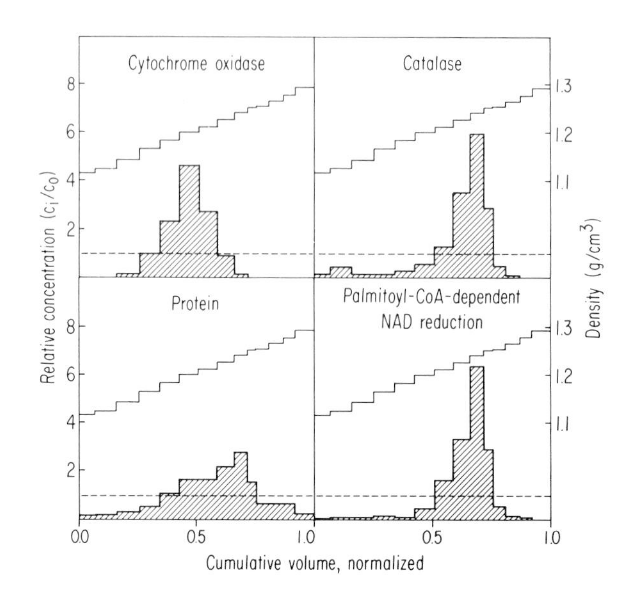

Fig. 6. Peroxisomal localization of palmitoyl-CoA oxidation demonstrated by equilibrium density centriguation of a peroxisome-rich fraction obtained from the livers of clofibrate-treated rats (from 17).

using specific optical techniques (22) and the appropriate 4-carbon substrates (P.B. Lazarow, unpublished observations).

Crotonase was assayed by following the disappearance upon hydration of the specific 263 nm absorption band of crotonyl-CoA. β-Hydroxybutyryl-CoA dehydrogenase was assayed in reverse, using acetoacetyl-CoA as substrate, and measured by following the doubling of the thioester band absorbance at 233 upon thiolysis of acetoacetyl-CoA to two acetyl-CoA's. Purified peroxisomes were found to catalyze each of these three reactions. The results obtained with thiolase are illustrated in Fig. 7. Major peaks of each of these enzyme activities were found coincident with catalase in the equilibrium density fractionation experiment of Fig. 6. These results provide further support for the pathway of Fig. 3.

III. ROLE OF PEROXISOMES IN ETHANOL OXIDATION

A. Peroxidatic Reaction

Peroxisomes oxidize ethanol by means of the well-known

peroxidatic reaction of catalase (top half of Fig. 2). This reaction is limited by the rate of H_2O_2 production, which in turn depends on the availability of substrates for the oxidases. Three types of experiments have provided information on the functioning of this pathway in liver (Table 1). The activities of the peroxisomal oxidases have been measured on purified peroxisomes (column 1) by Leighton et al. (21); their combined activities exceed the liver's usual rates of ethanol oxidation of 1-2 μmol/min/g.

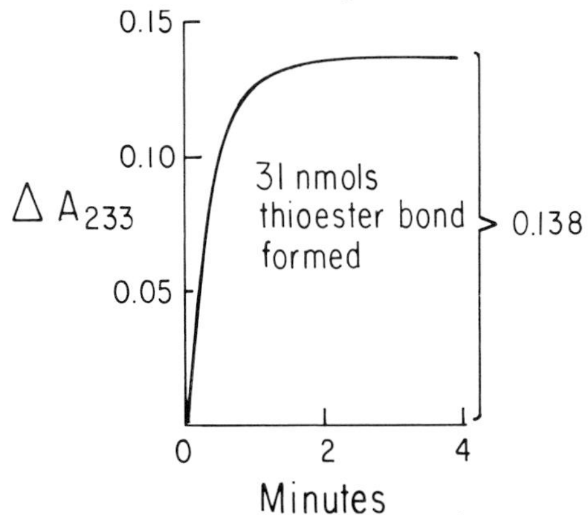

Fig. 7. Thiolase activity of peroxisomes from clofibrate-treated rats. The reaction mixture contained 30 nmol of acetoacetyl-CoA and 100 nmol of CoASH in 30 mM phosphate buffer, pH 7.4 at 37°.

Sies and Chance (23) have measured H_2O_2 in perfused liver by virtue of its reaction with catalase to form Compound I, identified by dual-wavelength spectrophotometry, and have quantitated the rate of H_2O_2 production by methanol titration. Using this technique, Oshino et al. (24) have measured H_2O_2 production with various substrates, as shown in column 2. In the absence of any substrate, H_2O_2 production is very small, because the flow-through perfusion system depletes the liver of endogenous substrates. When glycolate or urate is added, substantial increments in H_2O_2 formation are seen, although the rates are somewhat smaller than observed by Leighton et al. (21). This might reflect the presence of O_2 gradients within the perfused liver lobule, since the activities of these oxidases are proportional to O_2 tension.

Third, Thurman and McKenna (25) have directly measured ethanol uptake by perfused livers (column 3). They find that the addition of glycolate, urate, or D-amino acids produces substantial increases in ethanol uptake.

In addition to the substrates listed in Table 1, the

TABLE I. H_2O_2 METABOLIC RATES (μMOL/MIN/G LIVER)

SUBSTRATE	OXIDASE ACTIVITY PURIFIED PEROXISOMES LEIGHTON ET AL. (21)		H_2O_2 PRODUCTION PERFUSED LIVER OSHINO ET AL. (24)		ETHANOL UPTAKE PERFUSED LIVER* THURMAN & MCKENNA (25)		
	SUBSTRATE CONC. MM	RATE U/G	SUBSTRATE CONC. MM	RATE[#] U/G	SUBSTRATE CONC. MM	ETOH MM	RATE[#] U/G
NONE				.05		10 OR 30	0.3
						60	0.6
GLYCOLATE	7	1.1	3	+0.4	15	10	+0.3
URATE	0.04	3.1	1	+0.7	0.5	10	+0.3
					0.5	30	+2.6
D-ALANINE	55	1.4			40	60	+2.0

[#]PLUS SIGN DENOTES INCREASE ABOVE NO SUBSTRATE VALUE
*ALCOHOL DEHYDROGENASE INHIBITED BY 4-METHYLPYRAZOLE

experiments described above in section II demonstrate that palmitoyl-CoA is an excellent substrate for H_2O_2 production. Results consistent with these were obtained by Oshino et al. (24): addition of octanoate to their liver perfusate resulted in H_2O_2 production. These facts would suggest that fatty acids could stimulate ethanol oxidation by the catalase peroxidation mechanism. The available evidence, however, does not support this point of view: Thurman and Scholz (26) and Williamson et al. (27) have reported inhibitions of ethanol uptake by octanoate or oleate in perfusion experiments. However, the following facts must be kept in mind in interpreting these experiments. Both ethanol and fatty acids bring reducing equivalents to the liver, and at similar concentrations may compete for available NAD. Fatty acid oxidation is active at 10 μM palmitoyl-CoA (17) whereas ethanol peroxidation increases from 10 to 80 mM ethanol (25). It would be of interest to measure the effects of small amounts of palmitoyl-CoA on the oxidation of large amounts of ethanol in liver treated with 4-methylpyrazole to block the alcohol dehydrogenase pathway.

It may also be remembered that the striking effects of clofibrate in increasing fatty acid oxidation also involve an increased capacity to produce H_2O_2 from palmitoyl-CoA (17). Clofibrate treatment might therefore be expected to enhance ethanol detoxification. Indeed, Kahonen et al. (28) have reported that ethanol clearance *in vivo* is increased 2-3 fold in clofibrate-treated rats. However, Hawkins et al. (29) found only a 50% increase which could be explained by the drug induced increase in liver mass.

What then is the physiological contribution of the peroxidatic pathway to ethanol oxidation? Based on liver perfusion experiments using 4-methylpyrazole and/or aminotriazole, (to inhibit alcohol dehydrogenase and catalase, respectively), Thurman and McKenna (25) concluded that ethanol peroxidation is small at < 20 mM ethanol, but accounts for as much as 50% of the clearance of 50-80 mM ethanol. These are presumably minimal values, since the concentrations of endogenous oxidase substrates are probably lower than normal in the perfused liver, even with the recirculating perfusate method. Thurman and McKenna (25) have also demonstrated that with the addition of adequate oxidase substrates, the peroxidatic pathway alone is capable of oxidizing ethanol as fast as, or faster than, all the liver pathways normally do. Thus the question is: what are the physiological concentrations of oxidase substrates and what rate of H_2O_2 formation occurs *in vivo*?

According to Eggleston and Krebs (30), freeze-clamped rat liver contains 0.07 mM urate, sufficient to support as much as 3.1 μmol H_2O_2 production/min/g liver (21). In addition, D-amino acids are supplied to the liver via the portal circulation, and these may also support H_2O_2 synthesis. It appears, however, that the actual rate of H_2O_2 formation is smaller than this. Oshino et al. (31) have applied their spectroscopic method to liver *in situ* in an anaesthetized rat, and have obtained a value of 0.4 μmol/min/g liver. Even this lower rate of H_2O_2 production, were it to be used entirely for ethanol oxidation, would make a substantial contribution to ethanol detoxification.

B. Peroxisomal NADH Reoxidation?

The second mechanism for ethanol detoxification is the alcohol dehydrogenase pathway. However, as has been pointed out by several authors (e.g. 25), the rate-limiting step in this pathway, at least at high ethanol concentrations, is not the ADH reaction itself, but rather the reoxidation of NADH. Furthermore, many of the deleterious effects of chronic ethanol intoxication stem from the continual over-reduction of the NADH pool.

It is generally believed that the cell sap NADH is reoxidized by the mitochondria, via an electron shuttle, but it is unclear whether the mitochondria do the whole job. In fact, one might suppose that in the case of ethanol intoxication, the ADP of the cell would be largely phosphorylated to ATP, and the mitochondrial oxidation of NADH would be limited by the availability of ADP. That this is in fact the case was demonstrated by Videla and Israel (32), who showed that uncouplers of mitochondrial respiration increase the metabolism of ethanol.

It would be very useful to the intoxicated liver cell, therefore, to have another NADH oxidation mechanism not requiring ADP. In fact (Fig. 1) the peroxisomes potentially provide such a mechanism with their lactate-mediated NADH oxidation system. However, this pathway appears to be inactive under normal physiological conditions, due to its high K_m for lactate (10).

Another, hypothetical, scheme for the peroxisomal oxidation of cell sap reducing equivalents is shown in Fig. 8. It involves coupling the peroxisomal acyl-CoA oxidase activity with

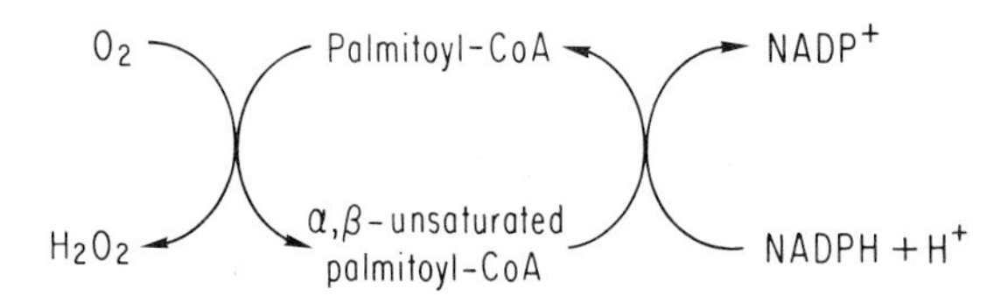

Fig. 8. Hypothetical mechanism for coupled NADPH oxidation.

the NADPH-driven enoyl-CoA reductase of the fatty acid elongation systems, reported to be present in microsomes (33) and mitochondria (34). Cell sap NADP may be indirectly reduced by NADH, even though the two nucleotide pools are not directly coupled. Hoek and Ernster (35) have suggested that this may occur by way of the mitochondrial transhydrogenation system, driven by ATP, and an isocitrate shuttle mechanism for exporting the NADPH back to the cell sap. Lundquist et al. (36), propose a transhydrogenation via a malic enzyme shuttle. The high NADH/NAD ratio characteristic of ethanol intoxication would inhibit the oxidation of fatty acids beyond the first desaturation reaction due to the scarcity of NAD. It would favor the cyclic oxidation depicted in Fig. 8 by means of either of the transhydrogenation shuttles that have been proposed. Furthermore, the pathway of Fig. 8 might account for some of the incorporation of ^3H_2O into fatty acids observed by Scholz et al. (37) in the absence of appreciable incorporation of ^{14}C-glucose.

These various pathways for ethanol metabolism are summarized in Fig. 9. The contributions of the mitochondrial and peroxisomal systems of NADH oxidation could be evaluated using aminotriazole and specific inhibitors of mitochondrial respiration in place of 4-methylpyrazole.

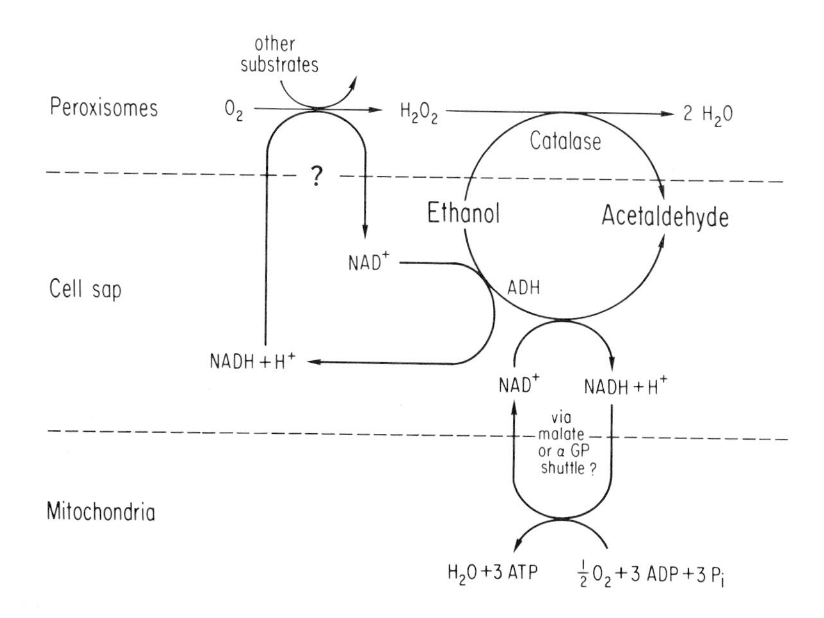

Fig. 9. Pathways of ethanol oxidation.

It might be pointed out that since clofibrate can cause striking increases in peroxisomal fatty acid oxidation and H_2O_2 production, it has the potential to increase both peroxisomal pathways of ethanol oxidation: the peroxidatic reaction of catalase, and the hypothetical oxidation of reducing equivalents via acyl-CoA desaturation. In addition, clofibrate, a clinically effective hypolipidemic agent, might be able to prevent or reverse the formation of a fatty liver. Indeed, evidence in favor of this possibility has been reported for both acute (38) and chronic (39) ethanol administration. Clofibrate would appear worthy of further investigation for its possible usefulness in alcoholism.

IV. ACKNOWLEDGEMENTS

The author appreciates many stimulating discussions with Dr. C. de Duve. This research was supported by NSF Grant BMS-72-0224 and NIH Grant AM19394.

V. REFERENCES

(1). C. de Duve, J. Cell Biol. 27, 25A (1965).
(2). C. de Duve, H. Beaufay, P. Jacques, Y. Rahman-Li, O.Z. Sellinger, R. Wattiaux and S. de Coninck, Biochim. Biophys. Acta 40, 186 (1960).
(3). Beaufay, H., Jacques, P., Baudhuin, P., Sellinger, O.Z., Berthet, J., and de Duve, C., Biochem. J. 92, 184 (1964).
(4). Hogg, J.F., Ann. N.Y. Acad. Sci. 168, 209 (1969).
(5). Hruban, Z., Vigil, E.L., Slesers, A., and Hopkins, E., Lab. Invest. 27, 184 (1972).
(6). Novikoff, A.B., Novikoff, P.M., Davis, C., and Quintana, N., J. Histochem. Cytochem. 21, 737 (1973).
(7). de Duve, C., and Baudhuin, P., Physiol. Rev. 46, 323 (1966).
(8). Baudhuin, P., Müller, M., Poole, B., and de Duve, C., Biochem. Biophys. Res. Commun. 20, 53 (1965).
(9). McGroarty, E., Hsieh, B., Wied, D.W., Gee, R., and Tolbert, N.E., Aruch. Biochem. Biophys. 161, 194 (1974).
(10). Poole, B., Thesis, The Rockefeller University (1968).
(11). Kielin, D., and Hartree, E.F., Proc. Roy. Soc. B. 119, 141 (1936).
(12). Chance, B., and Oshino, N., Biochem. J. 122, 225 (1971).
(13). Krebs, H.A., and Perkins, J.R., BIochem. J. 118, 635 (1970).
(14). McManus, I.R., Contag, A.O., and Olson, R.E., J. Biol. Chem. 241, 349 (1966).
(15). de Duve, C., in Alcohol and Aldehyde Metabolizing Systems (R.G. Thurman, T. Yonetani, J.R. Williamson, and B. Chance, eds.), Vol. I, 161. Academic Press, New York, 1974.
(16). Herzog, V. and Fahimi, H.D., J. Cell Biol. 60, 303 (1974).
(17). Lazarow, P.B., and de Duve, C., Proc. Natl. Acad. Sci. 73, 2043 (1976).
(18). Cooper, T.G., and Beevers, H., J. Biol. Chem. 244, 3514 (1969).
(19). Hess, R., Stäubli, W., and Riess, W., Nature 208, 856 (1965).
(20). Svoboda, D.J., and Azarnoff, D.L., J. Cell Biol. 30, 442 (1966).
(21). Leighton, F., Poole, B., Beaufay, H., Baudhuin, P., Coffey, J.W., Fowler, S., and de Duve, C., J. Cell Biol. 37, 482 (1968).
(22). Lynen, F., and Ochoa, S., Biochim. Biophys. Acta 12, 299 (1953).
(23). Sies, H., and Chance, B., FEBS Lett. 11, 172 (1970).
(24). Oshino, N., Chance, B., Sies, H., and Bücher, T., Arch. Biochem. Biophys. 154, 117 (1973).

(25). Thurman, R.G., and McKenna, W.R., in Biochemical Pharmacolgoy of Ethanol (E. Majchrowicz, ed.), 57, Plenum Press, New York, 1975.
(26). Thurman, R.G., and Scholz, R., Fed. Proc. 34, 634 (1975).
(27). Williamson, J.R., Scholz, R., Browning, E.T., Thurman, R.G., and Fukami, M.H., J. Biol. Chem. 244, 5044 (1969).
(28). Kähönen, M.T., Ylikahri, R.H., and Hassinen, I., Life Sciences 10, 661 (1971).
(29). Hawkins, R.A., Nielsen, R.C., and Veech, R.L., Biochem. J. 140, 117 (1974).
(30). Eggleston, L.V., and Krebs, H.A., Biochem. J. 138, 425 (1974).
(31). Oshino, N., Jamieson, D., Sugano, T., and Chance, B., Biochem. J. 146, 67 (1975).
(32). Videla, L., and Israel, Y., Biochem. J. 118, 275 (1970).
(33). Nugteren, D.H., Biochim. Biophys. Acta 106, 280 (1965).
(34). Harlan, W.R., Jr., and Wakil, S.J., J. Biol. Chem. 238, 3216 (1963).
(35). Hoek, J.B., and Ernster, L., in Alcohol and Aldehyde Metabolizing Systems (R.G. Thurman, T. Yonetani, J.R. Williamson, and B. Chance, Eds.), Vol. I, 351. Academic Press, New York, 1974.
(36). Lundquist, F., Damgaard, S.E., and Sestoft, L., *ibid.*, p. 405.
(37). Scholz, R., Kaltstein, A., Schwabe, U., and Thurman, R.G., *ibid.*, p. 315.
(38). Brown, D.F., Metabolism 15, 868 (1966).
(39). Spritz, N., and Lieber, C.S., Proc. Soc. Exp. Biol. Med. 121, 147 (1966).

SOME EFFECTS OF AGING AND LIPID-LOWERING DRUGS ON LIPID METABOLISM AND THE HEPATIC ENDOPLASMIC RETICULUM

E. Spring-Mills, D.L. Schmucker, and A.L. Jones

Cell Biology Section, VA Hospital
University of California

Certain effects of aging, reproductive status and lipid--lowering drugs on lipid metabolism, liver structure and hepatic drug metabolizing enzymes in male Sprague-Dawley, Holtzman strain rats are presented. Attention is focused on the following: 1) retired breeder rats have statistically higher serum cholesterol levels and a shorter life expectancy than age-matched virgin counterparts; 2) the retired breeder rat may be a suitable animal model for testing lipid-lowering agents and the pathogenesis of certain types of hyperlipoproteinemia; 3) oxandrolone may be an unusually potent hypocholesterolemic agent capable of inducing the proliferation of hepatic smooth endoplasmic reticulum without an associated increase in drug-metabolizing enzymes; 4) specific alterations in hepatic fine structure occur as a function of age; 5) certain age-related structural changes in the liver may be reversible; and, 6) subtle changes in cell fine structure may be detected by quantitative electron microscopy.

I. INTRODUCTION: AGING, REPRODUCTIVE ACTIVITY AND ARTERIOSCLEROSIS

Aging is a normal physiologic process which operates throughout the life cycle. It involves progressive, unfavorable changes in the potentials, adaptability and viability of the organism which ultimately lead to senescence and death. Aging proceeds at unique rates with different organ systems and the sequence of changes is often species specific. Nevertheless, in rats and man, the increase in serum lipids with aging follows remarkably similar patterns. Boberg *et al* (1) have shown that in male rats, serum phospholipids and cholesterol increase steadily from birth to 20 months of age, while triglycerides increase from birth to 10 months of age and then plateau. Certain populations of rats and humans with elevated serum lipids have a very high incidence of cardiovascular disease and show signs of premature aging. Wexler (2) discovered that breeding predisposes several strains of rats to the spontaneous development of arteriosclerosis. In susceptible male breeder rats, the incidence and severity of the disease is directly related to the frequency of matings and number of litters sired (3). The arterial changes appear in a specific anatomical sequence and even develop in breeder rats fed low fat diets (4). Male breeder rats die sooner than female breeder rats and at an earlier stage of the vessel disease (1-5). Male breeder and retired breeder rats (RB) which develop arteriosclerosis, diabetes and other degenerative disorders have a drastically reduced life expectancy. The majority die before they are a year and a half old, whereas their virgin litter mates normally live to be 2½ to 3 years old (6,7).

II. LIPID LEVELS IN MALE RATS

Although the severity of arteriosclerosis in the male retired breeder is not directly proportional to serum lipid levels, hyperlipidemia precedes and accompanies the disease (2,8). As shown in Table 1, we have found that 6-9 month old recently retired male breeder rats have approximately 1.7 times as much cholesterol, 1.5 times as much triglyceride and essentially the same concentration of free fatty acids as 3-month old virgin males of the same strain. Of particular interest is the finding that although the retired breeders have significantly higher cholesterol levels than virgin rats of the same age, their serum triglyceride levels are equivalent. These values were obtained from rats that were fasted 24 hr and bled between 8 and 11:00 a.m. Cholesterol was determined using the o-phthaladehyde method of Rudel and Morris (9), triglycerides according to the uv enzymatic method of Eggstein and Kreutz (10) using tripalmitin as standard and free fatty acids according to the combined procedures of Dole (11), Trout and Estes (12).

III. PLASMA LIPOPROTEINS OF YOUND ADULT AND RETIRED BREEDER RATS

Because cholesterol and triglycerides are carried in the blood within several different types of lipoprotein complexes, 2 years ago, we began to separate and characterize the plasma lipoproteins of young adult and retired breeder rats. The

TABLE 1
Serum Lipid Levels of Male Rats

GROUP	*CHOLESTEROL (mg/100 ml ± SEM)*	*TRIGLYCERIDE (mg/100 ml ± SEM)*
Young Virgins (3 - mo)	63.46 ± 2.54 (36)	66.04 ± 3.27 (36)
Mature Virgins (6 - mo)	81.65 ± 2.76 (32)	97.63 ± 7.17 (32)
Retired Breeders (6-9 mo)	105.64 ± 5.91 (47)	93.02 ± 7.22 (47)
	P	*P*
YV vs MV	< 0.0005	< 0.0005
YV vs RB	< 0.0005	< 0.0005
MV vs RB	< 0.01	ns

( ) = *number of animals*
ns = not statistically significant

lipoproteins were removed from the rat plasma at density 1.225 by ultracentrifugation and subsequently separated and purified by agarose column chromatography according to the procedure of Rudel, Morris and Felts (13).

Very low density (VLDL), low density (LDL) and high density (HDL) lipoproteins isolated by this technique are similar, if not identical, to the corresponding lipoprotein classes separated by conventional sequential ultracentrifugation (at densities 1.006, 1.063 and 1.21, respectively). Moreover, the column elution patterns obtained by monitoring the column eluate at 280 nm during lipoprotein separation provide a record of the efficiency of the separation and semi-quantitative information about plasma lipoprotein distribution since the % T at 280 nm is nearly proportional to the concentration of the material. The column separation is on the basis of size, the largest lipoprotein fraction is eluted first. VLDL, LDL and HDL, the 3 major classes of plasma lipoproteins which we separated, are similarly distributed in the two groups of rats; however, measurements of the areas under the peaks indicated that retired breeder rats have increased amount of HDLs. Similar increases in HDLs have been observed in geriatric human populations, but elevations in HDL or alpha lipoproteins *per se* are not presently associated with specific clinical disorders (14). The most unusual feature of the retired breeder pattern is the increased quantity of particles in the intermediate peak. Increased quantities of similar size particles are found in humans with Type III hyperlipoproteinemia (15). The particles are in the size range of very low density lipoproteins but on electrophoresis the particles have beta rather than pre-beta or VLDL electrophoretic mobility. These lipoproteins often migrate in a smear between normal LDL and VLDL, presumably as a result of loss of small apoprotein "C peptides" from very low density lipoproteins and a greatly increased content of arginine-rich protein which may be transferred to the surface as the C-proteins are lost (16). These remnant or intermediate particles (d < 1.019) are richer in cholesteryl esters than normal VLDLs of the same size. Cholesterol-rich particles are believed to be normal intermediates that are formed during the breakdown of VLDLs and chylomicrons to LDL. A defect in the final steps of VLDL and chylomicron catabolism is postulated to account for the accumulation of abnormally large concentrations of these particles in plasma of patients with Type III disease.

Figure 1 summarizes our electron microscopic analysis of the purity and particle sizes of the column lipoprotein fractions within each of the 4 major peaks. Beginning with the upper left hand quandrant, VLDL, intermediate, LDL and HDL

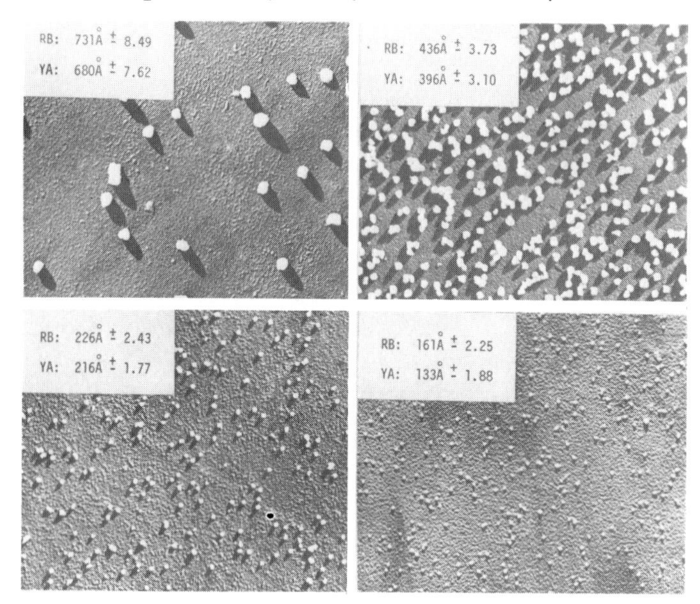

Fig. 1. Retired breeder lipoprotein particles from the 4 major peaks of the agarose column fractions are shown here. The mean particle size plus or minus the standard error of the mean is given for retired breeders and yound adults. Upper left = VLDL. Upper right = intermediate. Lower left = LDL. Lower right = HDL.

particles from retired breeder rats are shown. The mean particle size, plus or minus the standard error of the mean is given for both retired breeders and young adults. All peaks from the retired breeder rats contain particles of larger mean size than the particles in the corresponding peaks from the young animals. These differences are all statistically highly significant. The reason for the increase in size of the retired breeder particles is not known but may be related to impaired lipoprotein catabolism.

IV. AGE-RELATED CHANGES IN HEPATIC FINE STRUCTURE

Since the liver is a major site of lipoprotein metabolism and also is implicated in the pathogenesis of Type III disease, we have investigated the fine structure of this organ in our rats. In the first study (17), the livers from 3-month old virgin and 6-month old retired breeder male rats were perfused via the portal vein and fixed with paraformaldehyde-glutaraldehyde (18), post-fixed in osmium and embedded in Spurr's Epoxy resin (19). Morphometric analysis of the tissues was conducted according to the procedures of Weibel and his co-workers (20-22). A coherent double lattice test system was used over low magnification micrographs for determining the volume densities (i.e. the cm^3 of structure A per cm^3 of intralobular liver tissue) of the extrahepatic space, hepatocytes, hepatocyte cytoplasm, nuclei, mitochondria, biliary space, microbodies and lysosomes (dense bodies). A coherent multipurpose test system was used to evaluate the volume and surface densities of the rough endoplasmic reticulum (RER), smooth endoplasmic reticulum (SER) and the Golgi apparatus on the high magnification micrographs.

It was discovered that centrolobular parenchymal cells in livers of retired breeder rats are significantly larger and contain a relatively greater volume of lysosomes than similar cells from young adult livers. Nevertheless, young adult liver cells were found to have significantly greater surface area of SER than liver cells from retired breeders. Subsequently, the hepatic fine structural differences were quantitated between age-matched virgin and retired breeder rats that were 90, 180 and 480 days old (23).

Morphometric analysis revealed several distinct differences in liver structure between rats of different ages, as well as between age-matched virgin and RB animals. The volume densities and specific volumes of centrolobular hepatocytes as a function of age and breeding are shown in Figure 2. The data show that there is a net increase in the mean specific volume

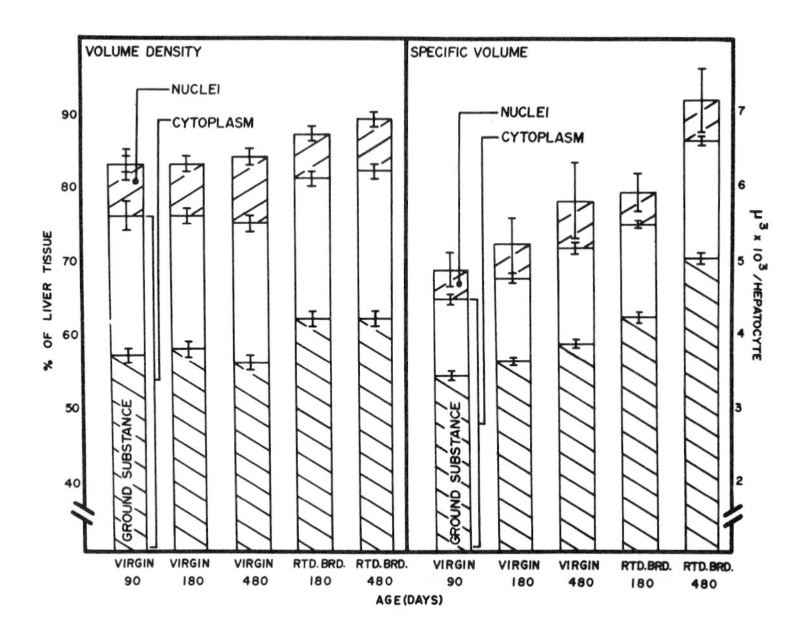

Fig. 2. Volume densities and specific volumes of centrolobular hepatocytes as a function of age and breeding. All values are expressed as the mean ± SEM. The left panel demonstrates the relative volumes (V_V) and is measured in cm^3 per cm^3 of intralobular liver tissue (% of liver tissue). The estimated mean volumes of individual hepatocytes are represented in the right panel and are measured in μ^3. [From Schmucker. Age-related changes in hepatic fine structure: A quantitative analysis. J. Gerontol 31, 135-143 (1976)].

of the hepatocytes up to 480 days of age in both virgin and RB rats. In addition, the mean cell volumes of the RB hepatocytes are significantly greater than those of age-matched virgin animals ($p < 0.005$). The increase in cell size is reflected in larger cytoplasmic and ground substance compartments in the hepatocytes of the virgin and RB animals (Tables 2 and 3). The volume densities of the mitochondria and the microbodies (peroxisomes) do not change appreciably. Both the relative and specific volumes of the lysosomes (dense bodies) increase steadily with age, but the differences are significant only between the 90 and 480-day old virgin rats.

A qualitative comparison of the relative amounts of endoplasmic reticulum (ER) in the hepatocytes of age-matched virgin and RB rats did not reveal any distinct differences.

TABLE 2
Hepatic Morphometric Parameters in Virgin Rats as a Function of Age

Tissue Components	*Age (Days)*		
	90	*180*	*480*
Extrahepatocyte (V_V)[b]	.18 ± .02	.17 ± .01	.16 ± .01
Hepatocyte (V_V)	.83 ± .02	.83 ± .01	.84 ± .01
(μ^3)[c]	*4860 ± 235*	*5213 ± 360*	*5792 ± 516*
Nucleus (V_V)	.07 ± .01	.08 ± .01	.09 ± .01
(μ^3)	*435 ± 53*	*477 ± 38*	*636 ± 45*
(n_V)[dε]	177 ± 8	161 ± 11	155 ± 12
Cytoplasm (V_V)	.76 ± .02	.76 ± .01	.75 ± .01
(μ^3)	*4483 ± 50*	*4736 ± 38*	*5156 ± 45*
Mitochondria (V_V)	.17 ± .01	.16 ± .01	.17 ± .01
(μ^3)	*1015 ± 54*	*1021 ± 46*	*1190 ± 39*
Microbodies (V_V)	.011 ± .001	.013 ± .001	.012 ± .001
(μ^3)	*62 ± 3*	*79 ± 4*	*86 ± 4*
Dense bodies (V_V)	.003 ± .000	.004 ± .000	.006 ± .001
(μ^3)	*17 ± 1*	*26 ± 1*	*41 ± 5*
SER (s_V)[e]	5.33 ± .13	6.71 ± .32	8.02 ± .30
(μ^2)[f]	*18145 ± 525*	*24230 ± 1178*	*31299 ± 1248*
RER (s_V)	2.47 ± .20	4.19 ± .24	3.58 ± .30
(μ^2)	*8397 ± 697*	*15152 ± 837*	*13816 ± 1174*
Golgi (s_V)	.36 ± .05	.39 ± .07	.31 ± .04
(μ^2)	*1246 ± 160*	*1385 ± 232*	*1205 ±143*

[b]*Volume per unit volume of intralobular tissue.*
[c]*Volume per average mononuclear hepatocyte.*
[d]*Number per* cm^3 *of intralobular tissue.*
[e]*Surface area per unit volume of hepatocyte ground substance (excludes mitochondria, microbodies, dense bodies).*
[fε] *Surface area per average mononuclear hepatocyte.*
Note: V_V ≃ *numerical density;* n_V ≃ *numerical density;* s_V ≃ *surface density.*
[From Schmucker. Age-related changes in hepatic fine structure: A quantitative analysis. J. Gerontol. 31, 135-143 (1976)].

However, morphometric analysis of the SER and RER membranes clearly demonstrated that:

1) the amount of SER continues to increase up to 480 days of age in both groups of rats - by 58% in the virgins and 61% in the RBS.

TABLE 3

Hepatic Morphometric Parameters in Retired Breeder Rats as a Function of Age[a]

Tissue Components	*Age (Days)*	
	180	*480*
Extrahepatocyte (V_V)[b]	.13 ± .01	.10 ± .01
Hepatocyte (V_V)	.87 ± .01	.89 ± .01
(μ^3)[c]	*5890 ± 258*	*7176 ± 438*
Nucleus (V_V)	.07 ± .01	.07 ± .01
(μ^3)	*431 ± 39*	*573 ± 63*
(n_V)[d]	145 ± 5	125 ± 8
Cytoplasm (V_V)	.81 ± .01	.82 ± .01
(μ^3)	*5471 ± 39*	*6604 ± 63*
Mitochondria (V_V)	.17 ± .01	.18 ± .01
(μ^3)	*1158 ± 53*	*1484 ± 65*
Microbodies (V_V)	.011 ± .001	.013 ± .001
(μ^3)	*73 ± 4*	*104 ± 6*
Dense bodies (V_V)	.005 ± .000	.005 ± .001
(μ^3)	*33 ± 3*	*43 ± 4*
SER (s_V)[e]	4.55 ± .23	6.23 ± .28
(μ^2)[f]	*19124 ± 1086*	*31198 ± 1473*
RER (s_V)	2.75 ± .16	3.29 ± .25
(μ^2)	*11541 ± 653*	*16500 ± 1317*
Golgi (s_V)	.26 ± .04	.16 ± .03
(μ^2)	*1111 ± 194*	*797 ± 151*

[a] *All values expressed as MEAN ± SEM.*
[b] *Volume per unit volume of intralobular tissue.*
[c] *Volume per average mononuclear hepatocyte.*
[d] *Number per cm*^{3} *of intralobular tissue.*
[e] *Surface area per unit volume of hepatocyte ground substance (excludes mitochondria, microbodies, dense bodies).*
[f] *Surface area per average mononuclear hepatocyte.*
Note: $V_V \simeq$ *numerical density;* $n_V \simeq$ *numerical density;* $s_V \simeq$ *surface density.*
[From Schmucker. Age-related changes in hepatic fine structure: A quantitative analysis. J. Gerontol. 31, 135-143 (1976)].

2) the RB rat livers contain less RER and SER ($p < 0.005$) than those of virgin animals at similar ages (Tables 2,3; Fig. 3).

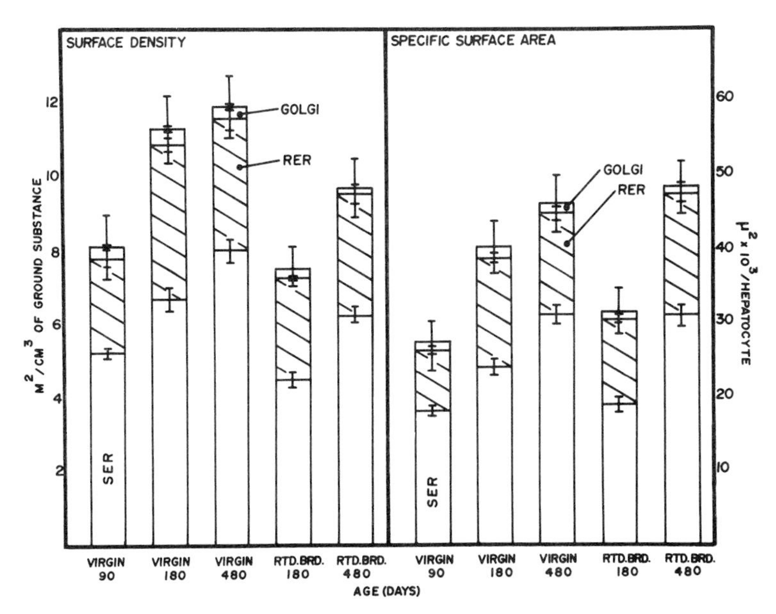

Fig. 3. Surface densities and specific surface areas of smooth surfaced endoplasmic reticulum (SER), rough surfaced endoplasmic reticulum (RER), and Golgi membranes as a function of age and breeding. All values are expressed as the mean ± SEM. The left panel (surface area to volume ratio) is expressed as m^2 per cm^3 of hepatocyte ground substance; whereas the specific surface areas are measured in μ^2 per average mononuclear hepatocyte. (23).

The surface densities of the Golgi membranes decrease by approximately 20 and 37% in the virgin and RB rat liver, respectively, between 180 and 480 days of age. Similarly, the specific surface areas decline during the same period in both groups of animals (Tables 2,3; Fig. 3).

The surface densities of both the SER and RER are greater in the livers of the virgin rats than in those of the RB animals at 480 days of age (Fig. 3). When the specific surface areas (that is, the μ^2 of SER/average mononuclear hepatocyte) are compared, there is no difference in the amount of SER. This, however, is only an apparent paradox: the hepatocytes are significantly larger and there are fewer parenchymal cells per volume of intralobular liver tissue in RB animals than in the livers of the age-matched virgin rats. The larger RB hepatocytes contain the same amount of membrane as the smaller parenchymal cells of the virgin animals. However, the greater number of hepatocytes per volume of liver tissue in the virgin rats results in significantly more membrane per cm^3 of tissue. Thus, the use of geometrical units, i.e. volume and surface densities, affords a more direct comparison of morphometric data with biochemical values obtained from tissue homogenates because the latter can be expressed as a function of tissue volume.

V. THE EFFECT OF OXANDROLONE ON SERUM AND LIVER OF MALE RATS

Oxandrolone (17β-hydroxy-17α methyl-2 oxa-5α androstan-3-one), a synthetic steroid-like lactone, has 6 times the anabolic and one-third the androgenic activity of methyl testosterone (24) Like certain other anabolic steroids, oxandrolone exerts a hypolipidemic effect. In several clinical trials oxandrolone effectively lowered plasma TG of most patients with Types III, IV and V hyperlipoproteinemia but produced little or highly variable effects on their plasma colesterol levels (25-28).

Recently, however, we discovered that oxandrolone administration for 21 days reduced the plasma cholesterol levels of RB rats (29). Subsequently, we measured the effect of this compound on serum and liver lipids of young adult (i.e. 2½ - 3-month old) virgins, 6-month old mature virgin, 6-month old recently retired breeder rats with the customary elevations of serum lipids and a small group of extremely old retired breeder rats (15 months) with cholesterol levels comparable to humans (Table 4). Oxandrolone produced no statistically significant effects on the blood or liver lipids of young virgin male rats and did not alter the body or liver weight of the treated animals (Tables 4-6). On the contrary, the drug reduced the serum cholesterol of all 3 of the other groups by a third or more of the initial value. In fact, oxandrolone actually lowered the serum cholesterol of the 6 - 9 month old MV and RB rats to a level significantly below that of the 3-month old young adults. Seven days after the last dose of oxandrolone, the serum cholesterol of the RB and MV rats returned to their pretreatment levels.

Morphometric analysis of the hepatic fine structure of young adult and retired breeder rats treated with oxandrolone for 21 days (30) revealed that following oxandrolone treatment, the livers of the young and old rats were no longer distinguishable on the basis of any morphometric parameters. In addition, oxandrolone, unlike many lipid-lowering drugs, does not induce a proliferation of hepatic microbodies.

The observation that oxandrolone increased the surface

TABLE 4

Effect of Oxandrolone on the Serum Lipid Levels of Male Rats

Group	pretreatment	21 days treatment	7 days post-treatment
	CHOLESTEROL (mg/100 ml ± SEM)		
YV Controls	70.18 ± 3.05(11)	78.90 ± 3.95(11)	71.72 ± 2.67(11)
YV Experimentals	75.57 ± 2.88(19)	68.68 ± 2.14(19)	68.00 ± 1.88(19)
MV Controls	83.66 ± 3.76(9)	75.55 ± 2.80(9)	75.88 ± 3.04(9)
MV Experimentals	82.72 ± 4.60(11)	55.54 ± 5.94(11)*	78.72 ± 6.29(11)
RB	94.5 ± 4.2 (8)	56.2 ± 3.0 (8)**	77.6 ± 5.5 (8)
OLD RB	222.7 ± 16.3(4)	134.1 ± 18.2(4)***	211.0 ± 16.0(4)

Group	pretreatment	21 days treatment	7 days post-treatment
	TRIGLYCERIDE (mg/100 ml ± SEM)		
YV Controls	58.35 ± 4.80(10)	72.04 ± 5.13(10)	75.68 ± 7.00(10)
YV Experimentals	66.67 ± 5.49(18)	80.01 ± 4.82(18)	75.55 ± 4.13(18)
MV Controls	96.30 ± 8.34(9)	93.53 ± 6.18(9)	92.28 ± 3.36 (9)
MV Experimentals	98.72 ± 11.48(11)	85.61 ± 10.15(11)	99.50 ± 9.52(11)
RB	85.88 ± 7.15(8)	81.41 ± 7.21(8)	98.95 ± 11.09(8)
OLD RB	---------------	111.8 ± 20.1(4)	154.8 ± 26.4 (4)

*Note: ( ) = number of animals; * = $p < 0.0025$
** = $p < 0.0005$; *** = $p < 0.005$.*

In these experiments, a sample of tail vein blood from each rat was collected & analyzed at 0,21 & 28 days. Control rats received 250 mg corn oil margarine/day. Experimental animals were treated for 21 days with 100 mg oxandrolone/Kg b.w./day mixed with 250 mg corn oil margarine. The 3 columns of numbers for each lipid show the animals' mean pretreatment value, the effect of 21 days of margarine or margarine + oxandrolone and the lipid levels 1 week after the cessation of oxandrolone treatment, respectively.

TABLE 5
Effect of Oxandrolone on Serum Free Fatty Acid Levels of Male Rats

Group	*Oxandrolone*	*Mean Serum Free Fatty Acid (meq/l ± SEM)*
Young Adults	0	0.605 ± 0.0390 (24)
Young Adults	+	0.582 ± 0.0296 (24)
Retired Breeders	0	0.518 ± 0.0360 (11)
Retired Breeders	+	0.444 ± 0.0365 (11)

( ) = number of animals.

Animals were divided into experimental and control groups. Each control animal received 250 mg corn oil margarine/day and each experimental animal was fed 250 mg corn oil margarine mixed with 100 mg oxandrolone/Kg body weight/day for 21 days. All animals were sacrificed by decapitation on the morning of Day 22, 12 hrs after the last dose and 24 hrs after the initiation of the fast. Pretreatment FFA levels were not determined on these animals.

TABLE 6
Effect of Oxandrolone on the Livers of Male Rats

	wet weight, gms	*Triglyceride mg%*	*Cholesterol mg%*
Young Adult			
Control	9.338±1.003(13)	10.090±0.980(7)	3.710±0.080(6)
Experimental	8.980±0.588(13)	9.130±0.890(7)	3.700±0.090(6)
Retired Breeder			
Control	13.341±2.097(22)	9.070±0.760(14)	3.320±0.070(14)
Experimental	13.681±1.830(18)	8.520±0.590(14)	3.330±0.080(15)

( ) = number of animals. Experimental protocol same as described in Table 5. Weight ± SEM.

density of smooth endoplasmic reticulum in livers of young adult rats by 35% and retired breeder rats by 68% prompted an investigation into the significance of the response by Dr. Luean Anthony, who proceeded to measure those modalities of SER function most often affected by drugs which induce SER proliferation (31). The concentration of cytochrome P-450 and cytochrome b_5 as well as the activities of NADPH-cytochrome *c* reductase and ethylmorphine N-demethylation were assayed in liver microsomes from young virgin and retired breeder control rats and those treated with oxandrolone for 21 days. The results showed that oxandrolone exerted no statistically significant effect on any parameter measured. Therefore, although oxandrolone increased the surface density of SER significantly as measured by quantitative EM, the change was not accompanied by concomitant increases in microsomal drug metabolism. This was surprising because most other lipid-lowering agents cause marked changes in the liver. Clofibrate, for example, causes hepatomegaly in rats, hepatic microbody proliferation, hypertrophy of the SER and enhanced microsomal metabolism of drugs (32-34).

Thus it appears that oxandrolone is similar to D-galactoseamine and dieldrin which, when administered to rats, result in a liver with a marked increase in the amount of SER but normal or reduced activity of drug metabolizing enzymes (35,36). The reason for the lack of coupling between membrane proliferation and enzyme induction is unclear and experiments examining the effects of short-term administration of oxandrolone at lower doses are necessary to resolve this problem.

VI. SUMMARY

In summary, we feel that the accumulated data indicates that the retired breeder rat is a potentially good animal model for

1) investigating the relationship between aging and lipid metabolism
2) studying the pathogenesis and control of certain types of hyperlipoproteinemia and
3) the evaluation of lipid-lowering drugs.

We would like to suggest that the mechanism of action and therapeutic efficacy of lipid-lowering drugs, as well as the pathogenesis of hyperlipoproteinemia are studied too often in animals rendered hyperlipidemic by unusual dietary manipulations or pharmacological agents that produce <u>extremely</u> high levels. Such experimental conditions probably do not simulate human hyperlipoproteinemias and obviate the chances for meaningful interpretations or applications of the results, especially when there appears to be no significant drug effect.

Although the RB rat may or may not turn out to be an "ideal" animal model to use for studies of this nature, it does afford at least two real advantages: 1) the hyperlipidemia and arteriosclerosis arise spontaneously in a predictable percentage of the breeders and 2) the serum TG and CHOL levels of the very old animals reach human levels.

The data we have presented on the effects of oxandrolone suggests that oxandrolone restores serum cholesterol and liver structure of old animals to young adult levels. Nevertheless, the rapidity and magnitude of the decrease in cholesterol also is a bit frightening because we do not yet know where the cholesterol goes upon removal from the blood. Chronic studies presumably will help to answer this question.

Finally, the data we are slowly accumulating on aging virgin and RB rats has important implications for many of us who work on the liver. They indicate that: 1) specific alterations in hepatic fine structure occur as a function of age which, in turn, may be related to age-related changes in hepatic capacity to metabolize drugs and lipids, synthesize plasma proteins, etc.; 2) certain structural changes which appear to accompany aging may be reversible; and, 3) subtle changes in cell fine structure may not be detected by routine qualitative electron microscopy. As a result, decisions regarding the amount and/or distribution of cells and organelles as well as the real or potential toxicity of a drug may be impossible to make unless the changes result in an all-or-none effect such as the phenobarbital induction of hepatic SER. Therefore, attempts to correlate qualitative structural results with quantitative functional data often may be unsatisfactory.

VII. ACKNOWLEDGEMENTS

The authors wish to thank Dr. James M. Felts, Mrs. Naomi Ross, Mrs. Anne Petersen, Mrs. Jill Mooney, Ms. Barbara Kriz, Ms. Joan Hahn, and Mr. Jack Quan for their assistance.

This work was supported by the Medical Research Service of the Veterans Administration and NIH grant HL-16363-01 MET, and NIH grant (NIAMDD) P17-AM18520.

VIII. REFERENCES

(1). Boberg, J., Carlson, L.A., and Fröberg, S., in "Nutrition in Old Age, Symposia of the Swedish Nutrition Foundation", (L.A. Carlson, Ed.), Vol. 10, p. 61. Almquist and Wiksell, Uppsala, Sweden.
(2). Wexler, B.C., J. Atheroscler. Res. 4, 57 (1964).
(3). Wexler, B.C., Circ. Res. 15, 485 (1964).
(4). Wexler, B.C., and Kittinger, G.W., J. Atheroscler Res. 5, 317 (1964).

(5). Wexler, B.C., Judd, J.T., and Kittinger, G.W., J. Atheroscler. Res. 4, 397 (1964).
(6). Wexler, B.C., Saroff, J., and Judd, J., Diabetes 19, 311 (1970).
(7). Wexler, B.C., Diabetes 19, 324 (1970).
(8). Wexler, B.C., Anthony, C.D., and Kittinger, G.W., J. Atheroscler Res. 4 (1964).
(9). Rudel, L.L. and Morris, M.D., J. Lipid Res. 14, 364 (1973).
(10). Eggstein, M., and Kreutz, F.H., Klin. Wschr. 44, 262 (1966).
(11). Dole, V.P., J. Clin. Invest. 35, 150 (1956).
(12). Trout, D.L., Estes, E.H., Jr., and Friedberg, S.J., J. Lipid Res. 1, 199 (1960).
(13). Rudel, L.L., Morris, M.D., and Felts, J.M., Biochem. J. 139, 89 (1974).
(14). Scanu, A. in "Structural and Functional Aspects of Lipoproteins in Living Systems" (E. Tria and A.M. Scanue, Eds.), Chpt. 3. Academic Press, New York, 1969.
(15). Fredrickson, D.S., Levy, R.I., and Lees, R.S., N. Engl. J Med. 276, 34 (1967).
(16). Shore, B., Shore, V., Salel, A. *et al.*, Biochem. Biophys. Res. Comm. 58, 1 (1974).
(17). Schmucker, D.L., Jones, A.L. and Spring-Mills, E., J. Gerontol. 29, 506 (1974).
(18). Wood, R.L. and Luft, J.H., J. Ultrastruc. Res. 12, 22 (1965).
(19). Spurr, A., J. Ultrastruc. Res. 26, 31 (1969).
(20). Weibel, E.R., Int. Rev. Cytol. 26, 235 (1969).
(21). Weibel, E.R., Stäubli, W., Gnagi, H.R., *et al.*, J. Cell Biol. 42, 68 (1969).
(22). Staubli, W., Hess, R., and Weibel, E.R., J. Cell Biol. 42, 92 (1969).
(23). Schmucker, D.L., J. Gerontol. 31, 135 (1976).
(24). Fox, M., Minot, A.S., and Liddle, G.W., J. Clin. Endocrinol. 22, (1962).
(25). Sachs, B.A. and Wolfman, L., Metabolism 17, 400 (1968).
(26). Sachs, B.A. and Wolfman, L., Circulation 36, 11 (1967).
(27). Glueck, C.J., Swanson, F., and Hutsell, T., Circulation 42 (3), 158 (1970).
(28). Glueck, C.J., Metabolism 20, 691 (1971).
(29). Mills, E.S., Kriz, B.M., and Jones, A.L., Proc. Soc. Exp. Biol. Med. 144, 861 (1973).
(30). Schmucker, D.L., and Jones, A.L., J. Lipid Res. 16, 143 (1975).
(31). Anthony, L.E. and Jones, A.L., Biochem Pharmacol. 25, 1549 (1976).
(32). Svoboda, D.J., and Azarnoff, D.L., J. Cell Biol. 30, 442 (1966).

(33). Hess, R., Stäubli, W., and Riess, W., Nature (London) 408, 856 (1965).
(34). Lewis, N.J., Witiak, D.T., and Feller, D.R., Proc. Soc. Exp. Biol. Med. 145, 281 (1969).
(35). Koff, R.S., Davidson, L.J., Gordon, G., *et al.* Exp. Mol. Pathol. 19, 168 (1973).
(36). Hutterer, F., Schaffner, F., Klion, F.M., *et al.*, Science (New York) 161, 1017 (1968).

ON THE QUESTION OF WHETHER CYTOCHROME P-450 CATALYZES ETHANOL OXIDATION: STUDIES WITH PURIFIED FORMS OF THE CYTOCHROME FROM RABBIT LIVER MICROSOMES[1]

Kostas P. Vatsis and Minor J. Coon

The University of Michigan

The ability of rabbit liver microsomal cytochrome P-450 to catalyze the oxidation of ethanol to acetaldehyde was examined in a reconstituted enzyme system containing electrophoretically homogenous phenobarbital-inducible P-450$_{LM2}$, β-naphtoflavone-inducible P-450$_{LM4}$, or a partially purified mixture of P-450$_{LM1}$ and LM$_7$. Only very low rates of ethanol oxidation were observed, about 3-5 nmol per min per nmol P-450$_{LM}$, with no apparent specificity of any of the P-450$_{LM}$ forms for ethanol. In the presence of ethanol, essentially all of the NADPH oxidation in the reconstituted system could be accounted for by the formation of H_2O_2. This finding and the inability of ethanol to stimulate NADPH and oxygen utilization by the reconstituted enzyme system are incompatible with a mixed function oxidation reaction leading to acetaldehyde formation. Ethanol had no effect on benzphetamine hydroxylation by P-450$_{LM2}$. The low rates of ethanol oxidation could possibly be due to a slight peroxidatic acivity of P-450$_{LM}$ in the presence of H_2O_2 generated in the reconstituted system.

I. INTRODUCTION

Following the resolution of the cytochrome P-450-containing enzyme system of liver microsomes and reconstitution of an active hydroxylation system from the components (1-3), this laboratory reported the isolation and characterization of multiple forms of rabbit P-450$_{LM}$ (4-6)[2]. The PB-inducible form,

[1]This research was supported by Grant PCM76-14947 from the National Science Foundation and Grant AM-10339 from the United States Public Health Service.

$P\text{-}450_{LM2}$, and BNF-inducible form, $P\text{-}450_{LM4}$, have been purified to electrophoretic homogeneity and have been shown to possess distinct physiochemical properties (4-7) as well as significant structural differences revealed by studies on amino acid composition (7) and by immunochemical methods, including Ouchterlony diffusion analysis and radioimmune assays (6,8,9). Additional forms of rabbit $P\text{-}450_{LM}$ of different subunit molecular weight and electrophoretic mobility, e.g., $P\text{-}450_{LM1}$, LM_{3a}, LM_{3b}, and LM_7, have been isolated in a partially purified state (5,7, 10), but their relationship to $P\text{-}450_{LM2}$ and LM_4 is not yet known.

The occurence of multiple forms of $P\text{-}450_{LM}$ with somewhat different catalytic activities (5,10) may account for the unusually broad substrate specificity of the hepatic microsomal mixed function oxidase system (11). For example, $P\text{-}450_{LM2}$ is more active toward drugs such as benzphetamine and ethylmorphine as well as toward the 4 positions of biphenyl and the 16α position of testosterone, whereas $LM_{1,7}$ acts preferentially on the 6β position of testosterone. The position-specific oxygenaction of benzpyrene by different forms of purified rabbit liver $P\text{-}450_{LM}$ has also been reported (12). More recently, electrophoretically homogenous $P\text{-}450_{LM2}$ has been shown to catalyze the peroxide-dependent hydroxylation of a host of compounds in the absence of NADPH and the reductase (13). The present studies were undertaken to determine whether highly purified $P\text{-}450_{LM}$ serves as a catalyst for the oxidation of ethanol to acetaldehyde in the reconstituted system, and to examine the effects of ethanol on $P\text{-}450_{LM2}$-mediated NADPH and oxygen utilization and H_2O_2 formation under basal and substrate-stimulated conditions.

II. RESULTS AND DISCUSSION

A. Studies on Ethanol as a Possible Substrate for $P\text{-}450_{LM}$ in The Reconstituted Enzyme System

Ethanol was incubated with electrophoretically homogenous $P\text{-}450_{LM2}$ or LM_4 or partially purified $LM_{1,7}$ (a mixture of about equal amounts of LM_1 and LM_7) in the presence of reductase,

---

[1]The abbreviations used are: $P\text{-}450_{LM}$, liver microsomal cytochrome P-450; PB, phenobarbital; BNF, β-naphtoflavone; dilauroyl-GPC, dilauroylglyceryl-3-phosphorylcholine; 3-MC, 3-methylcholanthrene; and TCA, trichloroacetic acid.

phospholipid, and NADPH. Table 1 shows that the various forms of rabbit liver P-450$_{LM}$ oxidized ethanol at only quite low rates, as judged by the spectral assays for acetaldehyde formation (15). With P-450$_{LM2}$, various phospholipid concentrations were tested, but these appeared to have no significant effect on the slight activity observed. Rates of NADPH oxidation were also determined in the experiments with P-450$_{LM2}$ under identical incubation conditions, and these were consistently 12-15% lower in the presence of ethanol.

Under basal conditions in the absence of a substrate for P-450$_{LM2}$, the total amount of NADPH and oxygen utilized in the reconstituted enzyme system can be accounted for entirely by the formation of H_2O_2; addition of substrate to this system gives rise to a stimulation of NADPH and oxygen utilization that is associated with the formation of hydroxylated substrate (16,17), as will be discussed in detail in the ensuing sections. Ethanol (0.1 M) did not cause a stimulation of NADPH utilization by the reconstituted enzyme system (Table 1), and , in

TABLE 1
Assay of Purified P-450$_{LM}$ Preparations for Ability to Oxidize Ethanol in the Presence of NADPH in the Reconstituted System

Purified P-450$_{LM}$ and type of microsome used as a source	Phospholipid concentration	Acetaldehyde formation[a]	NADPH oxidation[b]	
			Ethanol absent	Ethanol present
	(μg per ml)	nmol/min/nmol/ P-450$_{LM}$		
P-450$_{LM2}$				
(PB-induced)	0	1.6	33	31
"	5	3.4	102	90
"	10	5.1	102	88
"	30	4.8	90	80
P-450$_{LM4}$				
(PB-induced)	10	4.2		
P-450$_{LM4}$				
(BNF-induced)	10	5.5		
P-450$_{LM4}$				
(Untreated)	10	5.1		
P-450$_{LM1,7}$				
(PB-induced)	10	3.9		

a. Reaction mixtures contained P-450$_{LM}$ (0.1 μM), electrophoretically homogenous rabbit liver microsomal NADPH-cytochrome P-450 reductase (15 μg of protein per ml; specific activity *toward cytochrome c, 34.6 (14), dilauroyl-GPC as indicated, 15 mM $MgCl_2$, 80 mM ethanol, and 0.1 M potassium phosphate buffer, pH 7.4, in a total volume of 1.5 ml. Following preincubation at 37° for 5 minutes, reactions were started by adding NADPH (1.0 mM, final concentration) and were terminated after a 30-min incubation at the same temperature with 0.3 ml of 50% TCA. Zero-time controls contained all of the components, but TCA was added before NADPH and the flasks were kept at room temperature. All incubations, including zero-time, were carried out in a duplicate in the main chambers of stoppered 25-ml Erlenmeyer flasks with center wells containing 1.0 ml of 0.015 M semicarbazide in 0.16 M sodium phosphate buffer, pH 7.0. After an overnight equilibration period at room temperature, the flasks were opened and the absorbance at 224 nm of the solutions in the center wells was determined as a measure of acetaldehyde formed, according to Lieber and DeCarli (15). The values shown are the average of three different experiments. Glycerol, which is used as a protective agent throughout the purification and subsequent storage of both P-450$_{LM}$ and the reductase, was present in these incubations at a final concentration of 0.3 - 0.5%, depending on the preparation of P-450$_{LM}$ present. The total absorbance at 224 nm of zero-time controls (0.5-0.6) ranged between 70 and 88% of corresponding values obtained for the experimental incubations. Control experiments showed that glycerol is decomposed by TCA to give volatile products which contribute to the absorbance at 224 nm in the presence of semicarbazide.*

b. Reaction mixtures were as described above, except that the total incubation volume was 1.0 ml and NADPH was present at a final concentration of 0.15 mM. Ethanol, when present, was added at the start of the preincubation at a final concentration of 0.1 M. After temperature equilibration at 37° for 3 min, reactions were started with NADPH, and the decrease in absorbance at 340 nm was monitored at the same temperature using a Gilford multiple sample absorbance recorder. In each case, reactions were carried out in duplicate. Values shown are the average of two different experiments.

experiments not shown, did not affect significantly the NADPH-dependent formation of H_2O_2 assayed with varying amounts of phospholipid as shown in Table 1. Rates of H_2O_2 formation corresponded to NADPH oxidation rates, irrespective of phospholipid concentration. In the experiment in which the P-450$_{LM2}$-containing reconstituted system was incubated with ethanol in the presence of 30 μg/ml phospholipid, the contents in the main chamber were assayed for hydrogen peroxide when the flasks were opened for determination of acetaldehyde in the center wells. It was found that the amount of H_2O_2 present was identical to that obtained in corresponding incubations in the absence of ethanol. These results indicate that, whether ethanol is absent or present at a concentration of 0.1 M, all of the electrons from NADPH are directed toward the formation of H_2O_2. Consistent with these observation, others have shown that the total amount of H_2O_2 formed in a reconstituted system containing highly purified P-450_{LM} from PB-induced rats was directly proportional to the amount of NADPH oxidized, and was coupled quantitatively in the presence of excess catalase and ethanol to form acetaldehyde.[3] In view of these findings, the mere fact that any ethanol oxidation, however slight, took place at a time when the net amount of reducing equivalents from NADPH available for this process was zero (Table 1) is inconsistent with a typical mixed function oxidation reaction catalyzed by P-450_{LM}. Therefore, such results rule out a hydroxylation pathway according to the reaction.

$$NADPH + H^+ + O_2 + CH_3CH_2OH \rightarrow NADP^+ + H_2O + CH_3CH(OH)_2$$

where the expected product is the hydrated form of acetaldehyde. The very low rates of ethanol oxidation obtained with P-450_{LM2} should be contrasted to the 20-fold higher rates of NADPH oxidation (Tables 1 and 3), H_2O_2 production, or formaldehyde formation from benzphetamine (Table 4) using the same P-450_{LM2} preparation under similar incubation conditions.

Previous studies in this laboratory have shown that P-450_{LM4} is induced by BNF but is also present in microsomes from untreated and PB-treated rabbits (7). This form of the hemeprotein has the same properties regardless of the type of microsomes used as the source for its isolation (7). 3-MC also induces P-450_{LM4} in rabbits (10,18,19). The data in Table 1 show that the same low rates of ethanol oxidation obtained with P-450_{LM2} are also observed in the presence of P-450_{LM4}, whether isolated from BNF-induced, PB-induced, or untreated rabbits, as well as with P-$450_{LM1,7}$ isolated from PB-induced animals. A considerably greater activity would have been expected with P-450_{LM2} than LM_4 in view of reports that NADPH-dependent rates of ethanol oxidation were increased both by PB treatment and chronic ethanol feeding to rats (20,21), whereas 3-MC treatment was without effect (20). In addition, microsomes from rats fed ethanol exhibited CO-reduced difference spectra with maximum absorption at 450 nm, and the ratio of the 430 to 455 nm ethyl isocyanide peaks of similar preparations reduced with dithionite remained unchanged from that of control animals, indicating that the hemeprotein induced by chronic ethanol administration had spectral characteristics similar to those of the form induced by PB (22). It is well known, that the ratio of

[1] A.Y.H.Lu and G. Miwa, personal communication

the 430 to 455 nm ethylisocyanide peaks of dithionite-reduced microsomes decrease upon treatment of animals with carcinogens, and the absorption maximum of CO-reduced difference spectra of similar preparations is not at 450 nm (23). In this context, the observation that P-450_{LM4} exhibited an ethanol-oxidizing activity comparable to that seen with P-450_{LM2}, coupled with the slight ethanol oxidation rates obtained with any of the forms of the rabbit P-450_{LM}, whether highly or partially purified (Table 1), may be taken as evidence for a lack of catalytic specificity of P-450_{LM} for ethanol or for lack of specificity of the overall process. It is thus quite conceivable that a number of different hemeproteins, e.g., hemoglobin or myoglobin, or even hematin, could be capable of ethanol oxidation in the presence of NADPH and reductase at rates comparable to those observed in the present studies using the various purified forms of rabbit P-450_{LM}. Hemoglobin has been shown recently to catalyze the hydroxylation of aniline in the presence of NADPH, oxygen, and partially purified rat liver NADPH-cytochrome P-450 reductase; hemoglobin-catalyzed rates of aniline hydroxylation were similar to those obtained with rat P-450_{LM} in the presence of a reconstituted system (24). In contrast to the lack of specificity of rabbit P-450_{LM} for ethanol (Table 1), rates of benzphetamine hydroxylation by P-450_{LM4} and $LM_{1,7}$ in the reconstituted system were 5 and 10%, respectively, of those obtained with P-450_{LM2} (5,10).

The possibility was considered that, since H_2O_2 is formed in the reconstituted system by autoxidation of reduced P-450_{LM2} (Fig. 2 and Table 3), ethanol might be oxidized by a peroxidatic mechanism according to the following reaction:

$$CH_3CH_2OH + H_2O_2 \xrightarrow{\text{P-450}_{LM2}} CH_3CHO + 2H_2O.$$

This possibility was examined by incubating P-450_{LM2} with a high concentration of alcohol in the presence of H_2O_2 generated slowly by glucose-glucose oxidase, i.e., conditions which are expected to favor peroxidatic activity. In these experiments, shown in Table 2, the rate of H_2O_2 generation was 1.9 μ*M* min^{-1} at 0.2 μg/ml glucose oxidase and increased in direct proportion to the glucose oxidase concentrations employed. Under these conditions, methanol- and ethanol-oxidizing activities of P-450_{LM2} appeared to be quite low (one-tenth of the values shown in Table 1 for the reconstituted enzyme system) and independent of H_2O_2 generation rates. In experiments not shown in which catalase was substituted for P-450_{LM2} at the same molar concentration, methanol and ethanol were oxidized at rates corresponding to those of H_2O_2 generated by the various concentrations of glucose oxidase. It is still conceivable, however,

TABLE 2
Peroxidatic Acticity of P-450$_{LM2}$ Toward Methanol and Ethanol

Glucose Oxidase added	Methanol oxidation	Ethanol oxidation
μg/ml	nmol aldehyde formed/min/nmol P-450$_{LM2}$	
0.2	0.3	0.4
0.5	0.3	0.5
1.0	0.3	0.6
2.0	0.3	0.5

Reaction mixtures contained P-450$_{LM2}$ (0.7 μM), dilauroyl-GPC (30 μg per ml), glucose oxidase as indicated, 80 mM methanol or ethanol, 15 mM, $MgCl_2$, and 0.1 M potassium phosphate buffer, pH 7.4, in a total volume of 1.5 ml. After a 5-min preincubation at 37°, glucose was added (10 mM final concentration), and incubations were carried out in a duplicate for 30 min at 37° as described in Table 1 when ethanol was the substrate, and in open 25-ml Erlenmeyer flasks without center wells when methanol was the substrate. At the end of the incubation 0.3 ml of 50% TCA as added to stop the reaction, and acetaldehyde formed from ethanol was assayed following an overnight diffusion period, as described in Table 1. Formaldehyde formed from methanol was determined by the method of Nash (25) as modified by Cochin and Axelrod (26). Values shown are the average of two different experiments. The P-450$_{LM2}$ preparation was the same as that used in the experiments shown in Table 1.

that the low rates of ethanol oxidation in the reconstituted system represent a slight peroxidatic activity of purified P-450$_{LM}$. When expressed per ml of reaction mixture, rates of acetaldehyde formation from ethanol in the reconstituted system in the presence of P-450$_{LM2}$ (Table 1; 0.2-0.5 μM min^{-1}) are similar to those obtained with P-450$_{LM2}$ and H_2O_2 generated by glucose-glucose oxidase as shown in Table 2. In the experiments with the P-450$_{LM2}$-containing reconstituted system in the presence of ethanol, the ratio of H_2O_2-generation rate to P-450$_{L,2}$ heme concentration varied from 17 to 120 min^{-1}, whereas in the experiments depicted in Table 2 this ratio ranged between 2.7 and 27 min^{-1}. Rates of acetaldehyde formation from ethanol, expressed as μM min, were identical in both cases. It is thus quite possible that P-450$_{LM2}$ possesses a slight peroxidatic activity toward methanol or ethanol (0.2-0.5 μM min^{-1}) that remains constant over a 60-fold range in the ratio of H_2O_2-generation rate to P-450$_{LM2}$ heme concentration, regardless of the manner in which a shift in this ratio is brought about. In this regard, P-450$_{LM2}$ would not resemble catalase, which can peroxidatically metabolize these alcohols ar rates that are proportional to H_2O_2-generation rates up to a certain value of the ratio mentioned above. If H_2O_2 , generated from NADPH and oxygen in the reconstituted system, were utilized by P-450_{LM2}, in a peroxidatic fashion to convert ethanol to acetaldehyde, it should be present in a lesser amount at the end of the reaction as compared to corresponding incubations in the absence of ethanol. As mentioned previously in connection with the studies shown in Table 1, this was not the case, i.e., equal amounts of H_2O_2 were detected in the absence or presence of ethanol. This apparent discrepancy could presumably be attributed to the fact that ethanol oxidation rates were so low (0.2-0.5 μ$\underline{M}$ acetaldehyde min^{-1}) compared to H_2O_2 generation rates (10.5 μ$\underline{M}$ min^{-1}; Tables 1 and 3) that the utilization of H_2O_2 for such a slight peroxidatic activity would not have been detected as a decrease in the total amount of H_2O_2 formed. Alternatively, a slow ethanol oxidation could proceed independently of the very rapid NADPH utilization or H_2O_2 formation by the reconstituted system.

The low ethanol-oxidizing activity of P-450_{LM2} in the reconstituted system could conceivably be attributed to a contamination of the enzyme components by catalase, which would lead to ethanol oxidation in the presence of H_2O_2 generated under these conditions. Although a very unlikely possibility in view of the homogeneity of these proteins upon slab gel electrophoresis, it was demonstrated nonetheless that neither P-450_{LM2} (0.1 μ$\underline{M}$) nor the reductase (15 μg of protein per ml) decomposes H_2O_2, as judged by the spectrophotometric assay of Beers and Sizer (27) carried out at 30°. These results also show that the reductase and P-450_{LM2} do not possess 'catalatic' activity.

B. Effect of Ethanol and Catalase Inhibitors on NADPH Oxidation and NADPH-Linked H_2O_2 Generation Catalyzed by the P-450_{LM2}-Containing Reconstituted Enzyme System

1. *Oxidation of NADPH in the Reconstituted Enyzme System*

The oxidation of NADPH in the reconstituted enzyme system containing highly purified P-450_{LM2} is shown in Fig. 1. In the absence of benzphetamine, NADPH oxidation was decreased slightly (about 15%) by ethanol at concentrations ranging from 0.02$\underline{M}$ to 0.1 $\underline{M}$, whereas a greater inhibition (about 45%)was observed at 0.5-1.0 $\underline{M}$ ethanol (Fig. 1A). These findings are in agreement with those shown in Table 1; different P-450_{LM2} and reductase preparations were used in the two cases. Similar results have been obtained by other investigators using either a crude fraction containing P-450_{LM2} (0.92 nmol/mg protein) , NADPH-cytochrome P0450 reductase, and phospholipid, isolated by DEAE-cellulose column chromatography of detergent-solubilized liver microsomes from untreated rats (28), or electrophoretically homogenous rabbit P-450_{LM2} as well as highly purified PB-inducible P-450_{LM} from rats.[3] It has been reported that the rate of cytochrome P-450 reduction by NADPH in liver microsomes from untreated rats was inhibited 35% by 0.1 M ethanol (29), which may explain the slight inhibitory effect of a similar concentration of ethanol on NADPH oxidation in the reconstituted enzyme system (Fig. 1A, plot b).

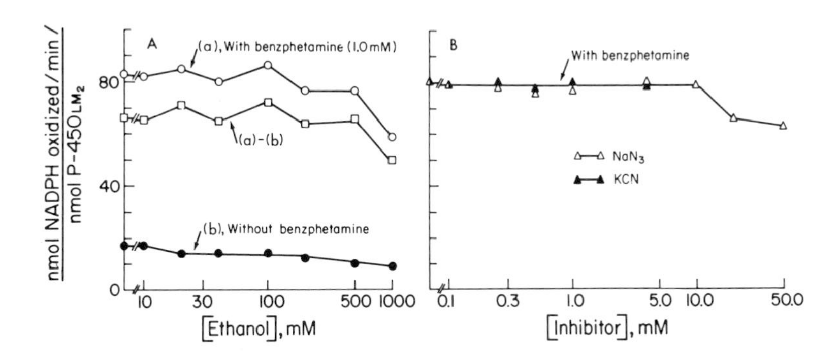

Fig.1. Effect of ethanol, azide, and cyanide on NADPH oxidation in the reconstituted enzyme system containing P-450_{LM2}. Expt. A: *Reaction mixtures contained P-450_{LM2} (0.1 μM), partially purified rabbit liver microsomal NADPH-cytochrome P-450 reductase (70 μg of protein per ml; specific activity towards cytochrome c, 7.5 (14), dilauroyl-GPC (30 μg per ml), 15 mM $MgCl_2$, 1.0 mM benzphetamine (when present), and increasing amounts of ethanol at the indicated final concentrations in a total volume of 1.0 ml of 0.1 M potassium phosphate buffer, pH 7.4. After preincubation at 37° for 3 min, NADPH (0.15 mM) was added and the decrease in absorbance at 340 nm was monitored at the same temperature using a Gilford multiple sample absorbance recorder. Reactions were linear for at least 5 min under these conditions.* Expt. B: *Increasing amounts of NaN_3 or KCN, at the indicated final concentrations, were added to the reconstituted system described in Expt. A in the presence of 1.0 mM benzphetamine and absence of ethanol. Azide or cyanide was preincubated at 37° with the other components prior to initiation of the reaction with NADPH. Each point on the curves (Expts. A and B0 represent the average of quadruplicate determinations.*

Unlike ethanol, benzphetamine when added to the reconstituted system gave rise to a 5-fold stimulation of NADPH utilization (Fig. 1A, plot a). Benzphetamine-stimulated rates of NADPH oxidation were unaffected by ethanol at concentrations up to 0.5 M, and even at 1.0 M ethanol the substrate-stimulated activity was decreased by only 30%. The same activity profile in the presence of ethanol was also seen for the net or benzphetamine-dependent NADPH oxidation rates, obtained by subtracting the basal from the overall rates (Fig. 1A, plot a minus b). In another experiment, the benzphetamine-stimulated oxidation of NADPH in the reconstituted system proceeded unimpaired in the presence of cyanide or azide at concentrations up to 4 mM and 10 mM, respectively (Fig. 1B), indicating that neither compound acts as an inhibitor of rabbit P-450_{LM2}-catalized drug hydroxylations.

2. *NADPH-Linked Hydrogen Peroxide Formation in the Reconstituted Enzyme System*

As mentioned previously, H_2O_2 formation accounts for all of the NADPH utilized or oxygen consumed by the reconstituted system in the absence of a substrate for P-450_{LM2}. Addition of benzphetamine to this system containing partially purified rabbit liver reductase results in a 4-fold stimulation of NADPH utilization or oxygen consumption with a concomitant 2-fold stimulation of H_2O_2 generation (NADPH oxidase activity)and the formation of hydroxylated substrate leading to the liberation of formaldehyde (hydroxylase activity). Under these conditions in the presence of benzphetamine, the amounts of H_2O_2 and formaldehyde formed account for 60% and 40% , respectively, of the overall NADPH or oxygen utilization by the reconstituted system; the observed stoichiometry of 1:1:1 for NADPH oxidation, oxygen consumption, and product (H_2O_2 + formaldehyde) formation is thus in accord with that expected for a mixed function oxidation reaction catalyzed by P-450_{LM} (16,17). Addition of benzphetamine to the reconstituted system containing highly purified rabbit liver reductase gives rise to a somewhat different profile with respect to the degree of stimulation of NADPH and oxygen utilization and the relative amounts of H_2O_2 and formaldehyde formed, but the overall stoichiometry is maintained at 1:1:1 (Table 3).

The time course of NADPH-linked H_2O_2 formation by the reconstituted enzyme system at two different concentrations of P-450_{LM2} is shown in Fig. 2A. Reactions were linear for 20 min at 37°, except at the higher concentration of P-450_{LM2} in the presence of substrate. From the linear portion of the curves it may be seen that H_2O_2 formation in the reconstituted system was proportional to the concentration of P-450_{LM2} whether

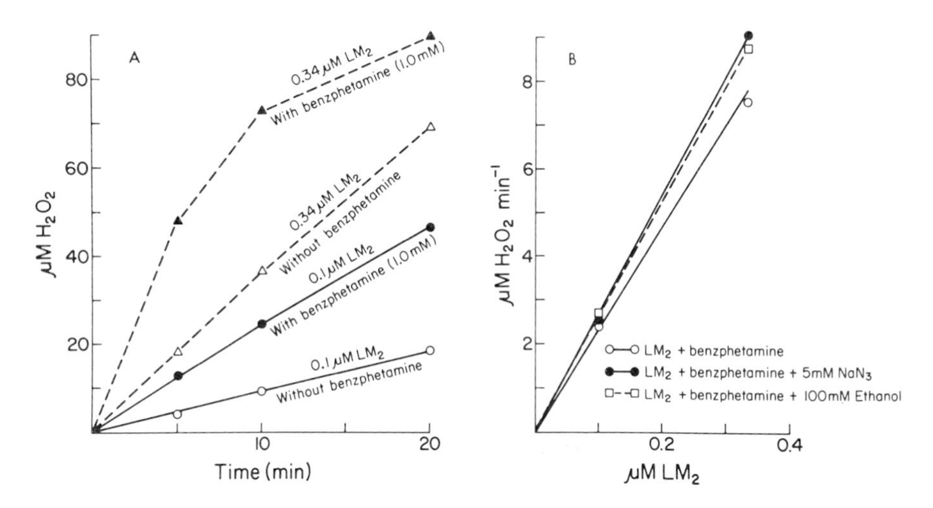

Fig. 2. P-450$_{LM2}$-dependent generation of hydrogen peroxide from NADPH in the reconsituted enzyme system. Expt. A: Time course of H_2O_2 formation. Reaction mixtures contained either 0.1 μM P-450$_{LM2}$ with reductase (70 μg of protein per ml) and dilauroyl-GPC (30 μg per ml), or 0.34 μM P-450$_{LM2}$ with reductase (240 μg of protein per ml) and dilauroyl-GPC (60 μg per ml) in 0.1 M potassium phosphate buffer, pH 7.4, containing 15 mM $MgCl_2$; the total volume of the reaction mixture was 1.5 ml. After preincubation at 37° for 3 min in the absence or presence of 1.0 mM benzphetamine, reactions were started with NADPH (1.0 mM) and were carried out at the same temperature for the time intervals indicated. Reactions were terminated by adding 0.3 ml of 50% TCA. Zero-time incubations contained all components, but received 0.3 ml of 50% TCA followed by NADPH and were placed in an ice bath. H_2O_2 was assayed by the ferrithiocyanate method as descrbied by Thurman et.al. (30). All incubations were in duplicate in open test tubes and the net values were averaged. Results are expressed as nmol H_2O_2 formed per ml of incubation mixture. Expt. B: Effect of ethanol and azide on H_2O_2 formation Conditions were identical to those described in Expt. A., except that, when present, ethanol (100 mM) or NaN_3 (5 mM) was added during preincubation to the reconstituted system containing 1.0 mM benzphetamine and 0.1 μM or 0.34 μM P-450$_{LM2}$. Incubations were at 37° for 10 min. The corresponding incubations in Expt. A, carried out for lo min in the absence of these compounds, served as controls for this experiment.

The P-450$_{LM2}$ and reductase preparations were the same in Expts. A and B as employed in the studies shown in Fig. 1.

substrate was present or absent, while addition of benzphetamine gave rise to a 2.5-fold stimulation of this activity at either concentration of the hemeprotein.

In complete agreement with the lack of effect of azide and ethanol on basal or benzphetamine-stimulated rates of NADPH utilization by the reconstituted system (Fig. 1) neither compound had a significant effect on benzphetamine-stimulated rates of H_2O_2 formation at either concentration of P-450$_{LM2}$ under similar experimental conditions (Fig. 2B). In the case of azide these findings indicate that, as expected from their purity upon gel electrophoresis, neither the reductase nor P-450$_{LM2}$ used in these experiments (Figs. 1 and 2; Table 4) was contaminated with catalase. In addition, this experiment indicates that P-450$_{LM2}$ does not possess 'catalatic' activity.

3. *NADPH Oxidase Activity, Oxygen Consumption and Benzphetamine N-Demethylation in the Reconstituted System with P-450$_{LM2}$*

The data in Table 3 were obtained in parallel experiments designed to show the effect of ethanol on NADPH oxidase activity, oxygen consumption, and substrate hydroxylation in the reconstituted enzyme system. In the absence of benzphetamine, NADPH oxidation and H_2O_2 formation were equimolar and neither parameter was affected appreciably by ethanol; these results are consistent with the data previously shown in Table 1 and Fig. 1. Ethanol was likewise without effect on rates of oxygen utilization by the reconstituted enzyme system, a finding which strengthens the conclusion that some reaction other than a mixed function oxidation (hydroxylation) is necessary to account for the very low rates of ethanol oxidation observed in the reconstituted system containing P-450$_{LM2}$ under similar experimental conditions (Table 1). Moreover, benzphetamine hydroxylation activity and benzphetamine-stimulated rates of NADPH oxidation and oxygen utilization in the reconstituted system were unaltered by ethanol, indicating that this compound is not an inhibitor of P-450$_{LM2}$-catalyzed drug hydroxylation. In accord with these observations, the K_s of binding of ethanol to P-450$_{LM2}$, determined from difference spectra recorded at 22° with 3 μM P-450$_{LM2}$ in 0.1 M phosphate buffer, pH 7.4, containing 20% glycerol, was shown to be 1.3 M and 0.5 M, respectively, in the absence and presence of dilauroyl-GPC (30 μg/ml), Using rabbit liver microsomes, Imai and Sato (31) showed that the K_s of a series of alipathic alcohols increased from 0.5 to 0.9 M with decreasing chain length of the alcohol.

In the presence of excess catalase to accomodate all of the H_2O_2 generated from NADPH in the reconstituted system, i.e., under conditions favoring the peroxidatic mode of action of

TABLE 3
Effect of Ethanol on NADPH Oxidase Activity, Oxygen Utilization, and Benzphetamine Demethylation in the Reconstituted Enzyme System Containing Electrophoretically Homogenous P-450$_{LM2}$

Substrate added	Ethanol present	Change in Components			
		NADPH loss	O_2 uptake	H_2O_2 formation	CH_2O formation
		nmol/min/nmol/ P-450$_{LM2}$			
None	-	103	76	105	
"	+	90	76	111	
Benzphetamine	-	225	213	77	115
"	+	224	212	96	115

The P-450$_{LM2}$ and reductase preparations used in these studies were the same as those employed in the experiments depicted in Tables 1 and 2. When present, ethanol and benzphetamine were at a final concentration of 100 mM and 1 mM, respectively. All reactions were initiated with NADPH and were carried out in duplicate at 37° following a 3-min equilibration period at the same temperature. In all such assays P-450$_{LM2}$ was the rate-limiting component. Rates of NADPH oxidation were determined in 1.0-ml reaction mixtures containing P-450$_{LM2}$ (0.1 μM), NADPH-cytochrome P-450 reductase (15 μg of protein per ml), dilauroyl-GPC (30 μg per ml), 14 mM $MgCl_2$, 0.1 M potassium phosphate buffer, pH 7.4, and NADPH (0.15 mM) as the final addition. Oxygen consumption was measured under the same conditions, but in 3-ml reaction mixtures containing NADPH at a final concentration of 0.3 mM; a Clark type electrode was used in conjunction with a Heath EU-20B recorder. Identical reaction mixtures, but in a total volume of 2.0 ml, were incubated in open tubes with NADPH (1.0 mM) for 5 min after 37°; after termination of the reactions with 0.5 ml of 30% TCA, a 1.0-ml aliquot of the TCA supernatant fraction was removed and assayed for H_2O_2 as described in Fig.2, while a second 1.0-ml aliquot was assayed for formaldehyde as described in Table 2. Zero-time incubations contained all components but received 0.7 ml of 30% TCA followed ty NADPH and were placed in an ice bath. Rates of H_2O_2 and formaldehyde formation were linear for 5 min under these conditions. Values shown are the average of two different experiments.

catalase (32), ethanol oxidation was obtained at rates proportional to H_2O_2 generation rates (Table 4). These findings strongly suggest an indirect role of cytochrome P-450$_{LM}$ in ethanol oxidation as observed in the intact microsomal system, namely, the generation of H_2O_2 which is required for the peroxidatic activity of catalase, also found in microsomes. The NADPH-linked generation of H_2O_2 in liver microsomes is well established (11,30,33,34). In addition, our laboratory has demonstrated that the "NADPH oxidase" activity (35) of microsomes can be effectively reconstituted by the presence of both NADPH-cytochrome P-450 reductase and highly purified P-450_{LM} (16,17). In view of these results, inhibition of NADPH-linked ethanol oxidation in hepatic microsomes by inhibitors of cytochrome P450_{LM}-mediated hydroxylation, such as carbon monoxide (15,30) and antibodies to P-450_{LM}, does not necessarily imply that P-450_{LM} is involved as the direct catalyst in the oxidation of ethanol to acetaldehyde. Thus, the possibility must be considered that these effects are produced by inhibition of P-450_{LM}-dependent formation of H_2O_2 and its consequent unavailibility for the peroxidatic reaction of catalase. A similar conclusion has been reached by Estabrook and colleagues (36), who demonstrated that the inhibitory effect of metyrapone, a well-characterized inhibitor of P-450_{LM} function, on NADPH-dependent methanol oxidation in rat liver microsomes was due to an inhibition of H_2O_2 formation rather than to a direct effect on methanol oxidation involving P-450_{LM}.

TABLE 4
Ethanol Oxidation by Crystalline Beef Liver Catalase and H_2O_2 Generated from NADPH and P-450_{LM2} in the Reconstituted Enzyme System

H_2O_2-Generating System	H_2O_2 Formation[a]	Ethanol oxidation[b]
	μM product	min^{-1}
0.1 μM P-450_{LM2} + 1.0 mM benzphetamine	2.7	3.1
0.3 μM P-450_{LM2} + 1.0 mM benzphetamine	7.8	8.5

a. NADPH-cytochrome P-450 reductase and dilauroyl-GPC were added as described in Fig. 2 for incubations containing 0.1μM P-450_{LM2} and 0.3 μM, LM_2, respectively, in a total volume of 1.5 ml of 0.1 M phosphate buffer, pH 7.4, with 15 mM $MgCl_2$ and 100 mM ethanol. NADPH (1.0 mM) was added to start the reaction and incubations were for 15 min at 37°. H_2O_2 was assayed as described in Fig. 2.

b. Reaction mixtures were identical to those used for the H_2O_2 generation studies above, except that the final incubation volume was 3.0 ml. Crystalline beef liver catalase was also added to a final concentration of 1.5 mg/ml (6.1 μM) and reactions were for 15 min at 37° *in the presence of NADPH (1.0 mM). Incubations were carried out in stoppered flasks, and acetaldehyde was trapped in semicarbazide and estimated after overnight diffusion at room temperature as described in Table 1. The absorbance at 224 nm zero-time incubations was 50% and 42% of corresponding experimental incubations containing 0.1 μM* $P\text{-}450_{LM2}$ *and 0.3 μM* LM_2*, respectively. The* $p\text{-}450_{LM2}$ *and reductase preparations were the same as those used in the studies described in Figs. 1 and 2.*

III. ACKNOWLEDGEMENT

We are grateful to Barbara M. Michniewicz for skillful assistance and to Dr. Gerald d. Nordblom for his help in experiments involving the oxygen electrode.

IV. REFERENCES

(1). Lu, A.Y.H., and Coon, M.J., J. Biol. Chem. 243, 1331(1968).
(2). Coon, M.J., and Lu, A.Y.H., in "Microsomes and Drug Oxidations" (J.R. Gillette, A.H. Conney, G.J. Cosmides, R.W. Estabrook, J.R. Fouts, and G.J. Mannering, Eds.), p. 151, Acad.Press, New York, (1969).
(3). Lu, A.Y.H., Junk, K.W., and Coon, M.J., J. Biol. Chem. 244, 3714(1969).
(4). van der Hoeven, T.A., Haugen, D.A., and Coon, M.J., J. Biochem.Biophys.Res.Comm, 60, 569 (1974).
(5). Haugen, D.A., van der Hoeven, T.A., and Coon, M.J., J. Biol.Chem. 250,3567, (1975).
(6). Coon, M.J. Haugen, D.A., Guengerich, F.P., Vermilion, J.L., and Dean, W.L., in "The Structural Basis of Membrane Function" (Y. Hatefi and L. Djavadi-Ohaniance, Eds.), p. 409, Acad. Press, New York (1976).
(7). Haugen, D.A., and Coon, M.J., J. Biol.Chem., In Press (1976).
(8). Dean, W.L., and Coon, M.J., Fed. Proc. 35, 1535 (1976).
(9). Dean, W.L., and Coon, M.J., J.Biol.Chem. , In press, (1977).
(10). Coon, M.J., Vermilion, J.L., Vatsis, K.P., French, J.S., Dean, W.L., and Haugen, D.A., Amer. Chem. Soc., Symposium Series, In press (1976).
(11). Gillette, J.R., Adv. Pharmacol. 4,219 (1966).
(12). Wiebel, F.J., Selkirk, J.K., Gelboin, H.V., Haugen, D.A., can der Hoeven, T.A., and Coon, M.J., Proc.Natl. Acad.Sci. USA 72 , 3917 (1975).

(13). Nordblom, G.D., White, R.E. and Coon, M.J., Arch. Biochem. Biophys.175, 524 (1976).
(14). Vermilion, J.L., and Coon, M.J., Biochem.Biophys.Res. Commun. 60, 1315 (1974).
(15). Lieber ,C.S. and DeCarli, L.M., J. Biol. Chem. 245, 2505 (1970).
(16). Nordblom, G.D., and Coon, M.J., Fed. Proc. 35, 281 (1976).
(17). Nordblom, G.D. and Coon, M.J., Arch.Biochem.Biophys., in press (1977).
(18). Kawalek, J.C., Levin, W., Ryan, D., Thomas, P.E., and Lu, A.Y.H., Mol.Pharmacol. 11, 874 (1975).
(19). Hashimoto, C., and Imai, Y., Biochem.Biophys.Res.Comm. 68, 821 (1976).
(20). Lieber, C.S., and DeCarli, L.M., Life Sci. 9 (Part II), 267 (1970).
(21). Khanna, J.M.,Kalant, H., and Lin, G., Biochem.Pharmacol. 21, 2215 (1972).
(22). Rubin, E., Lieber, C.S., Alvares, A.P., Levin, W., and Kuntzman, R., Biochem.Pharmacol. 20, 229 (1971).
(23). Mannering, G.J., in "Fundamentals of Drug Metabolism and Drug Disposition" (B.N. LaDu, H.G. Mandel, and E.L. Way, Eds.), p. 206, Williams and Wilkins Company, Baltimore (1971).
(24). Mieyal, J.J., Ackerman, R.S., Blumer, J.L., and Freeman, L.S., J.Biol.Chem. 251,3436 (1976).
(25). Nash, T., Biochem.J. 55, 416 (1953).
(26). Cochin, J., and Axelrod, J., J.Pharmacol.Exp.Ther. 125, 105 (1959).
(27). Beers, R.F., and Sizer, I.W., J. Biol. Chem. 195, 133 (1952).
(28). Teschke, R., Hasumura, Y., and Lieber, C.S., Arch.Biochem. Biophys. 163, 404 (1974).
(29). Rubin, E., Gang, H., Misra, P.S., and Lieber, C.S., Am. J. Med. 49, 801 (1970).
(30). Thurman, R.G., Ley, H.G., and Scholz, R., Eur.J.Biochem. 25, 420 (1972).
(31). Imai, Y., and Sato, R., J. Biochem. 62, 239 (1967).
(32). Oshino, N., Oshino, R., and Chance, B., Biochem. J. 131, 555 (1973).
(33). Hildebrandt, A.G., and Roots, I., Arch. Biochem.Biophys. 171, 385 (1975).
(34). Werringloer, J., and Estabrook ,R.W., Hoppe-Seyler's Z. Physiol.Chem.357, 1063 (1976).
(35). Gillette, J.R., Brodie, B.B., and LaDu, B.N., J. Pharmacol. Exp. Ther. 119, 532 (1957).
(36). Werringloer, J., Chacos, N., Estabrook, R.W., Roots, I., and Hildebrandt, A.G., in "Alcohol and Aldehyde Metabolizing Systems" Vol. II (R.G. Thurman, T. Yonetani, J.R. Williamson and B. Chance, Eds.) Acad.Press, New York, in press, (1977).

EVIDENCE FOR THE DIRECT INVOLVEMENT OF HEPATIC CYTOCHROME P450 IN ETHANOL METABOLISM

Gerald T. Miwa, Wayne Levin
Paul E. Thomas and Anthony Y. H. Lu

Hoffmann-LaRoche Inc.

A radiometric assay for the [1-^{14}C] -acetaldehyde produced by the oxidation of [1-^{14}C] ethanol has been developed to study ethanol metabolism by liver microsomes and by a purified, reconstituted system containing cytochrome P450, NADPH-cytochrome c reductase and phosphatidylcholine. The metabolism of ethanol to acetaldehyde by the reconstituted system has also been verified by gas chromatographic and spectrophotometric assays. The reconstituted system is free of contaminating catalase and alcohol dehydrogenase. Furthermore, antibodies specific for cytochrome P450 have been used to demonstrate a cytochrome P450-mediated pathway for ethanol metabolism in both the reconstituted and microsomal systems. It is concluded that ethanol is directly oxidized to acetaldehyde by the cytochrome P450-containing drug metabolizing system.

I. INTRODUCTION

Ethanol is metabolized primarily by alcohol dehydrogenase (ADH) *in vivo* (1). Ethanol metabolis,however, also occurs *in vitro* in microsomal preparations which are devoid of ADH. Orme-Johnson and Ziegler were the first to demonstrate the microsomal origin of this ADH-independent pathway of ethanol metabolism (2). Since the microsomal ethanol oxidizing system has some properties in common with the drug-metabolizing enzyme system, Lieber and DeCarli suggested that ethanol was primarily metabolized by the cytochrome P450-containing monooxygenase in microsomal preparations (3).

Catalase, however, is a known contaminant in microsomal preparations and both catalase-dependent (4-6) and catalase-independent (2,3,7) pathways of ethanol metabolism have been reported. The catalase-dependent route of ethanol metabolism requires the presence of H_2O_2 for the peroxidative metabolism of ethanol (8). Since microsomal NADPH-dependent H_2O_2 generation occurs at a rate greater than the rate of ethanol metabolism (9) and catalase can function peroxidatively in ethanol oxidation, the significance of a catalase-independent pathway of microsomal ethanol oxidation has been questioned (4-6,9).

The purpose of the present study was to examine more carefully the possible role of cytochrome P450 in hepatic microsomal ethanol metabolism. Antibodies, specific to purified rat liver cytochromes P450 and P448, have been used to demonstrate a cytochrome P450-dependent pathway of ethanol oxidation in microsomes. In addition, a catalase and ADH-free, reconstituted system composed of a homogenous preparation of NADPH-cytochrome c reductase, synthetic phospholipid and a highly purified preparation of cytochrome P450 has been used to demonstrate the direct involvement of cytochrome P450 in ethanol oxidation.

III. ACETALDEHYDE QUANTITATION

A simple radiometric assay was developed to measure the acetaldehyde formed during ethanol metabolism [1-^{14}C] -ethanol (1μCi, 46 mM) was incubated in closed reaction vials with either microsomes or the reconstituted system in addition to the approporiate cofactors. The incubations were done in scintillation vials which were sealed with rubber vial stoppers. A plastic center well containing a piece of filter paper, previously soaked in a solution of semicarbazide hydrochloride and air dried, was suspended over the reaction mixture from the rubber stopper. After incubating under the appropriate conditions, the reaction was terminated with either trichloroacetic acid or sodium hydroxide. Following an overnight incubation at 37°C to form the semicarbazone derivative, the filter papers were removed and air dried to remove traces of [1-^{14}C]-ethanol. The [1-^{14}C] -acetaldehyde semicarbazone derivative was then quantitated by liquid scintillation. The recovery [1-^{14}C]-acetaldehyde as the semicarbazone derivative was determined with authentic (1,2-^{14}C)-acetaldehyde (New England Nuclear Corp.,).

The identity of the [1-^{14}C] -acetaldehyde metabolite was confirmed by thin layer chromatography and mass spectral analysis of the semi-carbazone derivative. Thin layer chromatography of the ethanol metabolite semicarbazone revealed only a single peak with an R_f value (0.60)[1] identical to an acetaldehyde-semicarbazone standard. The metabolite and authentic acetaldehyde semicarbazones were extracted from the plates with ethanol and dried. The mass spectra of the residues are shown in Figures 1A and 1B. Electron impact ionization of the samples

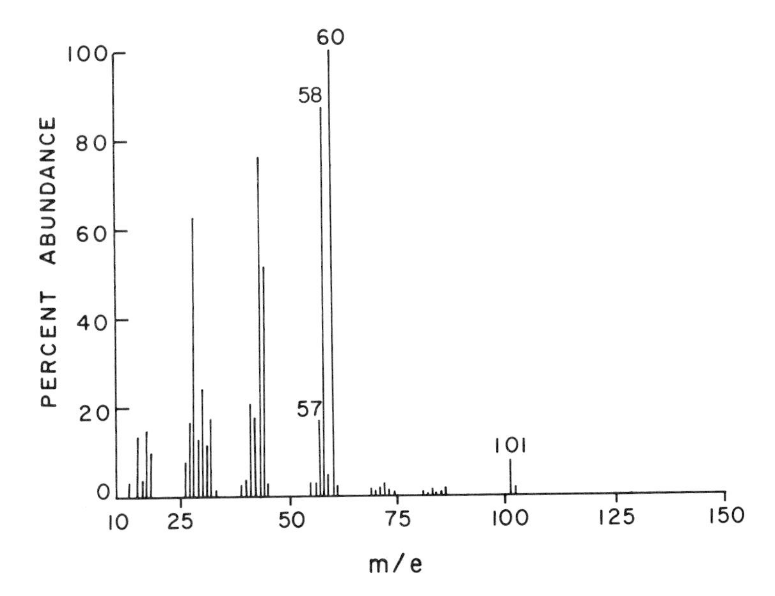

Fig. 1A.

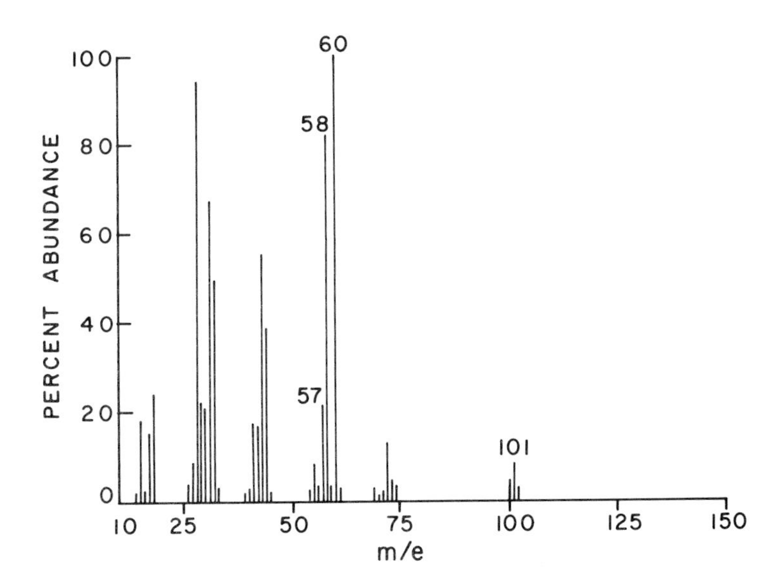

Fig. 1B.(Legend for both figures). Mass spectra of authentic acetaldehyde and ethanol metabolite semicarbazone derivatives. Samples of authentic acetaldehyde semicarbazone and the ethanol metabolite derivative were introduced into the mass spectrometer by a direct insertion probe. Mass spectra, without background subtraction, produced by electron impact ionization of the

authentic acetaldehyde semicarbazone (Fig. 1A) and ethanol metabolite semicarbazone (Fig. 1B) derivatives are illustrated.

gave molecular ions, m/e 101, base peaks, m/e 60, and fragment ions, m/e 58, 57 with relative intensities 8%, 8%; 100%, 100%; 88%; 82%; and 17%. 21% for the acetaldehyde and metabolite semicarbazones respectively.

IV. COMPARISON OF VARIOUS ASSAY METHODS FOR MEASURING ACETALDEHYDE

In addition to the radiometric assay for acetaldehyde form - ation, UV absorbance of the semicarbazone derivative (3) and gas chromatography (GC) of the free acetaldehyde have been used to confirm the production of acetaldehyde by the reconstituted system.We are indebted to Drs. Lieber and Ohnishi of the Bronx VA Hospital for measuring acetaldehyde by their HC assay. We have also employed incubation conditions similar to those used by Drs. Vatsis and Coon which are reported elsewhere in this Symposium. Under these conditions, (0.525 nmol rabbit cytochrome P450 LM_2, 150 μg dilaruoylglyceryl-3-phosphorylcholine, 7000 units NADPH-cytochrome c reductase and 50 mM ethanol incubated with an NADPH-generating system for 10 min at 37°C) only slight metabolism, corresponding to 3-6 nmol. min^{-1}. nmol P450 LM_2, is detectible by the UV assay. One unit of NADPH-cytochrome c re- ductase activity is defined as the amount of enzyme that will catalyze the reduction of cytochrome c at an initial rate of 1 $nmol.min^{-1}$ at 22°C with a saturating concentration of NADPH. The rate of metabolism of ethanol to acetaldehyde by the re- constituted system containing cytochrome P450 from phenobarbital- treated rats was determined to be 9.4, 22.5 and 21.7 $nmol.min^{-1}$. nmol $P450^{-1}$ for the radiometric, UV and GC assay respectively, when the reactions were terminated with NaOH. When the reaction was quenched with a solution of perchloric acid and thiourea and assayed by GC as routinely done by Drs. Lieber and Ohnishi, the turnover number was 19.2 nmol.$min.^{-1}$.nmol P450. The value for the radiometric assay has not been corrected for recovery since the samples of commercial (1,2-^{14}C)-acetaldehyde used to estimate recovery was not pure. Preliminary evidence indicates that the purity of the (1,2-^{14}C)-acetaldehyde standard was 55.5%. Based on this value for purity, the recovery would be 43.6% resulting in a calculated turnover number of 21.6 nmol.$min.^{-1}$ nmol $P450^{-1}$, in agreement with the values obtained by either GC or spectrophotometric analysis. Standard curves, using (1,2-^{14}C)-acetaldehyde are linear when either NaOH or TCA is used indicating a constant recovery factor over the concen- tration range investigated. Since the acetaldehyde purity is based on preliminary evidence, we have chosen to omit the recovery factor in calculations of acetaldehyde determined by the radiometric assay. All values reported in this communication were determined by this procedure and have not been corrected for recovery. It is clear, however, that irrespective of the assay method employed, ethanol is metabolized to acetaldehyde by the reconstituted system containing cytochrome P450.

V. THE EFFECT OF ANTIBODIES TO CYTOCHROME P450 AND CYTOCHROME P448 ON MICROSOMAL ETHANOL METABOLISM

Antibodies to highly purified preparations of cytochrome P450 from phenobarbital-treated rats and cytochrome P448 from 3-methylcholanthrene-treated rats have been recently described (10,11). These antibodies do not cross react with either purified beef catalase or partially purified rat liver catalase and rat microsomal electron transport components such as NADPH-cytochrome c reductase, cytochrome b_5 and NADH-cytochrome b_5 reductase.

The specificity of these antibodies for the cytochrome P450 component of microsomal electron transport systems permits a method for either implicating or excluding cytochrome P450 in ethanol metabolism. Table I summarizes the data obtained with microsomes from phenobarbital-treated rats and vaious antibody preparations. When microsomes are incubated with a NADPH-generating system and 1 mM azide, a catalase inhibitor, ethanol is oxidized to acetaldehyde at a rate of 4.7 $nmol.min^{-1}. mg^{-1}$ protein. When sodium azide is omitted, the rate increases to 8.5 $nmol.min^{-1}.mg^{-1}$. The difference 3.8 $nmol.min^{-1}. mg^{-1}$, is presumably due to catalase whereas the rate observed in the presence of sodium azide may be due to a cytochrome P450-dependent pathway. Thus, experiments involving the inhibition by cytochrome P450 and P448 antibodies of NADPH-dependent microsomal ethanol metabolism were conducted in the presence of 1 mM sodium azide.

Control γ-globulins isolated from the sera of untreated rabbits, do not significantly inhibit microsomal ethanol metabolism. In trast, antibodies isolated from the sera of rabbits treated with highly purified preparations of either cytochrome P450 or cytochrome P448 markedly inhibit ethanol metabolism.

Antibodies to both cytochrome P450 and cytochrome P448 do not cross react on Ouchterlony plates with rat liver catalase. To verify that the inhibition of ethanol metabolism by these antibody preparations does not result from the inhibition of catalase in the microsomal preparations, the effect of these antibodies on ethanol metabolism in the presence of an H_2O_2-generating system was also examined (Table I).

Ethanol is oxidized by the microsomal system in the presence of an H_2O_2-generating system via the Kielin-Hartree reaction (8). Sodium azide (1 mM) inhibits this H_2O_2-dependent route of ethanol oxidation by greater than 90%. Similar observations have also been reported by others (2,3,8) and demonstrate a catalase-dependent pathway of microsomal ethanol metabolism.

The lower rate of apparent catalase-dependent ethanol oxidation obtained in the presence of an H_2O_2-generating system relative to an NADPH-generating system probably reflects a lower rate of H_2O_2 production from the former system. The precise measurement of the catalase-dependent rate of ethanol metabolism in liver microsomes is difficult since it requires precise control over the H_2O_2-generation rate/catalase heme ratio. This has been demonstrated by Oshino *et.al.* (6).

In contrast to the inhibition obserbed with an NADPH-generating system, antibodies against cytochrome P450 are without effect on ethanol metabolism supported by an H_2O_2-generating system. These data confirm the absence of a significant cross reaction between these antibody preparations and catalase. Thus, the antibody experiments indicate that microsomal ethanol oxidation is at least partially mediated through a cytochrome P450-dependent pathway.

Since microsomes are capable of oxidizing NADPH to H_2O_2, several investigators have suggested that microsomal components, such as cytochrome P450, may function indirectly in ethanol metabolism by generating H_2O_2 from NADPH. Thus, sequential reactions involving the P450-dependent oxidation of NADPH to H_2O_2 followed by the catalase-dependent peroxidative metabolism of ethanol and H_2O_2 could explain an apparent cytochrome P450 catalyzed pathway. The results obtained with the microsomal system and cytochrome P450 specific antibodies are compatible with this scheme and do not differentiate between the direct or the indirect involvement of cytochrome P450 in microsomal ethanol oxidation.

VI. ETHANOL METABOLISM BY THE RECONSTITUTED P450 SYSTEM

Purified liver microsomal NADPH-cytochrome c reductase (12, 13), and cytochrome P450 purified from different species (14-17). were used to distinguish between the direct and indirect involvement of cytochrome P450 in ethanol oxidation. Table I summarizes data obtained with a reconstituted system composed of purified preparations of rat cytochrome P450, NADPH-cytochrome c reductase and synthetic phospholipid. Under optimal conditions, the rate of ethanol metabolism, uncorrected for acetaldehyde recovery, is about 10.5 $nmol.min.^{-1}.nmol$ cyto-

TABLE I

Effect of Cytochrome P450 specific antibodies on ethanol metabolism by microsomes and the reconstituted cytochrome P450 system. Microsomes (1 mg protein) from PB induces rats were incubated for 60 min at 37°C in sealed reaction vials with either an NADPH-generating system or an H_2O_2-generating system and with the antibody preparation (20 mg protein) indicated. Sodium azide (1 mM) was included, except where noted, in samples using the NADPH-generating system but was omitted from the samples using the H_2O_2-generating system. The reconstituted system was composed of 0.1 nmol purified cytochrome P450 from PB-treated rats, 1000 units of NADPH-cytochrome c reductase and 5 μg dilaurolyglyceryl-3-phosphorylcholine in addition to 1 mM EDTA, 0.1 M phosphate buffer, pH. 7.4 and 46 mM ($1^{-14}C$)-ethanol. Control and anti PB450 antibodies were added to a concentration equivalent to 5 mg antibody protein per nmol cytochrome P450. Sodium azide was omitted, except where noted, and the samples were incubated for 30 min at 37°C. The total incubation volume for either the microsomal or reconstituted system was 1.5 ml. All values are uncorrected for acetaldehyde recovery.

	Microsomes						Reconstituted Sytem		
	NADPH-Generating System			H_2O_2-Generating system			NADPH-Generating System		
Antibody	NaN_3	V^a	% Control	NaN_3	V^a	% Control	NaN_3	V^b	% Control
None	+	4.7	100	-	2.23	100	-	10.5	100
	-	8.5	181	+	0.14	6.3	+	10.8	103
Control IgG	+	4.6	98.0	-	2.06	92.3	-	9.75	92.8
PB P450 IgG	+	1.12	23.8	-	2.14	96.0	-	2.52	24.0
MC P448 IgG	+	1.73	36.8	-	2.15	96.4			

a. *The rate of acetaldehyde formation during the incubation with microsomes is expressed as $nmol.min.^{-1}.mg$ protein.*

b. *The rate of acetaldehyde formation during the incubation with the reconstituted system is expressed as $nmol.min.^{-1}.nmol\ P450^{-1}$.*

chrome $P450^{-1}$. This rate is greater than the azide insensitive rate observed in microsomes when normalized to the cytochrome P450 content. This is not surprising, however, since the reconstituted system offers the advantage of manipulating the

TABLE 2

Properties of the reconstituted cytochrome P450-dependent ethanol oxidation system. The reconsituted system was the same as described in Table 1. The samples were incubated for 30 min at 37°C with the components indicated and the resulting (1-^{14}C)-acetaldehyde assayed as described in Section III. The NADPH-generating system was composed of 1.5 mol NADP, 7.5 mol glucose-6-phosphate and 0.7 units of glucose-6-phoshphate dehydrogenase. The H_2O_2-generating system was composed of 15 mol glucose and 1.5 μg glucose oxidase.

Incubation Mixture	% Control
I. Complete = NADPH Generating System	100
- Rat PB P450	2.3
- Reductase	6.5
- Phosphatidylcholine	48.9
+ NaN_3 (1 mM)	93.1
+ Benzphetamine (1 mM)	19.4
II. Complete - NADPH Generating System	0.0
+ NADPH (0.3 mM)	103
+ NAD (0.3 mM)	0.0
+ H_2O_2-Generating System	15.2
+ H_2O_2-Generating System + NaN_3 (1 mM)	15.6

reaction components to provide optimal reaction conditions under which the rate is determined only by the cytochrome component. Thus, the incubation conditions used in the present study result in reaction rates which depend only on the cytochrome P450 concentration since the other components are in excess. A comparable situation may not exist in microsomes.

In contrast to the microsomal system, ethanol metabolism in the reconstituted system is unaffected by 1 mM sodium azide indicating the absence of contaminating catalase. Antibody to cytochrome P450, however, is equally effective in inhibiting ethanol oxidation by either system.

Ethanol metabolism in the reconstituted system is not supported by cofactors for either ADH or catalse (Table 2). For example, neither NAD^+ nor an H_2O_2-generating system is significantly effective in supporting ethanol oxidation. Furthermore, sodium azide is without effect on either NADPH or H_2O_2-dependent ethanol oxidation. In contrast, benzphetamine (1 mM), a typical

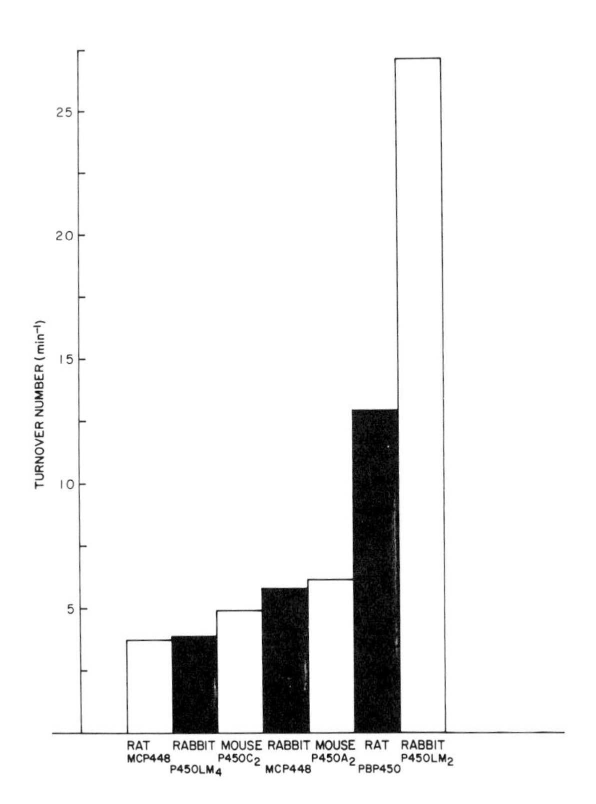

Fig. 2. Oxidation of ethanol by various cytochrome P450 preparations. The incubation mixture contained 0.1 nmol of various purified cytochrome P450 preparations, 1000 units of NADPH-cytochrome c reductase, 5 g phospholipid, 1 mM EDTA, an NADPH-generating system, and 0.1 M phosphate buffer, pH. 7.4 in a total volume of 1.5 ml. The reactions were initiated with (1-^{14}C)-ethanol (46 mM) and terminated after a 3o min incubation at 37°C with sodium hydroxide. The rate of (1 -^{14}C)-acetaldehyde formation, uncorrected for recovery is expressed as nmol. min^{-1}.nmol P450.

cytochrome P450 substrate, effectively inhibits ethanol metabolism. Furthermore, maximal catalytic activity requires all three components of the cytochrome P450-dependent monooxygenase system and NADPH as originally described by Lu *et.al.* (18).

The turnover numbers for cytochrome P450 preparations purified from a variety of sources are summarized in Fig. 2. The turnover numbers, uncorrected for acetaldehyde recovery and obtained under identical incubation conditions, range over nearly

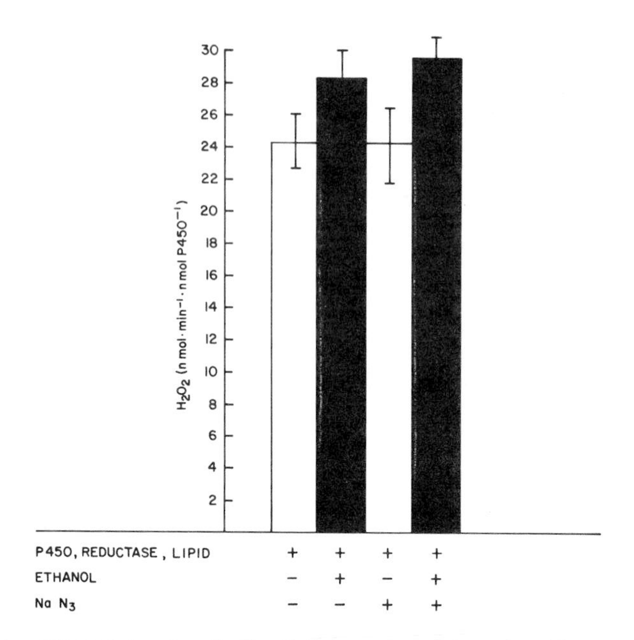

Fig. 3. NADPH-depdendent H_2O_2 generation by the reconstituted system. The incubation system contained 0.1 nmol cytochrome P450 from PB-treated rats, 1000 units NADPH-cytochrome c reductase, 5 µg phospholipid, 1 mM EDTA and 0.1 M phosphate buffer pH 7.4 in a total volume of 1.5 ml. Ethanol (46 mM) and/or sodium azide (1 mM) were added as indicated. The reactions were initiated by adding an NADPH-generating system and were terminated after intervals of 10 and 20 min with TCA. The H_2O_2 generated was assayed by the ferrithiocyanate method (19) and the mean and standard deviation for duplicate samples are indicated.

an order of magnitude and are generally greatest for heme proteins derived from phenobarbital-treated animals and lowest for heme proteins derived from 3-methylcholanthrene-treated animals.

Rabbit P450 LM_4 is a notable exception, however, since it was obtained from rabbits pretreated with phenobarbital. The spectral properties and molecular weight of this heme protein are, however, equivalent to a form of cytochrome P448 isolated from untreated, β-naphtoflavone-treated and 3-methylcholanthrene-treated rabbits and, therefore, Haugen and Coon (14) have concluded that these heme proteins are identical. The similarity between the turnover number for ethanol metabolism observed for rabbit P450 LM_4 and cytochrome MC P448 from 3-methylcholanthrene-treated rabbits is consistent with this conclusion.

In the absence of ethanol, NADPH is oxidized by the reconstituted system resulting in the production of H_2O_2 (Fig.3). The H_2O_2-generated *in situ* (24 $nmol.min^{-1}$. nmol $P450^{-1}$) does not result in the oxidation of ethanol by a Kielin-Hartree-like mechanism since the rate of H_2O_2 production is not decreased when ethanol is added to the system. In addition, the lack of effect of sodium azide on H_2O_2 production is consistent with the absence of catalase in the reconstituted system.

The generation of H_2O_2 from NADPH has also been reported by others (9) for the microsomal system and has served as a basis for excluding a catalase-independent pathway of microsomal ethanol metabolism. Our data, however, indicate that antibodies specific to cytochrome P450 selectively inhibit the cytochrome P450 component of microsomal ethanol oxidation. In addition, results obtained with a highly purified and well defined reconstituted cytochrome P450 system substantiate the experiments with antibodies in the microsomal system and indicate that cytochrome P450 is capable of directly catalyzing the oxidation of ethanol to acetaldehyde.

VII. ACKNOWLEDGEMENTS

We would like to thank Drs. M.J. Coon and K.P. Vatsis, University of Michigan, for many helpful discussions and samples of purified rabbit liver cytochromes P450 LM_2 and LM_4, Drs. C. Lieber and K. Ohnishi of the Bronx VA Hospital for measuring acetaldehyde by their GV method and Ms. D. Ryan, Ms. S. West and Dr. M. T. Huang for providing many of the enzymes used in this study. We wish to also thank Ms. C. Caso for the typing of this manuscript.

VIII. REFERENCES

(1). Mezey, E.,Biochem.Pharmacol. 25, 869 (1976).

(2). Orme-Johnson, W.H., and Ziegler, D.M., Biochem.Biophys. Res.Commun. 21, 78 (1965).

(3). Lieber, C.S., and DeCarli, L.M., J.Biol.Chem. 245,2505 (1970).

(4). Roach, M.K., Reese, W.N., and Creaven, P.J., Biochem.Biophys. Res. Commun. 36,596 (1969).

(5). Isselbacher, K.J., and Carter, E.A., Biochem.Biophys. Res. Commun. 39, 530 (1970).

(6). Oshino, N., Oshino, R., and Chance, B., Biochem.J. 131, 555 (1973).

(7). Teschke, R., Hasumura, Y., and Lieber ,C.S., J. Biol. Chem. 250, 7397 (1975).

(8). Keilin, D., and Hartree, E.F., Biochem. J. 39, 293 (1945).

(9). Thurman, R.G., Ley, H.G., and Scholz, R., Eur.J.Biochem. 25, 420 (1972).
(10). Thomas, P.E., Lu, A.Y., Ryan, D., West, S., Kawalek, J., and Levin, W., Mol.Pharmacol.12, 746 (1976).
(11). Thomas, P.E., Lu, A.Y., Ryan, D., West, S., Kawalek, J., and Levin, W., J. Biol. Chem. 251,1385 (1976).
(12). Dignam, J.D., and Strobel, H.W., Biochem.Biophys.Res.Commun. 63, 845 (1975).
(13). Yasukochi, Y., and Masters, B.S.S., J. Biol. Chem., 251, 5337 (1976).
(14). Haugen, D.A.,and Coon, M.J., J.Biol.Chem. in press
(15). Ryan, D., Lu, A.Y.H., Kawalek, J., Est, S., and Levin, W., Biochem.Biophys.Res. Commun, 64, 1134 (1975).
(16). Kawalek, J., Levin, W., Ryan, D., Thomas, P.E., and Lu,A.Y., H., Mol.Pharmacol. 11, 874 (1975).
(17). Huang, M.T., West, S., and Lu, A.Y.H., J.Biol. Chem. 251, 4659 (1976).
(18). Lu, A.Y.H., Junk, K.W., and Coon, M.J., J. Biol. Chem. 244, 3719 (1969).
(19). Hildebrandt, A.G., and Roots, I., Arch.Biochem.Biophys. 171, 385 (1975).

ADVANTAGES OF GAS CHROMATOGRAPHIC DETERMINATION OF ACETALDEHYDE IN STUDIES WITH ETHANOL AND A RECONSTITUTED ENZYME SYSTEM CONTAINING ELECTROPHORETICALLY HOMOGENEOUS PHENOBARBITAL-INDUCIBLE CYTOCHROME P-450 FROM RABBIT LIVER MICROSOMES

K.P. Vatsis and R.G. Thurman

University of Michigan
University of Pennsylvania

High purified rabbit liver cytochrom P-450 incubated in a reconstituted mixed function oxidase system failed to oxidize ethanol.

The original observation by Orme-Johnson and Ziegler (1) that hepatic microsomes catalyze the O_2- and NADPH-dependent oxidation of ethanol to acetaldehyde led to numerous suggestions from subsequent studies with inhibitors that catalase (2-4) and alcohol dehydrogenase (ADH) (3) are responsible for this activity. Based on the differential effects of azide on ethanol oxidation supported by NADPH as compared to an H_2O_2-generating system, Lieber and DeCarli (5) proposed that the NADPH-linked system operates independently of catalase but is instead akin to the liver microsomal cytochrome P-450 (P-450_{LM})-containing drug-metabolizing system. However, Thurman et al. (6) demonstrated that H_2O_2 produced from NADPH and O_2 by P-450_{LM} is utilized further for the peroxidatic reaction of catalase, while detailed studies by Oshino et al. (7) established that statements concerning the involvement or lack of involvement of catalase in the NADPH-dependent oxidation of ethanol cannot be made (5) without a precise knowledge of the ratio of H_2O_2 generation rates to catalase heme concentration. Subsequently, some investigators detected ethanol oxidation in reconstituted enzyme systems containing crude solubilized preparations of P-450_{LM} which were free of catalase and ADH (8-10), whereas others failed to obtain similar results with identical preparations (11). Since the purity of those preparations (7-11) was at best questionable, the controversy remained unresolved. Furthermore, though Teschke et al. (12) have concluded that the NADPH-dependent oxidation of higher aliphatic alcohols by microsomes may be taken as evidence for the participation of P-450_{LM} in the oxidation of all alcohols, several papers presented at this symposium (13-15) and elsewhere (16,17,18) strongly suggest that "microsomal" ADH (19) may be solely responsible for the metabolism of propanol, butanol, and pentanol.

The question of whether $P\text{-}450_{LM}$ is capable of catalyzing the oxidation of ethanol to acetaldehyde directly can be answered satisfactorily only with the use of highly purified $P\text{-}450_{LM}$ preparations isolated according to procedures recently published by both Coon's (20,21) and Lu's (22) groups. The purpose of these experiments, therefore, was to examine the activity of electrophoretically homogeneous rabbit $P\text{-}450_{LM2}$ (20,21) toward ethanol in a reconstituted enzyme system containing highly purified NADPH-cytochrome P-450 reductase, synthetic phospholipid, and NADPH; acetaldehyde formation was quantitated using the highly specific head-space gas chromatographic technique. That the $P\text{-}450_{LM2}$ and reductase preparations employed in these studies were functional catalytically was established by examining the activity of the reconstituted enzyme system with classical mixed function oxidase substrates. As may be seen in the Table, $P\text{-}450_{LM2}$ catalyzed the N-demethylation of benzphetamine at very high rates in agreement with previous reports (28). It is noteworthy that the turnover number for benzphetamine at 30° (Table) is proportionately less than that previously obtained at 37° under similar incubation conditions but with different $P\text{-}450_{LM2}$ and reductase preparations (see Table 3 of ref. 29), indicating that different $P\text{-}450_{LM2}$ preparations do not differ significantly in their catalytic activities. Moreover, rates of cyclohexane hydroxylation by $P\text{-}450_{LM2}$ (Table) are similar to those previously reported by Nordblom and Coon (26). In sharp contrast, the same $P\text{-}450_{LM2}$ preparation was unable to catalyze the conversion of ethanol to acetaldehyde in the reconstituted enzyme system incubated for 30 min at 37° in the presence of NADPH or an NADPH-generating system. The amount of acetaldehyde formed from ethanol under these conditions was only two-fold higher than that seen in corresponding zero-time controls, and, most importantly, was at the limit of detection by the technique employed. These findings confirm those of Vatsis and Coon (29) who used the semicarbazone method for the estimation of acetaldehyde formed from ethanol by $P\text{-}450_{LM2}$ in the reconstituted enzyme system under identical incubation conditions. The slightly higher turnover numbers (4 to 5 nmol acetaldehyde/min/nmol $P\text{-}450_{LM2}$) reported by the latter investigators (29) most likely represent an overestimate of the ethanol-oxidizing capacity of the rabbit reconstituted

TABLE
*Catalytic Activity of Electrophoretically Homogeneous Rabbit P-450$_{LM2}$ in the Reconstituted Enzyme System Toward Various Substrates and Ethanol**

Compound Added	Activity[†]
	nmol/min/nmol P-450$_{LM2}$
Benzphetamine	59.0
Cyclohexane	42.5
Ethanol	1.0

**P-450$_{LM2}$, the phenobarbital-inducible form of rabbit liver microsomal cytochrome P-450, was purified to homogeneity (specific content, 18.0 nmol per mg of protein) according to previously published procedures (23). NADPH-cytochrome P-450 reductase was purified from phenobarbital-induced rabbit liver microsomes by affinity chromatography on 2',5'-ADP-Agarose (24); the preparation appeared to be homogeneous by SDS-polyacrylamide gel electrophoresis, and catalyzed the reduction of 38 μmol of cytochrome c per min per mg of protein at 30° when assayed as described by Vermilion and Coon (25). One unit of reductase is defined as the amount which catalyzes the reduction of 1.0 μmol of cytochrome c per min under these conditions.*

†*In all cases reactions were initiated by the addition of NADPH and were carried out in duplicate at 30° (benzphetamine, cyclohexane) or 37° (ethanol) following a 3-min equilibration period at the respective temperature; P-450 $_{LM2}$ was always the rate-limiting component. In experiments with benzphetamine and ethanol, duplicate zero-time control incubations contained all of the components but were quenched before the addition of NADPH. All assays were carried out using the same P-450$_{LM2}$ and reductase preparations, and the activities shown represent the average of two separate experiments. Benzphetamine N-demethylation was determined in 2.0-ml reaction mixtures containing P-450$_{LM2}$ (0.1 μM), electrophoretically homogeneous NADPH-cytochrome P-450 reductase (1.1 units per ml), dilauroylglyceryl-3-phosphorylcholine (dilauroyl-GPC; 30 μg per ml), 0.1 M potassium phosphate buffer, pH 7.4, 15 mM $MgCl_2$, 1.0 mM benzphetamine, and 1.0 mM NADPH. The reaction mixtures were incubated in open test tubes with shaking for 5 min at 30°; after termination of the reactions with 0.7 ml of 30% trichloroacetic acid, a 1.0-ml aliquot of the supernatant fraction was removed and assayed for formaldehyde as*

described previously (Vatsis and Coon, this volume). Cyclohexane hydroxylation was measured under the same conditions, but in 3.0-ml reaction mixtures containing 10 m$\underline{M}$ cyclohexane (added in 0.03 ml of methanol) and 0.17 m$\underline{M}$ NADPH. The increase in fluorescence (excitation, 366 nm; emission, 450 nm) was monitored at 30° using an Eppendorf recording spectrofluorometer. The value shown represents the cyclohexane-dependent NADPH oxidation which was obtained by subtracting the rate of NADPH oxidation in the absence of cyclohexane from that seen in the presence of this substrate; the cyclohexane-dependent oxidation of NADPH in the P-450$_{LM2}$-containing reconstituted system corresponds to the amount of cyclohexanol formed from cyclohexane under these conditions (26). Reaction mixtures of similar composition, except that dilauroyl-GPC was present at a final concentration of 10 μg per ml, were incubated in stoppered 25-ml Erlenmeyer flasks with 80 m$\underline{M}$ ethanol and 1.0 m$\underline{M}$ NADPH for 30 min at 37° with continuous shaking. Reactions were terminated by injecting 0.5 ml of 7% perchloric acid through the rubber septa, and the flasks were placed in a water bath at 60° for 60 min. Acetaldehyde was assayed by head-space gas chromatography (27) using a Hewlett-Packard model 5720A gas chromatograph equipped with a Porapak Q column (6' x 1/8") and a flame ionization detector (operating parameters: carrier gas flow, 35 ml per min; inlet temperature, 200°; column temperature, 125°; and detector temperature, 300°). The amount of acetaldehyde detected in zero-time controls was approximately 55% of that obtained for corresponding experimental incubations. The amount of acetaldehyde formed in reaction mixtures incubated without ethanol was 25% of that seen in the presence of ethanol. Moreover, similar results were obtained when NADPH was replaced by an NADPH-generating system composed of 5.0 m$\underline{M}$ glucose-6-phosphate, 1.0 m$\underline{M}$ NADP, and 0.7 unit of glucose-6-phosphate dehydrogenase per ml of reaction mixture.

system, since, as was pointed out previously (see Table 1 of ref. 29), blanks as high as 80% of the values obtained for experimental incubations had to be subtracted before calculating the turnover numbers in those experiments. Clearly, a reliable and accurate estimate of any enzymatic activity is not possible with such high blanks. In addition, caution should be exercised in expressing this negligible activity in terms of turnover numbers, since a requirement for P-450$_{LM2}$ was not demonstrated in those studies (29).

Despite the insignificant activity of electrophoretically homogeneous P-450$_{LM2}$ toward ethanol, ascertained both by the semicarbazone (29) and gas chromatographic methods (Table) for acetaldehyde detection in our laboratories, Miwa et al. (30) have reported turnover numbers of 60 for this form of rabbit P-450_{LM} incubated with ^{14}C-ethanol at 37° in the reconstituted enzyme system based on 45% recovery of ^{14}C-acetaldehyde (see ref. 30). This disparity in turnover numbers between our laboratories cannot be attributed to methodological differences, since Miwa et al. (30) obtained identical turnover numbers for ethanol metabolism by rat phenobarbital-inducible P-450_{LM} whether acetaldehyde formation was measured by the radiometric, semicarbazone, or gas chromatographic assay. Suffice it to say here that such high rates of ethanol oxidation by rabbit P-450_{LM_2} (30) are totally incompatible not only with the data presented here but also with the inability of ethanol to cause a corresponding increase in either NADPH utilization or O_2 consumption by the P-450_{LM_2}-containing reconstituted enzyme system under similar conditions (29). In other words, the apparent oxidation of ethanol to acetaldehyde (30) proceeds without the utilization of NADPH or O_2 by the reconstituted enzyme system in the studies by Miwa et al. (29), a finding which is incompatible with the stoichiometry of classical mixed function oxidases. Furthermore, ethanol at a final concentration (100 m$\underline{M}$) two-fold higher than that used by Miwa et al. (30) had no effect on benzphetamine N-demethylation by P-450_{LM_2} in the reconstituted enzyme system (29), indicating that ethanol cannot act as an 'alternate substrate' for P-450_{LM_2} as would be expected from the very high turnover numbers obtained for this compound (30). Though the reasons for these discrepancies (Table, 29,30) are not well understood, several studies currently in progress are aimed at their clarification.

REFERENCES

(1). Orme-Johnson, W.H., and Ziegler, D.M., Biochem. Biophys. Res. Commun. 21, 78 (1965).

(2). Roach, M.K., Reese, W.N., Jr., and Creaven, P.J., Biochem. Biophys. Res. Commun. 36, 596 (1969).

(3). Isselbacher, K.J., and Carter, E.A., Biochem. Biophys. Res. Commun. 39, 530 (1970).

(4). Khanna, J.M., Kalant, H., and Lin, G., Biochem. Pharmacol. 19, 2493 (1970).

(5). Lieber, C.S., and DeCarli, L.M., J. Biol. Chem. 245, 2505 (1970).

(6). Thurman, R.G., Ley, H.G., and Scholz, R., Eur. J. Biochem. 25, 420 (1972).

(7). Oshino, N., Oshino, R., and Chance, B., Biochem. J. 131, 555 (1973).
(8). Teschke, R., Hasumura, Y., Joly, J.G., Ishii, H., and Lieber, C.S., Biochem. Biophys. Res. Commun. 49, 1187 (1972).
(9). Mezey, E., Potter, J.J., and Reed, W.D., J. Biol. Chem. 248, 1183 (1973).
(10). Teschke, R., Hasumura, Y., and Lieber, C.S., Arch. Biochem. Biophys. 163, 404 (1974).
(11). Thurman, R.G., and Scholz, R., Drug Metab. Disp. 1, 441 (1973).
(12). Teschke, R., Hasumura, Y., and Lieber, C.S., J. Biol. Chem. 250, 7397 (1975).
(13). Vatsis, K.P., Chance, B., and Schulman, M.P., this volume.
(14). Vatsis, K.P., and Schulman, M.P., this volume.
(15). Brentzel, H.J., and Thurman, R.G., this volume.
(16). Carter, E.A., and Isselbacher, K.J., Ann. N.Y. Acad. Sci. 179, 282 (1971).
(17). Thurman, R.G., Federation Proceedings 36, 1640 (1977).
(18). Thurman, R.G., and Brentzel, H.J., Alcoholism - Clinical and Experimental Research 1, 33 (1977).
(19). Bechtold, M.M., Delwiche, C.V., Comai, K., and Gaylor, J.L., J. Biol. Chem. 247, 7650 (1972).
(20). van der Hoeven, T.A., Haugen, D.A., and Coon, M.J., Biochem. Biophys. Res. Commun. 60, 569 (1974).
(21). Haugen, D.A., van der Hoeven, T.A., and Coon, M.J., J. Biol. Chem. 250, 3567 (1975).
(22). Ryan, D., Lu, A.Y.H., Kawalek, J., West, S.B., and Levin, W., Biochem. Biophys. Res. Commun. 64, 1134 (1975).
(23). Haugen, D.A., and Coon, M.J., J. Biol. Chem. 251, 7929 (1976).
(24). Yasukochi, Y., and Masters, B.S.S., J. Biol. Chem. 251, 5337 (1976).
(25). Vermilion, J.L., and Coon, M.J., Biochem. Biophys. Res. Commun. 60, 1315 (1974).
(26). Nordblom, G.D., and Coon, M.J., Arch. Biochem. Biophys. 180, 343 (1977).
(27). Eriksson, C.J.P., Sippel, H.W., and Forsander, O.A., Anal. Biochem. 80, 116 (1977).
(28). Coon, M.J., Vermilion, J.L., Vatsis, K.P., French, J.S., Dean, W.L., and Haugen, D.A., in: "Drug Metabolism Concepts" (D.M. Jerina, Ed.), p. 46. American Chemical Society, Washington, D.C., 1977.
(29). Vatsis, K.P., and Coon, M.J., this volume.
(30). Miwa, G.T., Levin, W., Thomas, P., and Lu, A.Y.H., this volume.

RECONSTITUTION OF THE HEPATIC MICROSOMAL ETHANOL OXIDIZING SYSTEM (MEOS) IN CONTROL RATS AND AFTER ETHANOL FEEDING

Kunihiko Ohnishi and Charles S. Lieber

Sinai School of Medicine (CUNY), N Y.

1. An ethanol oxidizing system was reconstituted with cytochrome P-450, NADPH-cytochrome c reductase and phospholipid. This system also metabolizes prpoanol, butanol and benzphetamine. Neither catalase nor ADH play a role in this system. Characteristics of this system are similar to those of the microsomal ethanol oxidizing system (MEOS).

2. Chronic ethanol feeding of either male or female rats for about 4-12 weeks resulted in a significant quantitative and qualitative change in hepatic cytochrome P-450. The rise in cytochrome P-450 content was associated with preferential increase of Form I cytochrome P-450 by cyanide titration. SDS-polyacrylamide gel electrophoresis showed an induction of microsomal protein of 53,400 molecular weight, presumably an apoprotein of cytochrome P-450. The activity of azide insensitive microsomal alcohol (ethanol, propanol, butanol) oxidizing system (MEOS) strikingly increased in ethanol fed rats whether expressed as specific activity (nmoles aldehyde/min/mg protein) or turnover number (nmoles aldehyde/min/nmole of cytochrome P-450). The partially purified cytochrome P-450 from ethanol fed rats was more active for alcohol oxidation than the control preparation in the presence of an excess of NADPH-cytochrome c reductase and L-α-dioleoyl lecithin. There was no significant difference in the capacity of partially purified NADPH-cytochrome c reductase from either ethanol fed rats or controls to promote ethanol oxidation in the presence of cytochrome P-450 and L-α-dioleoyl lecithin. Thus, the adaptive increase of azide insensitive MEOS activity after chronic ethanol consumption can be explained, at least in part, by quantitative and qualitative changes of cytochrome P-450.

I. INTRODUCTION

We have shown that hepatic microsomes of rats oxidize ethanol, propanol and butanol to acetaldehyde, propionaldehyde and butyraldehyde (1-3). This microsomal ethanol (and higher alcohol) oxidizing system (MEOS) has characteristics similar to those of the mixed function oxidases (1,2). Although some attributed MEOS activity to catalase contaminating microsomes (4), to catalase plus ADH (5) or to unidentified enzymes (6), others ascribed a major part of the ethanol oxidizing activity of microsomes to a catalase and ADH-independent pathway (2,3,7,8). This was substantiated by solubilization and isolation of MEOS from catalase and ADH (9-12) and by differences in substrate specificities (3). After chronic ethanol feeding the activity of MEOS rose together with cytochrome P-450, NADPH-cytochrome c reductase and phospholipid (13,14). The present study showed that MEOS can be reconstituted with partially purified cytochrome P-450, NADPH-cytochrome c reductase and phospolipid and that the adaptive increase of azide insensitive MEOS can be explained by quantitative and qualitative change of cytochrome P-450 after chronic ethanol consumption.

II. RECONSTITUTION OF MEOS WITH CYTOCHROME P-450,NADPH-CYTOCHROME C REDUCTASE AND PHOSPHOLIPID

To define the role of cytochrome P-450, NADPH-cytochrome c reductase and phospholipid in MEOS activity, constituents of microsomal membranes were tested in their capacity to oxidize alcohols (ethanol, propanol, butanol). Cytochrome P-450 was partially purified by protease treatment and subsequent column chromatography on DEAE-cellulose using a stepwise elution of KCl gradient (15). A cytochrome P-450 rich fraction was obtained by the application of the eluting buffer with 0.3 m KCl after washing out catalase in the void volume of the eluting buffer (16). NADPH-cytochrome c reductase was partially purified according to the method of Levin *et.al.* (17) with the omission of the Emulgen 911 step. The effect of each of the two microsomal components and synthetic phospholipid on ethanol, propanol, butanol oxidation is shown in Fig. 1. Whereas each of the components sustained no or negligible oxidation, the rates of acetaldehyde, propionaldehyde and butyraldehyde production were strikingly increased when the three components were combined. The acetaldehyde production was dependent on the concentration of cytochrome P-450 and NADPH-cytochrome c reductase (16). This system also metabolized benzphetamine (commonly used for the study of mixed function oxidases). About 54-62% of the activity of MEOS was restored in the reconstituted system when the azide (0.1mM) insensitive ethanol, propanol and butanol oxidizing activity of total microsomes was used for comparison with the activity of the reconstituted system.

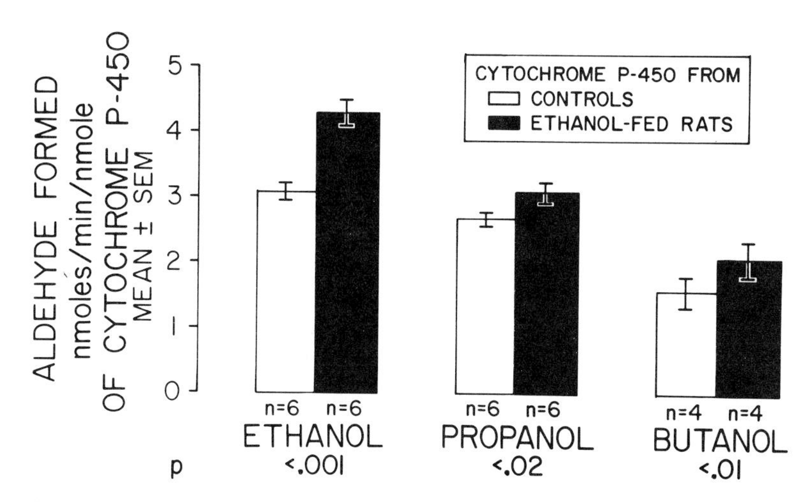

Fig.1. Ethanol, propanol and butanol oxidation by the reconstituted MEOS: Active oxidation of alcohols is achieved by the combination of the three components. Acetaldehyde, propionaldehyde and butyraldehyde in incubates were directly measured by gas-liquid chromatography.

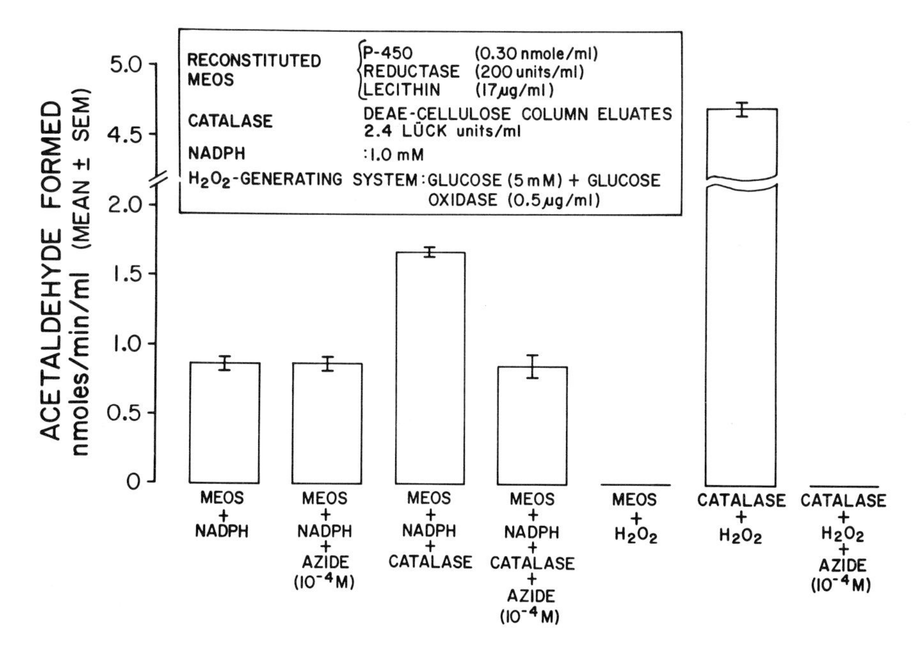

Fig. 2. Effect of Na-azide, H_2O_2-generating system and NADPH on the activity of the reconstituted MEOS: Differentiation of catalase from the reconstituted MEOS.

III. DIFFERENTIATION OF CATALASE ACTIVITY FROM THE RECONSTITUTED MEOS

To rule out the possibility that catalase might contaminate the reconstituted system, Na-azide (a potent catalase inhibitor) was used. Na-azide (0.1 mM) did not inhibit the ethanol oxidizing activity of the reconstituted system (Figure 2). Moreover, the activity of the reconstituted system was not sustained when NADPH was replaced by an H_2O_2-generating system, which produced more H_2O_2 than the reconstituted MEOS. The fact that no acetaldehyde was produced under these conditions not only shows the catalase free nature of the cytochrome P-450, NADPH-cytochrome c reductase and synthetic phospholipid used for the reconstituted system, but it also demonstrates the inability of H_2O_2 to replace NADPH for ethanol oxidation, thereby further differentiating the reconstituted MEOS from a catalase dependent system. This observation also shows that cytochrome P-450 does not act peroxidatically with H_2O_2 for ethanol oxidation in the reconstituted system. The fact that this system could oxidize propanol and butanol (poor or no substrate for catalase) (18, 19) further differentiates this system from catalase-dependent enzymes.

IV. DIFFERENTIATION OF ADH ACTIVITY FROM THE RECONSTITUED MEOS

The ADH inhibitor pyrazole (2 mM) (20) did not significantly inhibit the activity of the reconstituted MEOS (16). This system did not oxidize ethanol when NADPH was replaced by β-NAD^+ (optimal cofactor for ADH) (Figure 3). Other characteristics of the reconstituted MEOS, including pH optimum (7.4 for MEOS vs greater than 9 for ADH), and K_m value for ethanol (11.4 mM for MEOS vs lower than 2 mM for ADH) differentiate ADH from the reconstituted MEOS.

V. EFFECT OF CHRONIC ETHANOL CONSUMPTION ON THE ACTIVITY OF THE AZIDE INSENSITIVE MICROSOMAL ETHANOL, PROPANOL AND BUTANOL OXIDIZING SYSTEM (MEOS).

The azide (0.1 mM) insensitive MEOS activity strikingly increased in both male and female rats after chronic ethanol feeding whether expressed as specific activity or turnover number (Figure 4). Although the specific activity of the azide insensitive MEOS was higher in male than in female ethanol fed rats, the turnover number of the azide insensitive MEOS was similar in both sexes, reflecting a similar qualitative change of cytochrome P-450 in both sexes after chronic ethanol consumption.

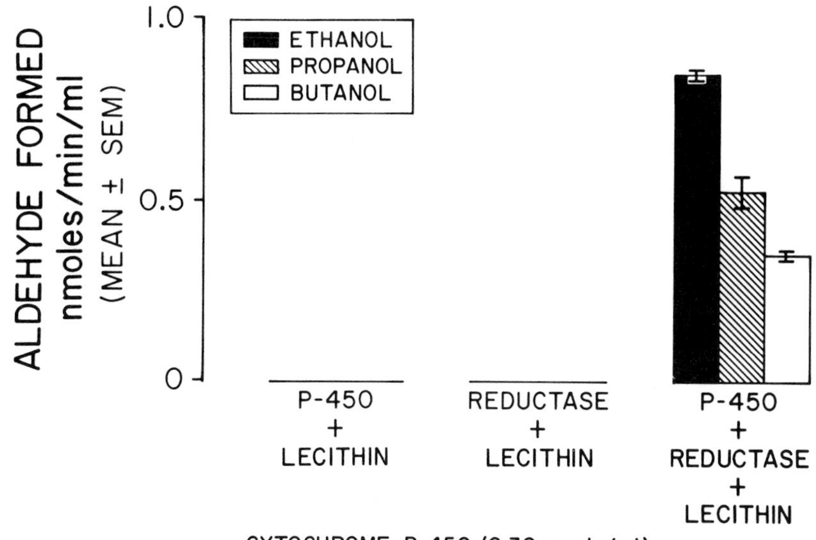

Figure 3. Effect of cofactor and pH on ethanol oxidation catalyzed by the reconstituted MEOS: The requirements for NADPH, the relative insensitivity to pyrazole and the optimal pH differentiate the reconstituted MEOS from ADH.

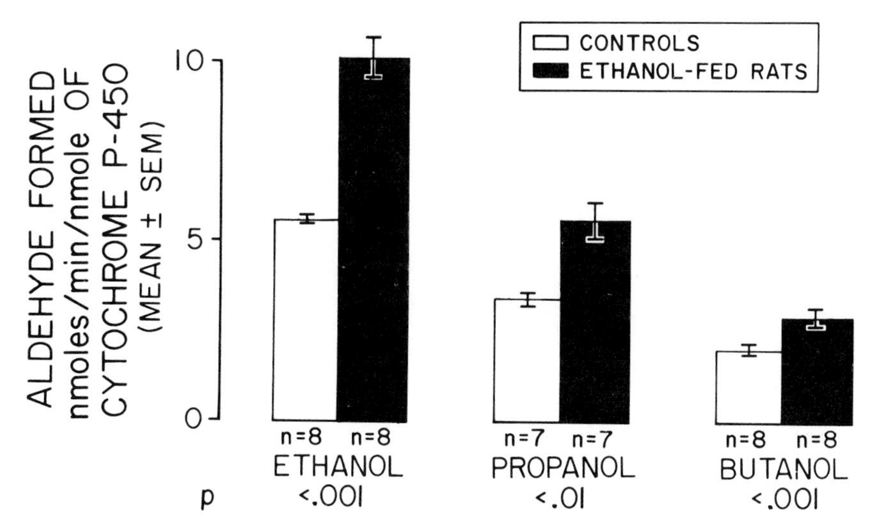

Fig. 4. Effect of Chronic Ethanol Consumption on the Activity of MEOS: Turnover rates are expressed as nmoles of product formed/min/nmole of cytochrome P-450. The reaction mixture, in a final volume of 3 ml, contained (per ml) 100 μmoles of phosphate buffer (ph 7.4), 5 μmoles of magnesium chloride, 1 μmole of NADPH,

and 50 μmoles of the respective alcohol (ethanol, propanol, butanol); 0.1 μmole of sodium azide was added when ethanol was used as a substrate. The reaction mixture was preincubated at 37°C for 5 min and the reaction was initiated by the addition of NADPH. The respective aldehyde formed in incubates were measured directly by gas-liquid chromatography.

VI. INDUCTION OF A NEW SPECIES OF MICROSOMAL PROTEIN (PRESUMABLY AN APO-PROTEIN OF CYTOCHROME P-450) AFTER CHRONIC ETHANOL CONSUMPTION

Liver microsomes prepared from ethanol fed rats and controls were subjected to protein electrophoresis on 0.1% SDS-polyacrylamide gel (21) (Figure 5a). Major differences were found in the distribution of the protein constituents with apparent molecular weights of 50,000. These differences are also shown by the scan of the protein band pattern in the 50,000 molecular weight region of these gels (Fig. 5b). After ethanol feeding a protein (presumably an apoprotein of cytochrome P-450) with apparent molecular weight of 53,400 (Band number 3) appeared.

VII. EFFECT OF CHRONIC ETHANOL CONSUMPTION OF THE SUBSPECIES OF CYTOCHROME P-450 AS DEFINED BY CYANIDE TITRATION

As found before (15,22,23), Form I of cytochrome P-450 was strikingly enhanced in both male and female rats after chronic ethanol consumption and Form II of cytochrome P-450 was also increased. The binding constant (K) for cyanide of three forms of cytochrome P-450 did not change after ethanol feeding. The degree of induction of Form I was similar in both sexes after chronic ethanol feeding although the specific content (A max 442-410 x 10^3/mg protein) of Form I of cytochrome P-450 was higher in males than in females.

VIII. RECONSTITUTION OF MEOS WITH CYTOCHROME P-450 FROM ETHANOL FED OR CONTROL RATS WITH CONTROL NADPH-CYTOCHROME C REDUCTASE AND SYNTHETIC L-α-DIOLEOYL LECITHIN

Cytochrome P-450 was partially purified from ethanol fed rats and controls and NADPH-cytochrome c reductase was partially purified from control animals as described before. The capacity of cytochrome P-450 from ethanol fed rats or controls to promote alcohol (ethanol, propanol, butanol) oxidation was compared. The partially purified cytochrome P-450 from ethanol fed rats was more active for alcohol oxidation than the control preparation (Figure 6).

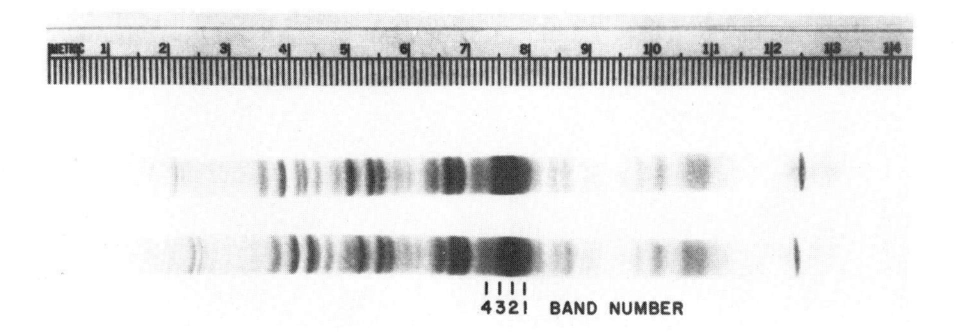

Fig. 5a. SDS-Polyacrylamide gel electrophoresis (21) of microsomes from ethanol fed rats (upper gel) or controls (lower gel); 40 μg of protein were used.

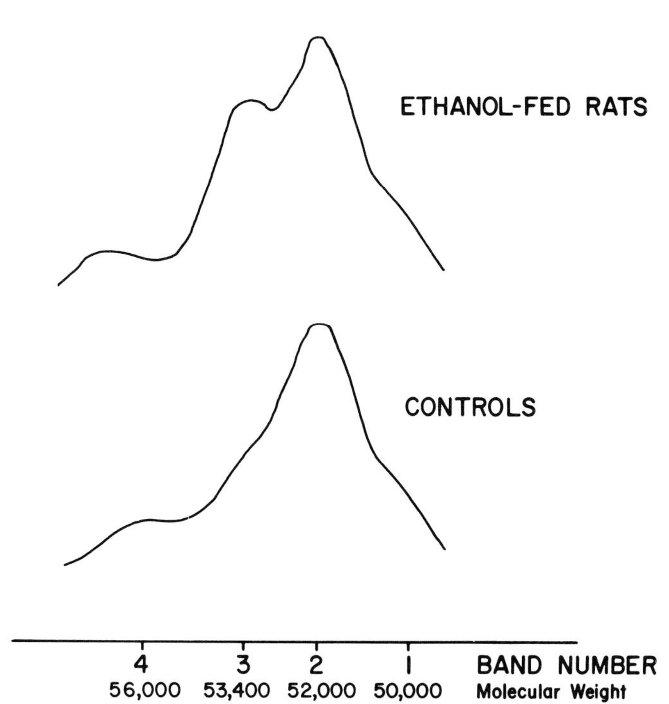

Fig. 5b. Scan of SDS-polacrylamide gel electrophoresis of hepatic microsomal protein (coomassie blue staining) from ethanol fed rats (upper scan) or controls (lower scan).

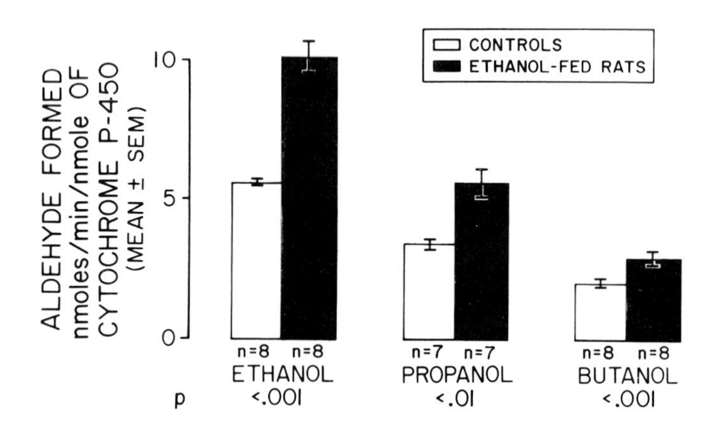

Fig. 6. Effect of chronic ethanol consumption on the capacity of purified cytochrome P-450 to promote alcohol oxidation in the reconstitued MEOS: The system consisted of NADPH-cytochrome c reductase (200 units/ml) from control rats, L-α-dioleoyl lecithin (17 μg/ml) and cytochrome P-450 (0.30 nmole/ml) from either controls or ethanol-fed rats.

IX. RECONSTITUTION OF MEOS WITH NADPH-CYTOCHROME C REDUCTASE FROM ETHANOL FED OR CONTROL RATS WITH CONTROL CYTOCHROME P-450 AND SYNTHETIC L-α-DIOLEOYL LECITHIN

There was no significant difference in the capacity of partially purified NADPH-cytochrome c reductase from either ethanol fed rats or controls to promote ethanol oxidation.

X. CONCLUSIONS

1. The microsomal ethanol oxidizing system (MEOS) can be reconstituted with cytochrome P-450, NADPH-cytochrome c reductase and phospholipid.

2. Neither catalase nor ADH play a role in the activity of the reconstituted MEOS.

3. Chronic ethanol consumption induces a cytochrome P-450

spectrophotometrically, electrophoretically and catalytically different from controls with:

(a) high affinity to cyanide (Form I)

(b) increase of protein (presumably an apoprotein) with molecular weight of 53,400

(c) greater alcohol oxidizing activity.

XI. ACKNOWLEDGEMENTS

The authors thank Ms L.M. DeCarli for generous advice throughout this study. The excellent technical assistance of Ms N. Lowe is gratefully acknowledged. This study was supported by the Medical Research Service of the Veterans Administration and USPHS Grants AA-00224 and AM-12511.

XII. REFERENCES

(1). Lieber, C.S., and DeCarli, L., Science 162, 917 (1968).
(2). Lieber, S.C., and DeCarli, L., J. Biol. Chem. 245, 2505 (1970).
(3). Teschke, R., Hasumura, Y., and Lieber, C.S., J. Biol. Chem. 250, 7397 (1975).
(4). Thurman, R.G., Ley, H.G., and Scholz, R., Eur.J.Biochem. 25, 420 (1972).
(5). Isselbacher, K.J., and Carter, E.A., Biochem.Biophys. Res. Commun. 39, 530 (1970).
(6). Roach, M.K., Reese, W.N., and Creaven, P.J., Biochem.Biophys. Res. Comm. 36, 596 (1969).
(7). Hildebrandt, A.G., and Speck, M., Naunyn Schmiedeberg's Arch.Pharmacol. 277, R31 (1973).
(8). Hildebrandt, A.G., Speck, M., and Roots, I., Naunyn Schmiedeberg's Arch. Pharmacol. 281, 371 (1974).
(9). Teschke, R., Hasumura, Y., Joly, J.-G., Ishii, H., and Lieber, C.S., Biochem.Biophys.Res.Commun. 49,1187 (1972).
(10). Mezey, E., Potter, J.J., and Reed, W.D., J. Biol. Chem. 248, 1183 (1973).
(11). Teschke, R., Hasumura, Y., and Lieber ,C.S., Arch.Biochem. Biophys. 163, 404 (1974).
(12). Corral, R.J.M., Yu, L.C., Rosner, B.A., Margolis, J.M., Rodman, H.M., Kam, W., and Landau, B.R., Biochem.Pharmacol. 24, 1825 (1975).
(13). Joly, J.-G., Ishii, H., Teschke, R., Hasumura, Y., and Lieber, C.S., Biochem.Pharmacol. 22, 1532 (1973).
(14). Ishii, H., Joly, J.-G., and Lieber ,C.S., Biochem.Biophys. Acta. 291, 411 (1973).
(15). Comai, K., and Gaylor, J.L., J.Biol.Chem. 248, 4947 (1973).
(16). Ohnishi, K., and Lieber ,C.S., Fed. Proc. 35, 706 (1976).

(17). Levin, W., Ryan, D., West, S., and Lu, A.Y.H., J.Biol. Chem. 249, 1747 (1974).
(18). Chance, B., Acta Chem.Scand. 1, 236 (1947).
(19). Chance, B., and Oshino, N., Biochem.J. 122, 225 (1971).
(20). Reynier, M., Acta Chem. Scand. 23, 1119 (1969).
(21). Laemmli, U.K., Nature 227, 680 (1970).
(22). Joly, J.-G., Ishii, H., and Lieber ,C.S., Gastroenterology 62, 174 (1972).
(23). Hasumura, Y., Teschke, R., and Lieber, C.S., J.Pharmacol. Exp.Ther. 194, 469 (1975).

MICROSOMAL ELECTRON TRANSPORT REACTIONS: THE FORMATION AND UTILIZATION OF HYDROGEN PEROXIDE AS RELATED TO ALCOHOL METABOLISM*

J. Werringloer, N. Chacos, and R.W. Estabrook

University of Texas Health Science Center;

I. Roots and A.G. Hildebrandt

Institut f. Klinische Pharmakologie

Cytochrome P-450 of liver microsomes is known to participate in the oxidative transformation of a wide variety of compounds. These reactions can be supported by NADPH in an oxygen dependent reaction or by hydrogen peroxide in an oxygen independent reaction. The present study reports on the formation of hydrogen peroxide during NADPH oxidation by liver microsomes and the utilization of hydrogen peroxide for the catalase dependent oxidation of methanol to formaldehyde. The stimulatory influence of compounds such as hexobarbital and benzphetamine, serving as "uncouplers" of cytochrome P-450 function, and the pattern of inhibition of hydrogen peroxide formation and methanol oxidation obtained with metyrapone and sodium azide respectively as well as as the inability to support methanol oxidation via cytochrome P-450 in the presence of hydrogen peroxide, lead to the conclusion that under the conditions reported here all of the oxidation of methanol can be attributed to the peroxidatic function of adventitious catalase present in the microsomal fraction of liver.

The oxidation of alcohols to aldehydes during the aerobic oxidation of NADPH by the microsomal fraction from liver has been attributed to the potential enzymatic functioning of three types of systems: a) the presence (1-3) of a unique MEOS (microsomal ethanol oxidizing system) where cytochrome P-450 presumably functions in alcohol oxidation in a manner

*Supported in part by a grant from the National Institutes of Health (5RO1-GM-16488).

analogous to its role in the oxidative transformation of a wide variety of drugs, steroids, and other compounds; b) the pyridine nucleotide dependent oxidation of alcohols (4,5) by an alcohol dehydrogenase bound to the membrane fragments derived from the endoplasmic reticulum; and c) the peroxidatic transformation of alcohols utilizing hydrogen peroxide (6-9) generated during NADPH oxidation by the microsomal electron transport system (10,11) in concert with adventitious catalase contaminating the preparations of microsomes. A significant amount of data has been presented supporting a role for each of these three types of systems although the relative contributions of each during the *in vivo* metabolism of alcohols by liver remains unresolved.

Critical to the consideration of a potential role of cytochrome P-450 in alcohol oxidation is a better understanding of the reactions of this pigment for the generation of utilization of hydrogen peroxide and the influence of the concomitant oxidation of other compounds such as drugs and steroids, known to be oxidatively metabolized by cytochrome P-450, during alcohol oxidation. The present report examines the influence of drugs found to be "uncouplers" of cytochrome P-450 function, i.e. oxygen utilization is stimulated with an associated increase in the rate of hydrogen peroxide formation. In addition, the effect of varying concentrations of the inhibitor of catalase, sodium azide, on the rate of methanol oxidation to formaldehyde by microsomes isolated from the livers of phenobarbital-treated male rats has been examined.

A. The Generation of Hydrogen Peroxide

The aerobic incubation of liver microsomes with NADPH results in a rather rapid rate of NADPH oxidation concomitant with the stoichiometric reduction of oxygen (Fig. 1). Approximately 50 to 60 percent of the reducing equivalents originating from NADPH are directed toward the formation of hydrogen peroxide; the remaining balance of reducing equivalents presumably are utilized by mixed function oxidation reactions of "endogenous substrates" where the stoichiometry of one mole of NADPH oxidized per one mole of oxygen reduced prevails. For experiments of this type, the ability to detect the formation of hydrogen peroxide by direct chemical analysis is dependent on the presence of an inhibitor of catalase, such as sodium azide. Repeated additions of NADPH to microsomal suspensions, in the absence of an NADPH generating system, shows an equivalent amount of hydrogen peroxide formed following each addition of NADPH (Fig. 1) although the initial rate of oxygen utilization, NADPH oxidation, and hydrogen peroxide formation

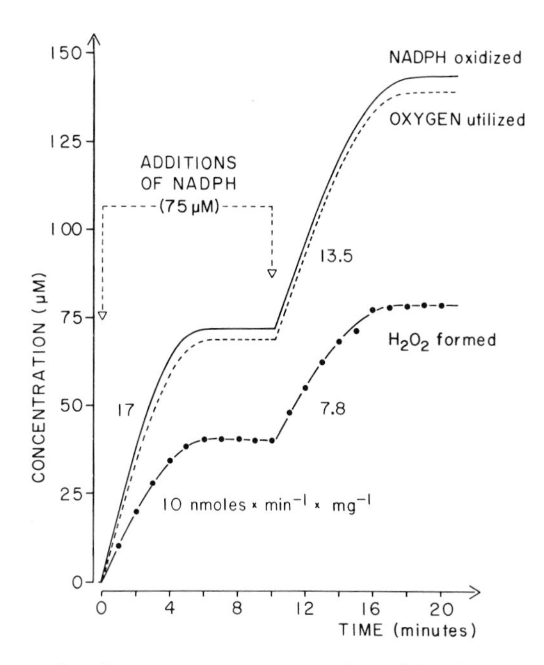

Fig. 1. The hepatic microsomal oxidation of NADPH and its relation to oxygen utilization and hydrogen peroxide formation. Rat liver microsomes from phenobarbital-treated animals were diluted to a protein concentration of 1 mg per ml in a reaction mixture containing 50 mM tris-chloride buffer, pH 7.5, 150 mM KCl, 10 mM $MgCl_2$, 2 mM 5'-AMP, and 1 mM sodium azide. Samples were removed at the times indicated for the determination of hydrogen peroxide colorimetrically using potassium thiocyanate and ferrous ammonium sulfate (13). Changes in oxygen concentration were determined polarographically while oxidation of NADPH was measured spectrophotometrically.

progressively decreases as the concentration of $NADP^+$ increases. $NADP^+$ is well recognized (12) as a competitive inhibitor for NADPH interaction with the required flavoprotein, NADPH-cytochrome P-450 reductase.

B. Effect of Hexobarbital

Studies have been carried out to determine the influence of hexobarbital, a substrate of cytochrome P-450 which is actively metabolized, on the rate of hydrogen peroxide formation during NADPH oxidation by liver microsomes. For these studies the oxidation of methanol to formaldehyde was determined by coupling the peroxidatic function of endogenous or exogenous catalase to hydrogen peroxide generated during the reaction (13). An NADPH generating system was also incorporated into the reaction medium to avoid the formation of $NADP^+$ resulting in a progressive inhibition during the course of the reaction (see above).

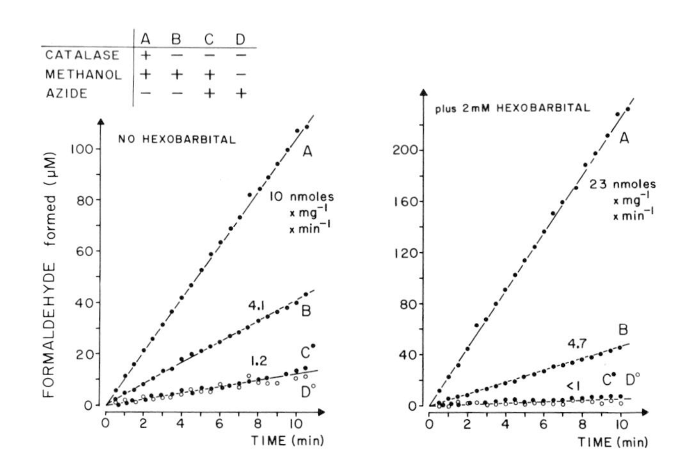

Fig. 2. The inhibition by sodium azide of methanol oxidation in the absence and presence of hexobarbital. Liver microsomes from phenobarbital treated rats were diluted in a reaction medium as described in Figure 1 supplemented with 5 mM sodium isocitrate, 0.5 units per ml of isocitrate dehydrogenase. Where indicated 50 mM methanol, 5000 units of catalase per ml., 2 mM hexobarbital, and 1 mM sodium azide were added. The reactions were initiated by addition of 0.2 mM NADPH. The concentration of formaldehyde formed was determined with the Nash reagent (24).

In the absence of hexobarbital, catalase associated with the microsomal preparation captured about 40 percent of the hydrogen peroxide formed for the oxidation of methanol to formaldehyde (Fig. 2, left side). Addition of an excess of catalase showed a rate of formaldehyde formation from methanol equivalent to that observed by the direct measurement of hydrogen peroxide (cf. Fig. 1). Determination of the content of endogenous catalase accompanying the preparation of microsomes - as measured by evaluating the rate of breakdown of hydrogen peroxide polarographically in an oxygen electrode apparatus under standard conditions (J. Werringloer, unpublished) - revealed the presence of approximately 2 picomoles of catalase heme per milligram of liver microsomal protein. It is apparent that the balance of peroxidatic and catalatic action of endogenous catalase associated with the preparation of liver microsomes favors the breakdown of hydrogen peroxide - a result consistent with the detailed analysis presented by Oshino *et al.* (14). It is of interest to note that 1 mM sodium azide reduced the rate of methanol oxidation in the presence or absence of added catalase to a level equivalent to the background rate of formaldehyde formation observed in the absence of methanol.

The presence of hexobarbital in the reaction mixture markedly stimulated the rate of hydrogen peroxide formation as reflected by the increased rate of methanol oxidation in the presence of excess catalase. In this way hexobarbital may be considered as an "uncoupler" of cytochrome P-450 function in a manner analogous to the action of aminopyrine during NADPH oxidation by liver microsomes from PCN (pregnenolone-α-carbonitrile) treated animals (11) but differing from the "uncoupling" influence of perfluorinated, aliphatic fluorocarbons as described by Ullrich and Diehl (15). Of interest is the marginal influence of hexobarbital on the rate of peroxidatic function of catalase originally associated with the preparation of microsomes and the nearly complete inhibition of the reaction by 1 mM sodium azide in the presence of hexobarbital.

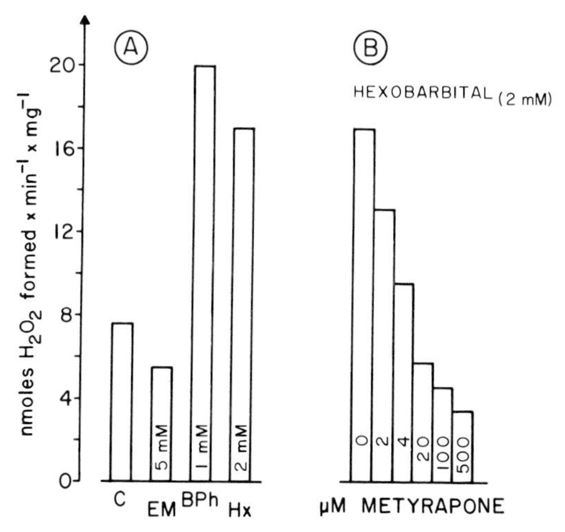

Fig. 3. The influence of various chemicals on the rate of hydrogen peroxide formation during NADPH oxidation by rat liver microsomes. A series of experiments were carried out as described in Figure 1 except for the addition of 5 mM ethylmorphine (EM), 1 mM benzphetamine (BPh), and 2 mM hexobarbital (Hx) as indicated. The data on the right side of the figure illustrate the inhibitory effect of varying concentrations of metyrapone on the rate of formation of hydrogen peroxide observed in the presence of 2 mM hexobarbital. The reaction mixtures contained 1 mg of microsomal protein per ml and the reactions were initiated by the addition of 0.3 mM NADPH.

The ability of various drugs to direct the function of cytochrome P-450 toward hydroxylation reactions or by serving as "uncouplers" resulting in a stimulation of hydrogen peroxide formation depends on the source of microsomes (e.g., from livers of various species of animals or from animals pretreated with different inducers). (16). The basis for this different action of drugs remains undefined. For the present series of experiments, where only liver microsomes from phenobarbital treated male rats were employed, ethylmorphine inhibited the rate of hydrogen peroxide formation while benzphetamine and hexobarbital markedly enhanced the generation of hydrogen peroxide (Figure 3). Thus, one cannot predict the actions of various drugs and the possible argument that substrates of cytochrome P-450 serve as inhibitors of alcohol oxidation or vice-versa cannot be used as a bias for concluding that cytochrome P-450 functions directly in the oxidation of alcohols to aldehydes.

C. Inhibitors of Alcohol Oxidation by Liver Microsomes

Previously we have presented data (17) confirming the results of Lieber and DeCarli (1) as well as Thurman *et al.* (7), showing the sensitivity of alcohol oxidation to inhibition by carbon monoxide. Direct comparison of the inhibitory effect of various ratios of carbon monoxide and oxygen revealed (17) a pattern of inhibition for hydrogen peroxide formation during NADPH oxidation *identical* to that observed for the oxidative transformation of ethylmorphine - a well characterized substrate for a cytochrome P-450 catalyzed reaction. As shown in Figure 3 (right side), metyrapone (2-methyl-1,2-bis [3-pyridyl]-1-propenone), another inhibitor of cytochrome P-450 function, inhibits the formation of hydrogen peroxide and therefore would be predicted to likewise inhibit the coupled peroxidatic reaction observed in the presence of catalase. Again, the pattern of inhibition observed corresponds to that obtained when varying concentrations of metyrapone are added to liver microsomes during the cytochrome P-450 catalyzed reaction for the N-demethylation of ethylmorphine. The fact that two inhibitors of cytochrome P-450 function influence the rate of hydrogen peroxide formation in a manner comparable to that observed for the oxidative metabolism of a drug substrate, ethylmorphine, does not permit one to distinguish between a direct or an indirect role for cytochrome P-450 in alcohol metabolism. As shown in Figure 4, sodium azide is a potent inhibitor of methanol oxidation during NADPH oxidation by liver microsomes. In contrast, 1 mM sodium azide has *no effect* on the cytochrome P-450 catalyzed N-demethylation of ethylmorphine. The effectiveness of sodium azide to inhibit the

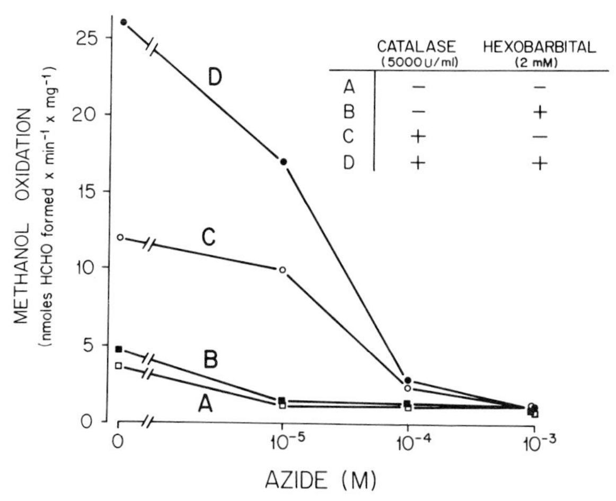

Fig. 4. The sensitivity of methanol oxidation during NADPH oxidation by rat liver microsomes to inhibition by various concentrations of sodium azide. Reactions were carried out as described in Figure 2 using 50 mM methanol and 5000 units per ml of catalase and 2 mM hexobarbital where indicated.

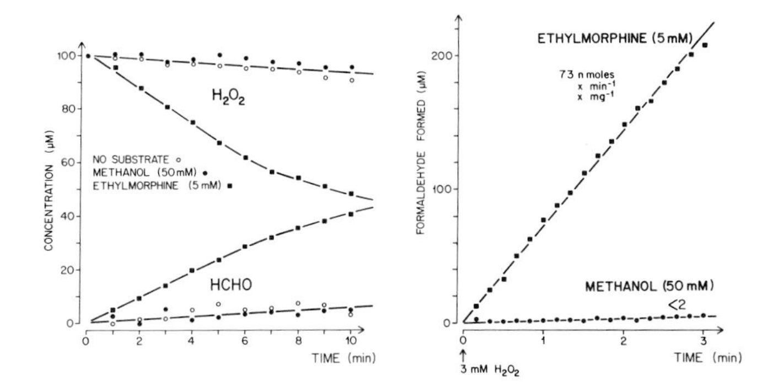

Fig. 5. Comparison of the rate of hydrogen peroxide utilization and the rate of formaldehyde formation as catalyzed by hepatic microsomes in the presence of ethylmorphine or methanol. Liver microsomes from phenobarbital treated rats (1 mg protein per ml) were diluted in the buffer mixture described in Figure 1 containing 1 mM sodium azide. The concentrations of hydrogen peroxide and formaldehyde were determined colorimetrically as described in Figures 1 and 2. The reactions on the left side of the Figure were initiated by the addition of 100 µM hydrogen peroxide while those on the right side of the Figure by addition of 3 mM H_2O_2.

oxidation of methanol to formaldehyde is comparable in the presence or absence of added catalase or in the presence or absence of an "uncoupler" of cytochrome P-450 function, hexobarbital. From these experiments it is concluded that the inhibition of alcohol oxidation observed when inhibitors of cytochrome P-450 function are employed merely reflects an interference in the rate of hydrogen peroxide formation during the cyclic function of cytochrome P-450 rather than a direct effect on alcohol oxidation involving cytochrome P-450.

D. Peroxidatic Function of Cytochrome P-450

Recent interest (18-22) has centered on the ability to catalyze a number of mixed function oxidation reactions by addition of hydrogen peroxide or organic peroxides to liver microsomes. As illustrated in Figure 5, addition of hydrogen peroxide, in the absence of NADPH, results in the rather rapid N-demethylation of ethylmorphine concomitant with the stoichiometric utilization of hydrogen peroxide (23). When the experiment was repeated using methanol rather than ethylmorphine as substrate, no significant rate of formaldehyde formation was observed and no increase in the rate of hydrogen peroxide utilization (above that observed in the absence of any addition of substrates to the microsomal suspension) was measurable. Since the peroxidatic function of cytochrome P-450 requires rather high concentrations of hydrogen peroxide (22,23) to obtain half-maximal rates of ethylmorphine N-demethylation, experiments were carried out using 3 mM H_2O_2 (Fig. 5; right side). Again, under conditions where a rapid rate of formaldehyde formation was observed associated with the N-demethylation of ethylmorphine, no formaldehyde was formed indicative of the conversion of methanol to formaldehyde. From these experiments it is concluded that methanol is not oxidized to formaldehyde in a manner analogous to that observed with a number of compounds recognized as substrates for cytochrome P-450 during the hydrogen peroxide-dependent peroxidatic function of cytochrome P-450.

E. Summary

The present studies indicate that the oxidation of methanol to formaldehyde, observed during NADPH oxidation by rat liver microsomes from phenobarbital treated animals, can be attributed solely to the function of catalase accompanying the preparation of microsomes. The influence of inhibitors or activators of cytochrome P-450 is explained by the role of cytochrome P-450 in generating hydrogen peroxide required for the peroxidatic action of catalase.

REFERENCES

(1). Lieber, C.S., and DeCarli, L.M., Science 162, 917 (1968).

(2). Lieber, C.S., and DeCarli, L.M., J. Biol. Chem. 245, 2505 (1970).

(3). Lieber, C.S., and DeCarli, L.M., Drug Metab. Disposition 1, 428 (1973).

(4). Isselbacher, K.J., and Carter, E.A., Biochem. Biophys. Res. Commun. 39, 530 (1970).

(5). Brentzel, H., and Thurman, R., Proc. Second Int. Symp. on Alcohol and Aldehyde Metabolizing Systems, p. 10 (1976).

(6). Roach, M.K., Reese, W.N., and Creaven, P.J., Biochem. Biophys. Res. Commun. 36, 596 (1969).

(7). Thurman, R.G., Ley, H.G., and Scholz, R., Eur. J. Biochem. 25, 420 (1972).

(8). Khanna, J.M., Kalant, H., and Lin, G., Biochem. Pharmacol. 19, 2493 (1970).

(9). Vatsis, K.P., and Schulman, M.P., in "Alcohol and Aldehyde Metabolizing Systems (R.G. Thurman, T. Yonetani, J.R. Williamson, and B. Chance, Eds.), p. 287. Academic Press, New York and London, 1974.

(10). Gillette, J.R., Brodie, B.B., and LaDu, B.N., J. Pharm. Exptl. Therap. 119, 532 (1957).

(11). Hildebrandt, A.G., Speck, M., and Roots, I., Biochem. Biophys. Res. Commun. 54, 968 (1973).

(12). Williams, C.H., Jr., and Kamin, H., J. Biol. Chem. 237, 587 (1962).

(13). Hildebrandt, A.G., and Roots, I., Arch. Biochem. Biophys. 171, 385 (1975).

(14). Oshino, N., Oshino, R., and Chance, B., Biochem. J. 131, 555 (1973).

(15). Ullrich, V., and Diehl, H., Eur. J. Biochem. 20, 509 (1971).

(16). Hildebrandt, A.G., Tjoe, M., and Roots, I., Z. Klin. Chem. Klin. Biochem. 13, 374 (1975).

(17). Werringloer, J., in "Proc. Third Int. Symposium on Microsomes and Drug Oxidations (V. Ullrich, I. Roots, A.G. Hildebrandt, R.W. Estabrook and A. Conney, Eds.). Pergamon Press, Oxford, in press (1977).

(18). Kadlubar, F.F., Morton, K.C., and Ziegler, D.M., Biochem. Biophys. Res. Commun. 54, 1255 (1973).

(19). Rahimtula, A.D., and O'Brien, P.J., Biochem. Biophys. Res. Commun. 60, 440 (1974).

(20). Rahimtula, A.D., O'Brien, P.J., Hrycay, E.G., Peterson, J.A., and Estabrook, R.W., Biochem. Biophys. Res. Commun. 60, 695 (1974).

(21). Hrycay, E.G., Gustafsson, J.A., Jungelman-Sundberg, M., and Ernster, L., Biochem. Biophys. Res. Commun. 66, 209 (1975).

(22). Nordblom, G.D., White, R.E., and Coon, M.J., Arch. Biochem. Biophys. 175, 524 (1976).

(23). Estabrook, R.W., and Werringloer, J., in "Proc. Third Int. Symposium on Microsomes and Drug Oxidations (V. Ullrich, I. Roots, A.G. Hildebrandt, R.W. Estabrook and A. Conney, Eds.). Pergamon Press, Oxford, in press (1977).

(24). Nash, T., Biochem. J. 55, 416 (1953).

STUDIES ON THE CHARACTERIZATION OF THE ENZYME COMPONENTS PARTICIPATING IN THE HEPATIC-MICROSOMAL OXIDATION OF ALIPHATIC ALCOHOLS

*K.P. Vatsis, +B. Chance and *M.P. Schulman

*University of Illinois
+University of Pennsylvania

The NADPH-dependent oxidation of methanol and ethanol in mouse liver microsomes was markedly inhibited by 0.1 mM cyanide or azide. However, although rates of ethanol oxidation with 3-acetylpyridine NAD were 50% of those seen with NADPH, 3-acetyl pyridine NAD was unable to substitute for NADPH in the microsomal oxidation of methanol. The methanol-oxidizing activity was not appreciably affected by the potent alcohol dehydrogenase inhibitor 4-methylpyrazole, while 4-n-propylpyrazole caused a small and non-specific decrease in this activity. Rates of ethanol oxidation in microsomes in the presence of NADPH were two-fold higher than corresponding values for methanol, but this difference in activity was almost totally abolished by 0.1 mM 4-n-propylpyrazole. Propylpyrazole-sensitive rates of ethanol oxidation were identical to cyanide- or azide-insensitive rates of acetaldehyde production and corresponded to the ethanol-oxidizing activity obtained with 3-acetylpyridine NAD as co-factor. Accordingly, only one K_m value (33 mM) was obtained with methanol while the kinetics with ethanol were clearly biphasic revealing the presence of a higher- (K_m = 3.3 mM) as well as a low-affinity (K_m = 20 mM) component, each contributing equally to the overall NADPH-linked metabolism of ethanol by microsomes. The activities with propanol and butanol remained unaltered in the presence of cyanide or azide, but were almost totally obliterated by 0.1 mM 4-n-propylpyrazole. 3-acetylpyridine NAD was as effective as NADPH in supporting the microsomal oxidation of propanol and butanol. It is concluded that the microsomal NADPH-dependent oxidation of methanol is mediated exclusively by catalase, whereas both catalase and alcohol dehydrogenase contribute equally to the metabolism of ethanol. In contrast, alcohol dehydrogenase is solely responsible for the oxidation of the higher alcohols.

I. INTRODUCTION

The specificity of the rat hepatic microsomal ethanol-oxidizing system for aliphatic alcohols has recently been examined in attempts toward the characterization of the enzymic component(s) participating in these reactions (1,2), especially in view of previous observations that propanol and butanol are largely ineffective in decomposing catalase Compound I peroxidatically (3-5). In particular, it was shown that NADPH supported the microsomal oxidation of methanol, ethanol, propanol, and butanol whereas only methanol and ethanol were metabolized in the presence of glucose and glucose oxidase. Based on these observations and on rather limited kinetic studies, it was concluded that liver microsomes contain an NADPH-linked alcohol-metabolizing system that operates independently of the peroxidatic reaction of catalase (1,2). However, studies by Oshino *et.al.* (6) have demonstrated conclusively that qualifying statements regarding the involvement or lack of involvement of catalase in alcohol oxidation cannot be made unless the ratio of H_2O_2 generation rates to catalase heme concentration is rigorously defined in each case.

Preliminary reports from our laboratory (7-9) indicate that the oxidation of alcohols in hepatic microsomes is mediated by the combined action of catalase and a propylpyrazole-sensitive component, most likely ADH[1]. The present studies were undertaken in order to investigate in detail the extent of participation of catalase and ADH in the overall metabolism of primary aliphatic alcohols. In addition, kinetic properties were defined for the microsomal system operating in the presence of NADPH. The results show unequivocally that catalase and ADH participate singly or in concert in the alcohol-oxidizing process, and that the relative contribution of each enzyme to overall metabolic rates depends strictly on the particular alcohol substrate.

II. RESULTS AND DISCUSSION

A. NADPH-Dependent-Alcohol-Oxidizing Activities of Mouse and Rat Hepatic Microsomes

Under our standard assay conditions, it was seen that KCl-washed liver microsomes from mice were capable of oxidizing all five primary alcohols in the presence of added NADPH (Table 1). Methanol oxidation rates were 50-60% of those seen with ethanol, while propanol was metabolized at rates comparable to those observed with methanol; activities decreased as the chain length of the alcohol increased above three carbon atoms. An identical pattern was observed with rat liver microsomes (KCl-washed) , though, dependening on the alcohol substrate, absolute activity levels were approximately 45-65% of the values obtained with similar preparations from mice. In contrast to the mouse, rat liver microsomes were unable to metabolize pentanol. With the exception of methanol, the alcohol-oxidizing activities shown for rat liver microsomes are in agreement with values reported by others (1,2). However, whereas in our experience methanol oxidation rates are consistently about 50-55% of those seen with ethanol, Teschke *et.al.* (1,2) obtained comparable activities for methanol and ethanol; these findings (1,2) are in direct conflict with a previous publication from the same laboratory which stated that the microsomal ethanol-oxidizing system in rats "has a rate of ethanol oxidation twice that of methanol" (13).

Washing mouse or rat liver microsomes twice with dilute Tris-acetate buffer did not alter the alcohol-oxidizing activity profile as compared to that seen with microsomal preparations which had been washed once with 0.15 M KCl (Table 1). The total amounts of protein recovered after two Tris-acetate washes were 95 and 90% of those obtained with KCl-washed microsomes from mice and rats, respectively. It was demonstrated several years ago that repeated washings with Tris-acetate buffer led to the removal from microsomal particles of "accessible" or loosely-bound ADH (11,12), and this was accompanied by severe losses of NAD-supported, NADH-dependent methyl sterol demethylase activity (11) as well as a 60% decrease in NADPH-dependent ethanol oxidation rates (12). Addition of the wash fraction or of purified ADH to these preparations resulted in complete restoration of both activitites (11,12,14). Since unwashed microsomes were not used in the present studies, the exact amount of ADH and alcohol-oxidizing activities lost by washing cannot be directly estimated. However, judging from the amounts of protein recovered in washed preparations (Vide supra), it is evident that isotonic KCl was as effective as dilute Tris-acetate in removing an equal amount of loosely-bound ADH from rat or mouse liver microsomes. Moreover, the results obtained upon gentle digestion of microsomes with crude snake venom or with purified phospholipase A (15) were both qualitatively and quantitatively similar to those seen following repeated washings with dilute Tris-acetate buffer (11). Of particular interest was the observation that NADH-dependent cytochrome b_5 reductase and methyl sterol demethylase activities of snake venom-digested microsomes could be restored to corresponding control values by the addition of NAD and 0.1 M butanol, thus revealing that "buried" ADH was present in these preparations in addition to the loosely-bound ADH which had been removed by the snake vermon treatment (15). The cytosolic and microsomal ADH enzymes were subsequently isolated and shown to be identical proteins as judged both by immunological criteria and substrate specificity studies (16). Thus, despite various claims to the contrary (13, 17), ample evidence exists in support of the conclusion that washed microsomes do, indeed, contain "microsomal-bound alcohol dehydrogenase" (15). The inability to detect ADH activity in washed microsomes (12,13,17) may be related to the insensitivity of the spectrophotometric assay commonly used for this purpose, i.e., the low amount of microsomal protein that must be employed to avoid light scattering and the fact that ethanol, instead of a higher alcohol, is used as substrate and the assay is carried out at room temperature. Soluble and microsomal alcohol dehydrogenase may resemble each other with respect to their relative activities toward primary alipatic alcohols of increasing carbon chain length, but the absolute level of activity with a given alcohol substrate may be drastically different for the two enzymes.

B. Effect of ADH and Catalase Inhibitors on Hepatic Microsomal NADPH-Dependent Oxidation of Aliphatic Alcohols

In agreement with the findings of others (2,17,18), the presence of 1.0 mM pyrazole did not alter the alcohol-oxidizing activity of mouse liver microsomes. Several studies have demonstrated that 4-Me-pyrazole is a potent inhibitor of ADH activity (19-21) and of ethanol metabolism in man and rats *in vivo* (22, 23) as well as by the perfused rat liver (24); its inhibitory potency is 1,000-fold greater than that of pyrazole (19), while its K_i value (0.013 μM) is approximately 20-fold less than that of pyrazole (21). As can be seen in Table 1 for mouse liver microsomes washed with either medium, 4-Me-pyrazole at a concentration of 1.0 mM had no effect on methanol oxidation and caused only a 25% decrease in the ethanol-oxidizing activity . In sharp contrast to methanol and ethanol, the oxidation of propanol, butanol, and pentanol was decreased to the same level by 4-Me-pyrazole thus resulting in similar residual or uninhibited activities for these alcohols; obviously, expressing these data on a percentage basis would give rise to variable inhibitions depending on the initial or uninhibited activity of the particular alcohol substrate. The pattern of inhibition of alcohol oxidation by 4-Me-pyrazole in rat liver microsomes was very similar to that seen with corresponding preparations from mice.

Propylpyrazole is among the most potent ADH inhibitors studied so far, its K_i being approximately 60-fold lower than that of pyrazole (21). Propylpyrazole, at a final concentration (0.02 mM) 50-fold lower than that employed with 4-Me-pyrazole, caused a precipitous decline in propanol- and butanol-oxidizing activities, whereas a more gradual decrease in activity was observed with methanol and ethanol; the inhibition seen for all

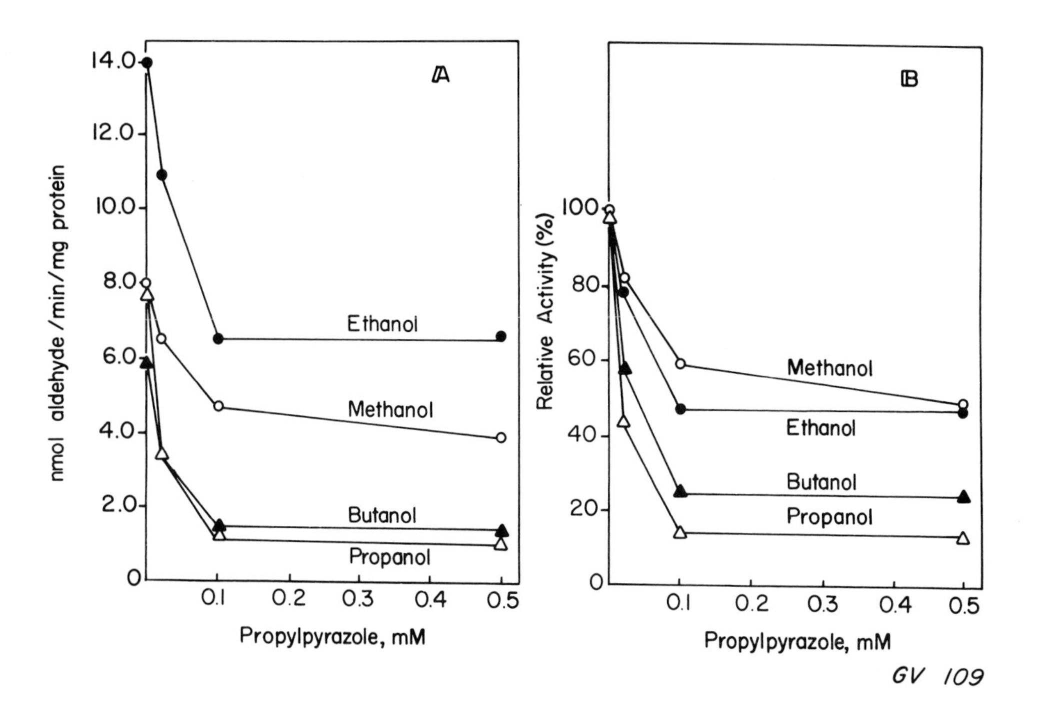

Fig. 1. Effect of propylpyrazole on NADPH-dependent alcohol oxidation by mouse liver microsomes. Reactions were carried out with KCl-washed microsomes (1.0 mg of protein per ml of reaction mixture) in the presence of 1.0 mM NADPH, 80 mM alcohol substrate, increasing concentration of propylpyrazole as shown, and Krebs-Ringer phosphate buffer, pH 7.4; the total volume of the reaction mixture was 1.5 ml. Incubations and product assays were done as described in Table 1. Values shown are the average of two separate experiments. Results are expressed in terms of absolute activity levels as nmol aldehyde formed per min per mg of protein (A), or as percent of the control activity obtained for each substrate in the absence of propylpyrazole (B).

TABLE 1

Effect of 4-Methylpyrazole on NADPH-Dependent Alcohol Oxidation in Mouse and Rat Hepatic Microsomes[a]

	Mouse Liver Microsomes				Rat Liver Microsomes			
	KCl-washed[b]		TrisAcetate-washed[c]		KCl-washed[b]		Tris-Acetate-washed[c]	
SUBSTRATE	Control	4-Methyl-pyrazole	Control	4-Methyl-pyrazole	Control	4-Methyl-pyrazole	Control	4-Methyl-pyrazole
	nmol Aldehyde/min/mg protein							
Methanol	7.2	6.1	7.6	6.9	4.8	3.2	3.9	2.7
Ethanol	15.3	11.7	16.9	12.4	8.3	6.6	8.2	6.4
Propanol	8.1	2.0	7.4	2.5	5.2	1.2	4.9	1.8
Butanol	6.5	3.8	6.3	3.5	3.0	1.2	2.8	1.1
Pentanol	4.6	3.2	4.7	3.2	N.D.[d]	- -	N.D.[d]	- -

a. Male Swiss mice (23-25 g body weight) and male Sprague-Dawley rats (190 g body weight) were used throughout these studies. After washing either in KCl or Tris-acetate as described below, microsomes were suspended in Krebs-Ringer-phosphate buffer, pH 7.4, to a concentrationof 7.5 - 8.0 mg protein/ml. Reaction mixtures consisted of microsomes from 100 mg of liver (1.5-1.7 mg protein), 1.0mM NADPH, 80mM alcohol substrate, and, when present, 1.0mM 4-methylpyrazole in a total volume of 1.5 ml of Krebs-Ringer-phosphate buffer, pH 7.4. Reactions were started by adding alcohol substrate and, following a 10-min incubation at 37° they were terminated with 0.3 ml of 50% trichloroacetic acid. Zero-time incubations contained all of the components, but trichloroacetic acid was added before the alcohol and the flasks were kept at room temperature. All incubations were carried out in duplicate in capped 25-ml Erlenmeyer flasks without center wells (methanol substrate), or with center wells (all other alcohols) containing 1.0 ml of 0.1 M semicarbazide in 0.16 M sodium phosphate buffer, pH 7.0. Formaldehyde formed from methanol was determined according to the method of Nash (10); all other aldehydes were assayed by reading *the aborbance at 224 nm of the solutions in the center wells following an overnight diffusion period at room temperature, as described by Teschke et.al.(1,2). Values shown are the average of two different experiments. b. Washed once by resuspending 1:5(w/v; 3 mg of protein/ml) in 1.15% KCl. c. Washed twice in 0.05 M Tris-acetate buffer, pH 7.4, containing 2 mM glutathione, and 30mM nicotinamide, by resuspending each time to a concentration of 3mg protein/ml as described by Moir et.al. (11) and by Isselbacher and Carter (12). d. D.D., not detectable.*

TABLE 2

Effect of Nucleotides and Inhibitors on Mouse Hepatic Microsomal Alcohol Oxidation

INCUBATION SYSTEM	Microsomal Alcohol-Oxidizing Activity			
	Methanol	Ethanol	Propanol	Butanol
Experiment 1[a]				
NADPH	7.9	13.9	7.7	5.9
+ Propylpyrazole (0.1 mM)	4.7	6.5	1.1	1.5
Propylpyrazole-sensitive	3.2	7.4	6.6	4.4
Experiment 2[b]				
NADPH	7.0	13.1	6.0	5.5
3-AP-NAD	0	6.1	6.9	6.9
"net NADPH-mediated"	7.0	7.0	- 0.9	- 1.4
Experiment 3[b]				
NADPH	7.2	13.8	6.7	5.6
+ Potassium cyanide (0.1mM)	2.8	6.8	5.3	5.5
+ Sodium azide (0.1mM)	2.2	5.7	8.1	6.4
Cyanide-sensitive	4.4	7.0	1.4	0.1
Azide-sensitive	5.0	8.1	- 1.4	- 0.8

a. The values shown for this experiment were taken from the studies depicted in Fig. 1.

b. KCl-washed mouse liver microsomes (1.0mg of protein per ml of reaction mixture) were incubated with 80 mM alcohol substrate and 1.0mM NADPH or 1.0mM 3-AP-NAD in a total volume of 1.5ml of Krebs-Ringer-phosphate buffer, pH 7.4. When present, potassium cyanide or sodium azide was added at the indicated final concentration to reaction mixtures containing NADPH as the cofactor. Incubations and product assays were carried out as described in the legend to Table 1. Values represent the average results of 3 (Experiment 2) or two (Experiment 3) separate studies.

the alcohols except methanol was complete at 0.1 mM propylpyrazole (Fig. 1.). Consistent with the inhibitory pattern previously observed in the presence of 4-Me-pyrazole (Table 1), 0.02 mM propylpyrazole gave rise to only a very small inhibition of methanol and ethanol oxidation (Fig. 1B), but inhibited the activities with propanol and butanol to the same level. In fact, residual or uninhibited rates for propanol and butanol were identical regardless of the propylpyrazole concentration employed (Fig. 1A). Unlike the pattern seen with the three higher alcohols, the activity with methanol decreased further as the propylpyrazole concentration was raised above 0.1 mM (Fig. 1.) suggesting a non-sepcific effect of this compound on the metabolism of methanol. Of particular interest is the observation that control rates were about two-fold higher with ethanol than with methanol, whereas this difference in activity was almost totally abolished by 0.1 mM propylpyrazole (Fig. 1A). Moreover, aniline p-hydroxylation was totally unaffected, while rates of N-demethylation of aminopyrine, ethylmorphine, or benzphetamine were decreased by only 15% in the presence of 0.03 mM propylpyrazole (25,26).

ADH activity is enhanced five-fold when 3-AP-NAD is substituted for NAD[2] (12,14), findings which are consistent with the relative binding of pyridine nucleotides to this enzyme (27). Table 2 shows that when NADPH was replaced by an equimolar concentration of 3-AP-NAD, the latter nucleotide was unable to support the oxidation of methanol and was almost 50% as effective as NADPH when ethanol was the substrate while propanol and butanol oxidation rates with 3-AP-NAD were 115 and 125%, respectively, of those obtained with NADPH. It is noteworthy that the portion of the total NADPH-linked ethanol oxidation which is blocked by 0.1 mM propylpyrazole (Fig. 1A, Table 2) corresponds to the rates of ethanol oxidation obtained with 3-AP-NAD as cofactor (Table 2). It may also be seen in Table 2 that the activities which emerge after subtraction of the 3-AP-NAD- from the NADPH-supported rates are identical for methanol and ethanol. These findings demonstrate directly the presence of ADH in microsomes as well as the participation of this enzyme in the oxidation of the higher aliphatic alcohols. Moreover, the two-fold higher rates with ethanol as opposed to methanol must be attributed to the involvement of two enzymatic components in the overall metabolism of ethanol. Additional evidence in favor of this view was obtained in subsequent experiments (results not shown) which revealed only one apparent K_m value (33 mM) for the NADPH-linked oxidation of methanol, whereas two apparent K_m values (3.3 and 20 mM) were seen with ethanol as substrate.

Repeated claims (1,2,13) that cyanide and azide do not cause an appreciable decrease in NADPH-dependent rates of ethanol oxidation by rat hepatic microsomes have been taken as evidence against the involvement of catalase in this reaction. However, as reported previously (7,8), cyanide and azide caused a marked inhibition of methanol and ethanol oxidation in mouse liver microsomes (Table 2). In the presence of these compounds the activity with methanol was decreased 60-70% while that with ethanol was inhibited 50-60%; residual rates for each alcohol were identical in the presence of either cyanide or azide. The cyanide- and azide-insensitive rates of acetaldehyde production were very similar both to propylpyrazole-sensitive rates of ethanol oxidation as well as to the ethanol-oxidizing activity supported by 3-AP-NAD; conversely, cyanide- and azide-sensitive rates of ethanol oxidation were very similar to propylpyrazole-insensitive rates of acetaldehyde formation (Table 2). These data support the conclusion that catalase and ADH contribute almost equally to the overall metabolism of ethanol by mouse liver microsomes. In contrast, cyanide gave rise to only a 20% decrease in propanol oxidation rates and left butanol oxidation unaffected, while azide actually caused a slight stimulation in propanol- and butanol-oxidizing activities (Table 2). Others have reported a similar effect of azide on the oxidation of the higher alcohols in rat liver microsomes (2).

III. SUMMARY

The profound difference in the response of the various alcohols to the widely different classes of inhibitor used is in sharp contradiction to the postulate that the metabolism of all primary alcohols in hepatic microsomes is mediated by a common catalyst. The observation that alcohol oxidation is markedly inhibited by the pyrazole derivatives provides strong presumptive evidence for the role of ADH in these reactions. The facts that the degree of inhibition exhibited by these compounds effectively increased in the same direction with the known specificity of ADH for aliphatic alcohols, as well as the ability of 3-AP-NAD to replace NADPH in supporting these reactions, reinforce the conclusion that the metabolism of higher alcohols in catalyzed almost exclusively by ADH. The data presented are also consistent with the interpretation that catalase and ADH participate almost to the same extent in the metabolism of ethanol, whereas it seems clear that the oxidation of methanol proceeds almost exclusively via the "peroxidatic" reaction of catalase.

IV. FOOTNOTES

[o]Present address: Department of Biological Chemistry, The

University of Michigan.

[1]The abbreviations used are: ADH, alcohol dehydrogenase; 4-Me-pyrazole, 4-methylpyrazole; Propylpyrazole, 4-n-propylpyrazole, 3-AP-NAD, 3-acetylpyridine NAD.

[2]K.P. Vatsis, and M.P. Schulman, unpublished observation.

V. ACKNOWLEDGEMENTS

We are grateful to Dr. H. Drott, Johnson Research Foundation, for his generous supply of propylpyrazole.

VI. REFERENCES

(1). Teschke, R., Hasumura, Y., and Lieber, C.S., Biochem. Biophys. Res. Commun. 60, 851 (1974).
(2). Teschke, R., Hasumura, Y., and Lieber, C.S., J. Biol. Chem. 250, 7397 (1975).
(3). Chance, B., Acta Chem. Scand. 1, 236 (1947).
(4). Deisseroth, A., and Dounce, A.L., Physiol. Rev. 50,319 (1970).
(5). Chance, B., and Oshino, N., Biochem. J.122, 225 (1971).
(6). Oshino, N., Oshino, R., and Chance, B., Biochem. J. 131, 555 (1973).
(7). Vatsis, K.P., and Schulman, M.P., Pharmacologist 17, 241 (1975).
(8). Vatsis, K.P., and Schulman, M.P., 10th Int. Congress Biochem., Hamburg, p. 667 (1976).
(9). Schulman, M.P., and Vatsis, K.P., Fed. Proc. 36, 333(1977).
(10). Nash, T., Biochem. J. 55, 416 (1953).
(11). Moir, N.J., Miller, W.L., and Gaylor, J.L., Biochem. Biophys. Res. Commun. 33, 916 (1968).
(12). Isselbacher, K.J., and Carter, E.A., Biochem. Biophys. Res. Commun. 39,530 (1970).
(13). Lieber ,C. S., and DeCarli, L.M., J. Biol. Chem. 245,2505 (1970).
(14). Carter, E.A., and Isselbacher, K.J., Ann. N.Y. Acad. Sci. 179, 282 (1971).
(15). Bechtold, M.M., Delwiche, C.V., Comai, K., and Gaylor, J. L., J. Biol. Chem. 247, 7650 (1972).
(16). Comai, K., Delwiche, C.V., Opar, G.E., and Gaylor, J.L., Intern. Res. Commun. System, March 1973.
(17). Lieber, C.S., and Rubin, E., and DeCarli, L.M., Biochem. Biophys. Res. Commun. 40, 858 (1970).
(18). Khanna, J.M., Kalant, H., and Lin, G., Biochem. Pharmacol. 19, 2493 (1970).
(19). Theorell, H., Yonetani, T., and Sjoberg, B., Acta Chem. Scand. 23, 255 (1969).
(20). Li, Ti-K., and Theorell, H., Acta Chem Scand. 23, 892 (1969).

(21). Dahlbom, R., Tolf, B.R., Akeson, A., Lundquist, G., and Theorell, H., Biochem. Biophys. Res. Commun. 57, 549 (1974).
(22). Blomstrand, R., and Theorell, H., Life Sci, 9 (Part II), 631 (1970).
(23). Blomstrand, R., and Foosell, L., Life Sci. 10, 523 (1971).
(24). Theorell, H., Chance, B., Yonetani, T., and Oshino, N., Arch. Biochem. Biophys. 151,434 (1972).
(25). Vatsis, K.P., and Schulman, M.P., in: "Alcohol Intoxication and Withdrawal: Experimental Studies III" (M.M. Gross, Ed.) Adv. Expl. Med. Biol., Plenum Press, New York, 1977, in press.
(26). Vatsis, K.P., and Schulman, M.P., in: "Alcohol and Aldehyde Metabolizing Systems" (R.G. Thurman, J.R. Williamson, H. Drott and B. Chance, Eds.) Vol. 2, Academic Press, New York, 1977, in press.
(27). Kaplan, N.O., Ciotti, M.M., and Stolzenbach, F.E., J. Biol. Chem. 221,833 (1956).

QUANTITATION OF PATHWAYS RESPONSIBLE FOR NADPH-DEPENDENT METHANOL, ETHANOL, AND BUTANOL OXIDATION BY HEPATIC MICROSOMES*

H.J. Brentzel and R.G. Thurman[x]

University of Pennsylvania

NADPH-dependent aldehyde formation from methanol (M), ethanol (E), and butanol (B) was studied in hepatic microsomes. A potent inhibitor of alcohol dehyrogenase (ADH), 4-n-propylpyrazole (1-10 μMolar), inhibited aldehyde production 80% from B, 40% from E, and 10-20% from M. Thus, the degree of inhibition correlated with the substrate specificity of the alcohols (B > E > M) for ADH. Under these conditions, cytochrome P-450-dependent oxidation of aminopyrine was unaffected by 4-n-propylpyrazole. A competitive inhibitor of ADH, trifluoroethanol (200 μM), also decreased rates of aldehyde production from E and B without inhibiting aminopyrine oxidation. Sodium azide (50 μM), an inhibitor of catalase-H_2O_2-dependent oxidations, produced greater than 50% inhibition of M and E oxidation; however, B oxidation was unaffected. The data are consistent with the hypothesis that ADH and catalase account for NADPH-dependent alcohol oxidation by hepatic microsomes. Postulation of a direct oxidation of alcohols via cytochrome P-450-dependent mechanisms (i.e., "MEOS") is unnecessary.

I. INTRODUCTION

It is generally accepted today that ethanol is oxidized by hepatic microsomes in the presence of NADPH and oxygen. However, the precise mechanism responsible for this phenomena is highly controversial. Lieber and his co-workers strongly support the hypothesis that an enzyme system similar to or identical with the drug metabolizing mixed function oxidase is responsible for ethanol oxidation by microsomes (1). However, a number of studies have shown that microsomal ethanol oxidation can be accounted for by H_2O_2 production via microsomal components subsequent to peroxidation of ethanol to acetaldehyde via catalase-H_2O_2 (2,3,4). Furthermore, Isselbacher and Carter (5) suggested a number of years ago that alcohol dehydrogenase was also a component of the microsomal ethanol oxidase, but this has been largely ignored in the recent debate over catalase-H_2O_2 vs. cytochrome P-450 dependent mechanisms. Recently, Teschke et al. (6) demonstrated that butanol oxidation by hepatic microsomes is insensitive to inhibition by sodium azide, an inhibitor of catalase-H_2O_2 and argue that this observation is evidence in favor of a unique microsomal ethanol oxidizing system distinct from catalase-H_2O_2 and presumably dependent upon cytochrome P-450. The purpose of this paper is to present experimental evidence which demonstrates that butanol oxidation both *in vitro* and in the intact perfused organ is catalyzed by alcohol dehydrogenase. The data indicate that microsomal ethanol oxidation involves an alcohol dehydrogenase plus a catalase-H_2O_2 component.

II. METHODS

Preparation of Microsomes

Once-washed microsomes were prepared in 0.15 M KCl as described by Teschke et al. (6). Protein was determined by the biuret reaction (7), employing appropriate controls for the turbidity of the microsomal suspension.

Hemoglobin-Free Liver Perfusion

A recirculating perfusion system described previously (8) was used with slight modifications. The perfusion fluid consisted of 65 ml of Krebs-Henseleit bicarbonate buffer, pH 7.4 (9) containing 2 g of bovine serum albumin per 100 ml. This solution was saturated in a temperature-regulated (37°) disc oxygenator with a mixture of oxygen and carbon dioxide (95:5). Details of the surgical techniques have been published elsewhere (10). The effluent perfusate flows past an oxygen electrode before returning to the oxygenator.

Microsomal Aldehyde Production from Ethanol and Butanol

Microsomal NADPH-dependent ethanol or butanol oxidation was measured as described previously (2). The reaction mixture contained 100 mM potassium phosphate buffer (pH 7.4), 20 mM nicotinamide, 10 mM magnesium chloride, 50 mM ethanol of butanol, and an NADPH-generating system. Incubations were performed in capped 25 ml Erlenmeyer flasks with center wells containing semicarbazide (15 mM in 100 mM phosphate buffer, pH 7.0) in a shaking water bath at 37°. Reactions were initiated by the addition of the NADPH-generating system, and were terminated after ten minutes by the addition of perchloric acid (final concentration - 1.4%). The aldehyde produced during the incubation was trapped as the semicarbazone and compared with aldehyde standards at 224 nm (11). Incubations were performed in duplicate.

Microsomal Formaldehyde Production from Methanol and Aminopyrine

Microsomes were incubated as described previously with 50 mM methanol or 5 mM aminopyrine as the substrate. The mixture was then heated for 8 min. at 55° with an equal volume of Nash reagent (12). Following centrifugation, the absorbance of the supernatant was determined at 405 nm and compared with formaldehyde standard solutions.

Ethanol Utilization by Perfused Livers

Samples of perfusate were collected every 15 min., deproteinized with perchloric acid and neutralized as above. Ethanol was then determined enzymatically by standard procedures (13).

III. RESULTS AND DISCUSSION

In the presence of the catalase inhibitor, sodium azide (Table I), methanol and ethanol oxidation was inhibited 43% and 34% respectively, confirming a catalase contribution to microsomal alcohol oxidation (1). On the other hand, butanol oxidation was not inhibited by azide indicating that it is a poor substrate for catalase-H_2O_2. The effect of azide upon the mixed function oxidation of animopyrine, a model substrate, was also examined. Control rates of drug metabolism as measured by formaldehyde production were unaffected by concentrations of azide employed in the experiments with alcohols, thus obviating a participation of cytochrome P-450 in the azide-sensitive rates of alcohol oxidation.

Table I
Effect of Sodium Azide on NADPH-Dependent Microsomal Alcohol Oxidation

	nmoles aldehyde/min/mg		
	Control	+50 μM N_3^-	% of control
Methanol	5.75	3.30	57%
Ethanol	11.28	7.41	66%
Butanol	7.79	10.67	137%
Aminopyrine	2.91	2.98	102%

Acetaldehyde and butyraldehyde production from ethanol (50 mM) and butanol (50 mM) were determined according to Teschke et al. (6). Formaldehyde production from methanol (50 mM) or aminopyrine (5 mM) was determined according to Nash (12). Data represent means from 5 microsomal preparations.

Table II
Effect of 4-n-Propylpyrazole on NADPH-Dependent Microsomal Alcohol Oxidation

	nmoles aldehyde/min/mg		
	Control	+50 μM Propylpyrazole	% of control
Methanol	8.48	7.01	83%
Ethanol	11.47	7.42	65%
Butanol	6.27	1.68	27%

Conditions as in Table 1. Data = average values from 4 microsomal preparations.

The inhibitory potency of the alkyl pyrazoles on alcohol dehydrogenase increases dramatically with increasing chain length (e.g., n-propylpyrazole has a Ki value in the nanomolar range (14). Therefore, 4-n-propylpyrazole was employed in order to determine what enzyme system(s) are responsible for the microsomal alcohol metabolism which remains in the presence of azide. The extent of inhibition of microsomal alcohol oxidation by n-propylpyrazole was as follows: methanol (17%), ethanol (35%), and butanol (73%) (Table II). This inhibition follows the substrate specificity of the alcohols for alcohol dehydrogenase (butanol > ethanol > methanol). When sodium azide and n-propylpyrazole were added together, microsomal ethanol oxidation was suppressed over 70% (results not shown). In another series of experiments, butyraldehyde production was inhibited up to 50% when a competitive substrate analogue for alcohol dehydrogenase, trifluoroethanol, was added to the microsomal incubation medium (Table III). Aminopyrine oxidation was minimally perturbed by either n-propylpyrazole or trifluoroethanol. Thus, we conclude that microsomal alcohol oxidation has an alcohol dehydrogenase component in addition to catalase-H_2O_2.

Table III
Effect of Trifluoroethanol on Microsomal Butanol and Aminopyrine Oxidation

μMolar Trifluoroethanol	μmoles Butyraldehyde/min/mg	% of control	Aminopyrine Oxidation % of control
0	4.5	100	100
75	3.3	73	93
150	2.3	51	91

Assays for butyraldehyde and formaldehyde as in Table I. Data represent means of duplicate determinations from one microsomal preparation.

Because microsomes in vitro may not reflect the physiological role of the endoplasmic reticulum in vivo, argument over the precise mechanism of alcohol oxidation in the subfraction may be of only academic interest. A more realistic model which closely approximates the actual metabolic environment in which alcohol metabolism occurs is the perfused liver. In the recirculating version of this model, ethanol and butanol were metabolized at similar rates at identical alcohol concentrations (30 mM); however, their sensitivity to 4-methylpyrazole varied (Table IV). The 4-methylpyrazole-insensitive rate of ethanol uptake by the perfused liver was 30% of the control value under these conditions. This residual rate was abolished by the addition of the catalase inhibitor, aminotriazole. Butanol oxidation, on the other hand, was entirely abolished by 4-methylpyrazole. These data support the hypothesis that butanol is oxidized entirely via an alcohol dehydrogenase-dependent pathway in the perfused liver and not via cytochrome P-450 as proposed by Teschke et al. (6). Furthermore, these experiments strongly suggest that alcohol dehydrogenase plus catalase account for ethanol uptake in the intact cell.

Table IV
Alcohol Oxidation by Livers from Normal Rats in Recirculating Perfusion System

	Ethanol		Butanol	
Normal	54.6 ± 5.1	(4)	56.9 ± 7.2	(7)
Normal + 4-Methyl-pyrazole	16.0 ± 3.0	(4)	1.9 ± 4.6	(7)

Livers from normal rats were perfused in the presence of 30 mM ethanol or butanol. Alcohol was determined enzymatically in deproteinized samples (13). 4-methylpyrazole concentration = 0.8 mM. Data represent mean ± S.E.M. (number of livers).

The data presented here clearly demonstrate that microsomal ethanol oxidation has a component which is azide-insensitive in confirmation of Teschke et al. (6). However, because microsomal ethanol oxidation is sensitive of ADH inhibitors (trifluoroethanol and n-propylpyrazole) and because propylpyrazole inhibition follows the homologous series butanol > ethanol > methanol, we conclude that alcohol dehydrogenase - not cytochrome P-450 - is the enzyme system involved. This conclusion is strengthened by the observation that inhibitor concentrations employed in these studies did not inhibit the cytochrome P-450-mediated metabolism of aminopyrine.

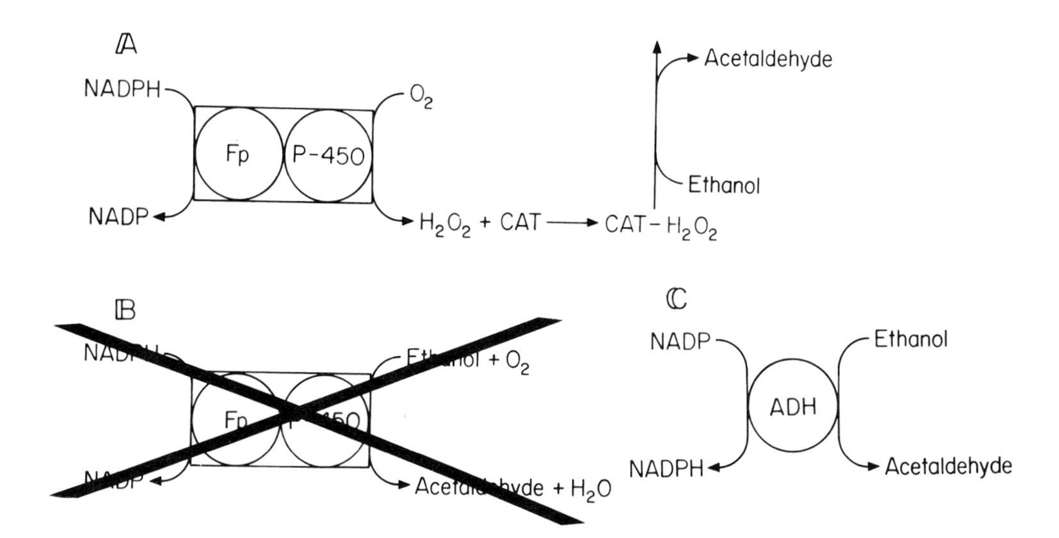

Fig. 1. Scheme depicting possible mechanisms of microsomal ethanol oxidation. Abbreviations used: Fp, cytochrome P-450 reductase; P-450, cytochrome P-450; CAT, catalase; ADH, alcohol dehydrogenase.

Fig. 1 depicts three pathways which could account for microsomal ethanol oxidation: peoxidatic activity of catalase-H_2O_2 (A), the microsomal ethanol oxidizing system (B), and alcohol dehydrogenase (C). The results presented here substantiate the role of catalase and alcohol dehydrogenase (schemes A and C) in both microsomal and whole organ alcohol oxidation. There is no need to postulate ethanol oxidation via cytochrome P-450.

*Supported, in part, by grants from the National Institute of Alcohol Abuse and Alcoholism, AA-00288 and K02 AA-70155.

[x]Present address: Department of Pharmacology, University of North Carolina at Chapel Hill, Chapel Hill, North Carolina, 27514

IV. REFERENCES

(1). Lieber, C.S., and DeCarli, L.M., J. Biol. Chem. 245, 2505-2512 (1970).
(2). Thurman, R.G., Ley, H.G., and Scholz, R., Eur. J. Bio-

chem. 25, 420-430 (1972).
(3). Lin, G., Kalant, H., and Khanna, J.M., Biochem. Pharmacol. 21, 3305-3307 (1972).
(4). Vatsis, K.P., and Schulman, M.P., In "Alcohol and Aldehyde Metabolizing Systems" (R.G. Thurman, T. Yonetani, J.R. Williamson, and B. Chanel, eds.), Academic Press, New York, pp. 287-298, 1974.
(5). Isselbacher, K.J., and Carter, E.A., Biochem. Biophys. Res. Commun. 39, 530-538 (1970).
(6). Teschke, R., Hasumura, Y., and Lieber, C.S., J. Biol. Chem. 250, 7397-7405 (1975).
(7). Gornall, A.G., Bardawill, C.J., and David, M.M., J. Biol. Chem. 177, 751-759 (1949).
(8). Scholz, R., Hansen, W., and Thurman, R.G., Eur. J. Biochem. 38, 64-72 (1973).
(9). Krebs, H.A., and Henseleit, K., Hoppe-Segler's Z. Physiol. Chem. 210, 33-64 (1932).
(10). Scholz, R., In "Stoffwechsel der Perfundierten Leber" (Staib, W. and Scholz, R., eds.), Springer, Berlin, pp. 25-47 (1968).
(11). Gupta, N.K., and Robinson, W.G., Biochem. Biophys. Acta 118, 431-439 (1966).
(12). Nash, T., Biochem. J. 90, 416-424 (1964).
(13). Bergmeyer, H.U. (editor) "Methoden der Enzymatrschen Analyse", Verlag Chemie, Weinheim, 1970.
(14). Theorell, H., Yonetani, T., and Sjöberg, B., Acta Chem. Scand. 23, 255-263 (1969).

CATALASE AND ALCOHOL DEHYDROGENASE-MEDIATED OXIDATION OF ETHANOL IN HEPATIC MICROSOMES OF ACATALASEMIC MICE

K.P. Vatsis and M.P. Schulman

University of Michigan
University of Illinois

Preincubation of hepatic microsomes from acatalasemic mice (C_sb) at 42-44° for 15 min resulted in a 97-99% inactivation of catalase and a concomitant 60-70% loss of ethanol oxidation (MEOS); neither enzymic activity was altered in microsomes of the genetic control mice (C_sa) by the same heat treatment. The residual MEOS activity in 43°-preincubated C_sb microsomes was unaffected by cyanide or azide but was totally abolished by propylpyrazole, a potent inhibitor of alcohol dehydrogenase. Neither propylpyrazole nor preincubation at 43° had any influence on NADPH-linked drug hydroxylations in either C_sa or C_sb microsomes. Added beef liver catalase restored MEOS activity of 43°-preincubated C_sb microsomes to levels that were slightly higher than those obtained with corresponding C_sa preparations under similar conditions. These results demonstrate that microsomal ethanol oxidation can be accounted for by catalase (60-70%) and alcohol dehydrogenase (30-40%) and that the postulation of MEOS, as proposed by Lieber and DeCarli (1), is unwarranted.

I. INTRODUCTION

The nature of the enzymic components responsible for the NADPH-linked oxidation of ethanol by hepatic microsomes has been the subject of controversy (1-4). Reports from this laboratory (5,6), using a mutant strain of mice (C_sb) with thermolabile hepatic catalase (7), demonstrated that microsomal ethanol metabolism, measured by ethanol disappearance, was abolished when catalase was heat-inactivated at 37°C. These findings were subsequently challenged (8,9) by the finding that, when microsomal ethanol metabolism was assayed by acetaldehyde formation, little difference was noted between C_sb mice and their genetic control counterparts, C_sa mice. Methodological differences have been responsible in part for the conflicting results and are discussed elsewhere (10). We have reexamined ethanol metabolism using the same methodology as Lieber and DeCarli (8). Again, the findings to be presented in this study demonstrate clearly that catalase is responsible for the major portion of ethanol metabolism in hepatic microsomes and do not support the claim (1,8) that this subcellular organelle contains a unique microsomal ethanol-oxidizing system, *i.e.*, MEOS.

II. RESULTS AND DISCUSSION

The catalase activity of hepatic microsomes of C_sb mice was not altered by preincubating the preparation up to 30°C; at 43°C, however, more than 96% of this enzymic activity was lost, while that of the control C_sa mice was unaltered. Similar results were obtained with two different assay procedures for catalase (Table 1).

Since catalase activity of the C_sb mice was destroyed during assay at 37°C, meaningful comparisons can only be made by comparing catalase and alcohol-oxidizing activities both assayed at 30° where catalase of the mutant mouse does not become inactivated. At this temperature, C_sa and C_sb microsomes exhibited equal rates for the NADPH-linked oxidations of methanol and ethanol (Table 2; Fig. 1). At 37°C, however, there was a loss of about 60 and 20%, respectively, of the oxidations of the two alcohols in non-preincubated C_sb microsomes as compared to corresponding C_sa preparations; catalase activity in the C_sb microsomes was decreased about 50% while that of the C_sa preparation was unaltered. By comparing control activities with those of microsomes heated at 43°C for 15 min before incubation at 30°C, it can readily be seen that the C_sb microsomes lost 83% and 64%, respectively, of their methanol- and ethanol-oxidizing activities; catalase loss in the C_sb preparation was 98% (Table 2). In contrast to methanol and ethanol, propanol oxidation rates decreased by only 20% in 43°-preincubated C_sb microsomes and the oxidation of butanol and pentanol proceeded unimpaired in these preparations (Fig. 1). Preincubation at 43°C had no effect on the alcohol-oxidizing activity of C_sa microsomes (Table 2; Fig. 1). It should be clear from the aforementioned that catalase is responsible for all of methanol as well as for the majority of ethanol oxidation, whereas only a very small portion of the propanol-oxidizing activity of hepatic microsomes can be accounted for by this enzyme. In complete accord with these findings, other studies from our laboratory (14-16) have shown that (a) the oxidation of methanol and ethanol in liver microsomes of Swiss mice was inhibited 60-70% and

TABLE 1

Effect of Preincubation at Different Temperatures on Catalase Activity. Comparison of Two Methods

Preincubation		Baudhuin et al.				Beers and Sizer			
		C_sa		C_sb		C_sa		C_sb	
Temp.	Time	Activity	%	Activity	%	Activity	%	Activity	%
0°	None	.317	100	.348	100	.618	100	.702	100
25°	10 min	.334	105	.334	96	.618	100	.744	106
30°	10 min	.343	108	.331	95	.588	95	.674	96
37°	10 min	.340	107	.230	66	.618	100	.478	68
43°	15 min	.326	103	.016	4	.588	95	.010	1

Liver microsomes from either strain were kept in an ice bath at 0°-4° or were preincubated in closed vessels as indicated above. Following preincubation, the vessels were immediately chilled in ice before assaying for catalase activity. When the method of Baudhuin et al. *(11) was used, aliquots from each suspension were incubated 10 minutes at room temperature in the presence of H_2O_2, and catalase activity was determined by measuring the residual H_2O_2 at 410 nm as the yellow peroxy titanium sulphate complex; results are expressed as ΔA_{410}/min/mg protein. The same preparations were assayed for catalase activity by following the decrease in absorbance at 240 nm as described by Beers and Sizer (12); the reaction was carried out for 3 minutes at room temperature, and results are expressed as ΔA_{240}/min/mg protein.*

50-60%, respectively, by 0.1 mM cyanide or azide, whereas propanol oxidation rates were decreased by only 20% in the presence of cyanide; (b) propylpyrazole, a potent inhibitor of alcohol dehydrogenase (17), did not appreciably affect methanol oxidation, but caused a 40% inhibition of ethanol oxidation rates and almost totally obliterated the oxidation of the higher alcohols (16); (c) the NADPH-dependent oxidation of propanol showed two K_m values one of which is consistent with catalase (16); and (d) crystalline beef liver catalase was capable of metabolizing propanol in the presence of glucose-glucose oxidase at rates comparable to those obtained with methanol and ethanol under similar conditions (14,16), though careful kinetic studies revealed that propanol was not nearly as good a substrate for this system as the two lower alcohols (16).

The effect of crystalline beef liver catalase on ethanol oxidation by C_sa and C_sb microsomes is shown in Fig. 2.

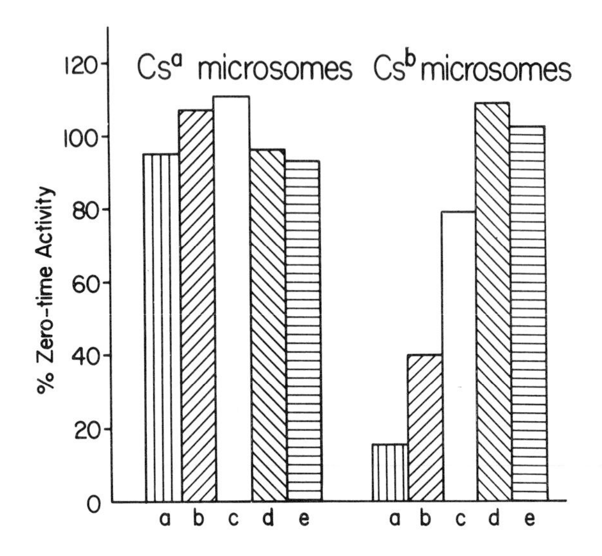

Fig. 1. Effect of preincubation temperature on NADPH-dependent alcohol oxidation in liver microsomes of C_sa and C_sb mice. Microsomes from either mouse strain were kept at 0°-4°C (zero-time controls) or were preincubated at 43°C for 15 min as described in Table 1. Alcohol-oxidizing activities were determined in reaction mixtures containing aliquots of the above microsomal suspensions (1.8 mg of protein from either condition or strain), 1.0 mM NADPH, and 80 mM alcohol substrate in a total volume of 1.5 ml of Krebs-Ringer-phosphate buffer, pH 7.4. Reactions were carried out for 10 min at 30° in closed 25-ml Erlenmeyer flasks without center wells (methanol substrate) or with center wells (all other alcohols) containing 1.0 ml of 0.1 M semicarbazide in 0.16 M sodium phosphate buffer, pH 7.0. Formaldehyde formed from methanol was determined according to the method of Nash (13). All other aldehydes were estimated after an overnight diffusion period by measuring the absorbance at 224 nm of the corresponding aldehyde semicarbazone, as described by Teschke et al. (9). Control or 'zero-time' activities (taken as 100% for microsomes kept at 0°-4°C) were as follows for C_sa and C_sb mice, respectively (in nmol aldehyde formed/min/mg protein): a. Methanol, 5.1 & 5.3; b. Ethanol, 12.7 & 11.9; c. n-Propanol, 8.4 and 10.2; d. n-Butanol, 7.7 & 7.5; & e. n-Pentanol, 5.1 & 5.3. Catalase activities (ΔA_{410}/min/mg protein) for C_sa and C_sb microsomes,

respectively, were: zero-time, 0.64 and 0.59; incubated at 30°C for 10 min, 0.67 and 0.64; incubated at 43°C for 15 min, 0.77 and 0.016.

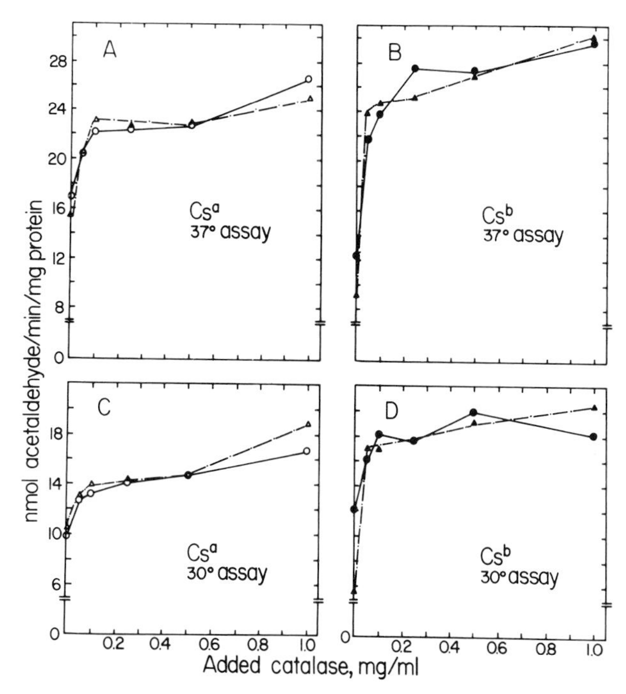

Fig. 2. Effect of added crystalline beef liver catalase on NADPH-dependent ethanol oxidation in liver microsomes of C_sa and C_sb mice. Microsomes from either strain were kept in an ice bath at 0°-4° C (o, ●) or were preincubated at 43°C for 15 min (△,▲) as described in the legend to Table 1. Reaction mixtures consisted of microsomes (1.8 mg of protein from either condition or strain), 1.0 mM NADPH, 80 mM ethanol, and increasing amounts of crystalline beef liver catalase (Sigma, 14,400 U/mg) at the indicated final concentrations (expressed per ml of reaction mixture) in a total volume of 1.5 ml of Krebs-Ringer-phosphate buffer, pH 7.4. Reactions were for 10 min at 37°C (panels A & B) or 30°C (panels C & D), and were carried out in the main chambers of stoppered 25-ml Erlenmeyer flasks with center wells containing 1.0 ml of 0.1 M semicarbazide in 0.16 M sodium phosphate buffer, pH 7.0. Reactions were stopped with 0.3 ml of 50% trichloracetic acid. After an overnight diffusion period at room temperature, the concentration of acetaldehyde semicarbazone was determined at 224 nm as described by Teschke et al. (9). Catalase activities (ΔA_{410}/min mg protein) for C_sa and C_sb microsomes, respectively, were: (a) zero-time, 0.53 & 0.50; (b) incubated at 30°C for 10 min, 0.50 and 0.61; (c) incubated at 37°C for 10 min, 0.52 & 0.35 and (d) incubated at 43°C for 15 min, 0.54 & 0.017.

TABLE 2
Effect of Preincubation of Microsomes on NADPH-linked Alcohol Oxidations at 37°C and 30°C

Substrate	Preincubation	37° Assay $C_s a$	37° Assay $C_s b$	30° Assay $C_s a$	30° Assay $C_s b$
		nmol aldehyde/min/mg protein			
Methanol	None	10.1	3.9	5.8	5.4
	43°, 15 min	10.5	2.1	5.5	0.9
Ethanol	None	14.0	11.0	10.8	10.2
	43°, 15 min	13.8	5.5	10.6	3.7

Microsomes from either strain were kept in an ice bath or preincubated at 43°C for 15 minutes as described in the legend to Table 1. Aliquots (1.5 mg protein) were subsequently assayed for alcohol oxidation in a total volume of 1.5 ml of Krebs-Ringer-phosphate buffer, pH 7.4, in the presence of 1.0 mM NADPH and 80 mM alcohol substrate. Incubations were for 10 minutes at 37°C or 30°C, and acetaldehyde formation was measured as described by Teschke et al. (9). Formaldehyde formation was assayed by the method of Nash (13). Catalase activities (ΔA_{410}/min/mg protein) for $C_s a$ and $C_s b$ microsomes, respectively, were: (a) zero-time, 0.50 and 0.49; (b) incubated at 30°C for 10 minutes, 0.51 and 0.50; (c) incubated at 37°C for 10 minutes, 0.48 and 0.26; and (d) incubated at 43°C for 15 minutes, 0.51 and 0.01.

Addition of small amounts of catalase (0.05-0.1 mg/ml) resulted in an enhancement of the ethanol-oxidizing activity of $C_s a$ microsomes (Fig. 2A, 2C), indicating that H_2O_2 formation from NADPH is in excess of the catalase concentration in these preparations under the conditions employed. Since heat pretreatment has no effect on the alcohol-metabolizing capacity of these preparations, rates of ethanol oxidation in both non-preincubated and 43°-preincubated $C_s a$ microsomes were increased by the same percent in the presence of catalase irrespective of assay temperature (Fig. 2A, 2C). Addition of catalase at concentrations greater than 0.1 mg/ml did not result significantly in a further enhancement of ethanol oxidation by the $C_s a$ preparations, *i.e.*, rates were approximately 1.2-fold higher at 1.0 mg/ml as opposed to 0.1 mg catalase/ml. When catalase was added to non-preincubated $C_s b$ microsomes assayed for activity at 37°C, ethanol oxidation rates were increased to levels that were slightly higher than those obtained with corresponding preparations from $C_s a$ mice under identical conditions (Fig. 2B). Thus, loss of catalase activity during assay

TABLE 3
Effect of Preincubation Temperature on Microsomal NADPH-linked Drug=metabolizing Enzyme Activities in C_sa and Csb Mice

Preincubation	C_sa –	C_sa +	C_sb –	C_sb +
	nmol product/min/mg protein			
Substrate				
Aniline	1.9	2.1	1.5	1.8
Ethylmorphine	10.7	10.1	7.8	7.4
Benzphetamine	8.4	9.1	8.9	9.3
Aminopyrine	16.1	14.8	14.2	14.0

Once-washed liver microsomes from either strain were suspended in 0.1 M sodium phosphate buffer, pH 7.4, to a protein concentration of 7.0 mg per ml. One-half of this suspension from each strain was preincubated for 15 minutes at 43°C, while the other half was kept in an ice bath at 0°-4°C. Rates of drug metabolism were determined in reaction mixtures consisting of microsomes (2.0 mg protein), 1.0 mM NADPH, 5.0 mM $MgCl_2$, and drug substrate (8 mM aniline or aminopyrine; 5 mM ethylmorphine or benzphetamine) in a total volume of 1.0 ml of 0.1 M sodium phosphate buffer, pH 7.4. All incubations were in duplicate and were carried out for 10 minutes at 37°C, except for aminopyrine which was for 5 minutes. Product formation was assayed as described previously (19).

of non-preincubated C_sb microsomes at 37°C gives rise to a higher H_2O_2/catalase ratio resulting in a consistently lower ethanol-oxidizing activity of these microsomes (Table 2; Fig. 2A, 2B). Furthermore, added catalase restored ethanol oxidation in 43°-preincubated microsomes to levels obtained with non-preincubated preparations from this strain regardless of the assay temperature (Fig. 2B, 2D). These results demonstrate clearly that the loss of ethanol oxidation in heat-pretreated C_sb microsomes is due to inactivation of the thermolabile catalase in these preparations, and not to an impairment of any other microsomal component or to a decrease in the NADPH-linked H_2O_2-producing enzymes.

Since it has been claimed that MEOS shares properties in common with the microsomal drug-metabolizing system (18), it was of interest to determine if drug metabolism in the C_sb preparation was decreased as was alcohol oxidation. That this was not the case can be seen in Table 3, in which rates of hydroxylation of four mixed function oxidase substrates were unaltered by preincubation at 43°C for 15 min.

It is doubtful that the remaining catalase in 43°-preincubated C_sb microsomes (1-4%) could account for the residual methanol- or ethanol-oxidizing activities in these preparations, since, under these conditions, the profound drop in catalase would result in such a marked increase in the H_2O_2/catalase ratio that the system would be essentially uncoupled peroxidatically. This statement presupposes equal rates of H_2O_2 generation from NADPH in C_sb microsomes before or after preincubation at 43°C for 15 min, *i.e.*, preincubation at this temperature only affects catalase and not the rates of NADPH-dependent H_2O_2 formation. That this is indeed the case was shown in the experiments illustrated in Fig. 2 in which catalase addition to 43°-preincubated microsomes resulted in ethanol-oxidizing activities that were slightly higher than those obtained with corresponding C_sa preparations under identical conditions. Clearly, if the NADPH-linked H_2O_2-producing enzymes had been adversely affected by exposure of C_sb microsomes to elevated temperatures, added catalase would not have given rise to any increase in ethanol oxidation by these preparations. Moreover, the experiments depicted in Table 3 demonstrated that preincubation at 43°C for 15 min had no effect on the drug-metabolizing activity of C_sb microsomes, indicating that cytochrome P-450 remained intact following heat pretreatment of these preparations. Cytochrome P-450 has been shown to catalyze H_2O_2 formation from NADPH both in rat liver microsomal suspensions (20-23) and in a reconstituted enzyme system containing electrophoretically homogeneous cytochrome P-450 from phenobarbital-induced rabbit liver microsomes (24,25).

As expected from a peroxidatically-uncoupled system, *i.e.*, a system in which catalase is functioning almost exclusively in its catalatic mode, addition of 0.1 mM cyanide or azide to 43°-preincubated C_sb microsomes was without effect on the residual alcohol-oxidizing activities. These findings suggest that some enzymic component other than catalase was responsible for alcohol oxidation in these preparations. Alcohol dehydrogenase activity has been reported previously in hepatic microsomes of the rat (26,27) and may account for a part of the ethanol oxidation in this organelle. Using propylpyrazole, a portion of MEOS was inhibited in mouse (15,16) and rat (28) microsomal preparations. In addition, kinetic studies with mouse hepatic microsomes have revealed two K_m values for ethanol and propanol, one consistent with alcohol dehydrogenase and the other with catalase (16).

TABLE 4

Effect of Propylpyrazole on NADPH-linked Alcohol and Drug Oxidations by Liver Microsomes of C_sa and C_sb Mice

Pre-incubation	Propyl pyrazole	% Control Activity* Ethanol Oxidation C_sa	C_sb	Methanol Oxidation C_sa	C_sb	Ethylmorphine Demethylation C_sa	C_sb
+	-	120	42	98	19	n.d.	n.d.
-	+	63	61	78	71	90	91
+	+	73	4	82	0	n.d.	n.d.

**Control activities for ethanol and methanol oxidations were determined by incubating microsomes at 30° for 10 minutes in the absence of 0.03 mM propylpyrazole and measuring acetaldehyde and formaldehyde, respectively, as described in legend to Table 2; for ethylmorphine demethylation control assays were carried out for 5 minutes at 37°. Preincubations were at 43° for 15 minutes. Catalase activities (ΔA_{410}/min/mg protein) for C_sa and C_sb microsomes, respectively, were (a) zero-time, 0.5 and 0.58; (b) incubated at 30° for 10 minutes, 0.49 and 0.55; (c) incubated at 43° for 15 minutes, 0.55 and 0.005.*

When propylpyrazole[1] was incubated with the microsomal preparations, the residual oxidation of methanol and ethanol in heat-pretreated C_sb microsomes was obliterated. There was little effect on ethylmorphine demethylation (Table 4) as well as on aniline hydroxylation and on benzphetamine and aminopyrine N-demethylation (results not presented). It seems clear that microsomal alcohol oxidation can be satisfactorily accounted for by the combined activities of catalase and alcohol dehydrogenase.

III. REFERENCES

(1). Lieber, C.S., and De Carli, L.M., J. Biol. Chem. 245, 2505 (1970).

(2). Isselbacher, K.J., and Carter, E.A., Biochem. Biophys. Res. Commun. 39, 530 (1970).

---

[1]We are grateful to Drs. H. Drott and B. Chance, Johnson Research Foundation, for their generous supply of propylpyrazole.

(3). Thurman, R.G., Ley, H.G., and Scholz, R., Eur. J. Biochem. 25, 420 (1972).
(4). Lin, G., Kalant, H., and Khanna, J.M., Biochem. Pharmacol. 21, 3305 (1972).
(5). Vatsis, K.P., and Schulman, M.P., Biochem. Biophys. Res. Commun. 52, 588 (1973).
(6). Vatsis, K.P., and Schulman, M.P., in "Alcohol and Aldehyde Metabolizing Systems" (R.G. Thurman, T. Yonetani, J.R. Williamson, and B. Chance, eds.), Vol. 1, p. 287. Acedemic Press, New York, 1974.
(7). Feinstein, R.N., Braun, J.T., and Howard, J.B., Arch. Biochem. Biophys. 120, 165 (1967).
(8). Lieber, C.S., and De Carli, L.M., Biochem. Biophys. Res. Commun. 60, 1187 (1974).
(9). Teschke, R., Hasumura, Y., and Lieber, C.S., Mol. Pharmacol. 11, 841 (1975).
(10). Vatsis, K.P., and Schulman, M.P., in "Cytochromes P-450 and b_5", (D.Y. Cooper, Rosenthal, O., Synder, R., and Witmer, C., eds.), Adv. Exper. Med. Biol. Vol. 58, p. 369. Plenum Press, New York, 1975.
(11). Baudhuin, P., Beaufay, H., Rahman-Li, Y., Sellinger, O. Z., Wattiaux, R., Jacques, P., and DeDuve, C., Biochem. J. 92, 179 (1964).
(12). Beers, R.F., and Sizer, I.W., J. Biol. Chem. 195, 133 (1952).
(13). Nash, T., Biochem. J. 55, 416 (1953).
(14). Vatsis, K.P., and Schulman, M.P., The Pharmacologist 17, 241 (1975).
(15). Vatsis, K.P., and Schulman, M.P., Tenth Internl. Congress Biochem., Hamburg, p. 667 (1976).
(16). Chance, B., Vatsis, K.P., and Schulman, M.P., in "Alcohol and Aldehyde Metabolizing Systems" (R.G. Thurman, J.R. Williamson, H. Drott, and B. Chance, Eds.), Vol. 2. Academic Press, New York, 1977, in press.
(17). Dahlbom, R., Tolf, B.R., Akeson, A., Lundquist, G., and Theorell, H., Biochem. Biophys. Res. Commun. 57, 549 (1974).
(18). Rubin, E., and Lieber, C.S., Science 172, 1097 (1971).
(19). Vatsis, K.P., Kowalchyk, J.A., and Schulman, M.P., Biochem. Biophys. Res. Commun. 61, 258 (1974).
(20). Hildebrandt, A.G., and Roots, I., Arch. Biochem. Biophys. 171, 385 (1975).
(21). Werringloer, J., and Estabrook, R.W., Hoppe-Seyler's Z. Physiol. Chem. 357, 1063 (1976).
(22). Werringloer, J., in "Proceedings of the Third International Symposium on Microsomes and Drug Oxidations" (V. Ullrich, I. Roots, A.G. Hildebrandt, R.W. Estabrook, and A.H. Conney, Eds.). Pergamon Press, Oxford, 1977, in press.

(23). Werringloer, J. Chacos, N., Estabrook, R.W., Roots, I., and Hildebrandt, A.G., in "Alcohol and Aldehyde Metabolizing Systems" (R.G. Thurman, J.R. Williamson, H. Drott, and B. Chance, Eds.), Vol. 2. Academic Press, New York, 1977, in press.

(24). Nordblom, G.D., and Coon, M.J., Fed. Proc. 35, 281 (1976).

(25). Vatsis, K.P., and Coon, M.J., in "Alcohol and Aldehyde Metabolizing Systems"(R.G. Thurman, J.R. Williamson, H. Drott, and B. Chance, Eds.), Vol. 2. Academic Press, New York, 1977, in press.

(26). Carter, E.A., and Isselbacher, K.J., Ann. N.Y. Acad. Sci. 179, 282 (1971).

(27). Comai, K., Delwiche, C.V., Opar, G.E., and Gaylor, J.L., Internl. Res. Commun. System, March, 1973.

(28). Brentzel, H.J., and Thurman, R.G., in "Alcohol and Aldehyde Metabolizing Systems" (R.G. Thurman, J.R. Williamson, H. Drott, and B. Chance, Eds.), Vol. 2, Academic Press, New York, 1977, in press.

ANALYSIS OF MITOCHONDRIAL AND MICROSOMAL CYTOCHROMES IN HUMAN LIVER BIOPSY

N. Sato, T. Kamada, H. Abe, B. Hagihara & B. Chance

Osaka University
University of Pennsylvania

Studies on mitochondrial and microsomal cytochromes in human livers have been carried out in relation to ethanol metabolism. To determine the cytochrome content in small biopsy samples, a micromethod was developed, in which different CO reassociation behavior among cytochromes aa_3 and P-450 and hemoglobin after photolysis of the CO-treated sample at low temperature was utilized. The results with 62 cases with a variety of liver diseases have revealed as follows: 1) the liver cytochrome aa_3 content did not significantly change in drinker with chronic hepatitis (CH) as compared to that of the non-drinkers, while it slightly decreased in the case of liver cirrhosis (LC); 2) cytochrome P-450 content significantly increased in the drinkers with CH ($p < 0.05$), while no change was observed in the drinkers with LC; 3) the P-450/b_5 ratio of non-drinkers' livers was constant, while this ratio was variable in the drinkers.

I. INTRODUCTION

It has been shown that the respiratory function in liver mitochondria of alcohol-treated animals has a close relation to the rate of reoxidation of NADH generated in the alcohol dehydrogenase pathway (e.g. 1-3), as well as to acetaldehyde metabolism in mitochondria (4,5). It has been suggested that the increased rate of ethanol consumption (e.g. 6) results from the increased mitochondrial reoxidation of the NADH (7). Thus, the determination of the mitochondrial function in patients' livers seems to be of clinical and biochemical importance, since it may yield information about the possible toxicity of ethanol on the underlining disease(s). Cytochrome P-450, which is presumably implicated in NADPH-dependent ethanol oxidation (8), has been shown to increase following chronic ethanol feeding in rats (9), a result suggesting that the adaptive increase in activity of MEOS may at least in part be involved in the increased rate of ethanol metabolism. Thus the determination of microsomal and mitochondrial cytochromes in the livers of patients with chronic liver disease seems appropriate.

In this paper we describe a new method for measuring simultaneously the contents of the cytochromes aa_3, P-450 and b_5 in very small sample obtained by needle biopsy. The result obtained shows no significant change of cytochrome aa_3 content in drinkers' livers, while a significant increase in the liver cytochrome P-450 content was observed with concomitant change of the ratio of P-450/b_5 contents by chronic alcohol intake.

II. MATERIALS AND METHODS

The liver specimens were obtained from patients of the 1st Department of Internal Medicine, Osaka University Hospital (Prof. H. Abe), and Internal Medicine, Osaka Prefectural Hospital (Director, Dr. A. Takaoka), and Gastroenterological Division, Center for Adult Disease (Director, Dr. J. Kojima), Osaka, Japan. Liver biopsies were performed on 62 subjects (46 males and 16 females) for clinical diagnosis. The specimens were classified by clinical and histological criteria into the following groups: a) normal (normal liver function test, no increase of transaminases, normal histology or mild fatty infiltration); b) chronic hepatitis; c) cirrhosis; d) primary hepatoma. All patients gave their consent to the procedure. Every patient received 100 mg Ca-pentobarbital 30 min before the laparotomy. The liver biopsy was taken with a Silverman needle. Since a great variety of drugs such as phenobarbital, meprobamate, phenylbutazone, griseofluvin induce the drug-hydroxylating enzymes in the endoplastic reticulum, and several drugs have been claimed to affect ethanol metabolism, the patients who had been taking such drugs including corticosteroid, contraceptives and analgesics were omitted from the present investigation. Patients who habitually drink alcohol beverages more than 40 g/day as ethanol for more than ten years (average 18 years) but stop drinking 5 to 30 days before biopsy were taken as drinkers. Severe alcoholics and patients with diabetes mellitus were excluded from this study.

Liver samples approximately 5 to 10 mg wet weight (1/3 - 1/4 of fragment biopsied) was washed with cold saline and then stored at -30°C. Low temperature spectra were recorded usually on the following day, but at most within one month. The thawed sample was homogenized in a glass-glass homogenizer with a medium containing 1.0 M sucrose, 1.0 mM EDTA and 20 mM Tris-HCl (pH 7.4).

The cytochrome aa_3 contents were determined according to the method of Sato *et al.* (10). Cytochrome P-450 content was determined by the absorbance difference between 449 and 490 nm in the same low temperature spectra obtained for measuring the cytochrome aa_3 content (10), after the absorbance contribution of cytochrome a at these wavelengths difference had been calculated. The value obtained with purified cytochrome aa_3 from pigeon heart mitochondria was 0.70. Cytochrome b_5 was determined by the NADH (0.1 mM) reduced minus oxidized low temperature spectrum of the homogenate in the presence of rotenone (5 μM) (see the legend of Fig. 1). The following equations are used for the calculation of the mitochondrial as well as microsomal cytochromes (m mol, g^{-1}) from the low temperature spectra I through III. The numerals I, II and III are shown in the legend of Fig. 1.

$$[aa_3] = \Delta E_{II(602-590\sim625)}/\Delta\varepsilon \times K \times SW$$

$$[P\text{-}450] = \frac{\Delta E_{II(449-490)} - 0.70 \times [\Delta E_{III(443-455)} - \Delta E_{II(443-455)}]}{\Delta\varepsilon \times K' \times SW}$$

$$[b_5] = \Delta E_{I(552-542\sim568.5)}/\ \Delta\varepsilon \times K'' \times SW$$

K, K' and K'' are conversion factors in the low temperature spectrophotometry of cytochromes aa_3(8.02), P-450(3.84) and b_5(8.02), respectively. Difference extinction coefficients ($\Delta\varepsilon$) are 13.1 for cytochrome aa_3 (11), 91.0 for cytochrome P-450 (12), and 21.0 for cytochrome b_5 (13). *SW* is the weight of the sample (g).

A typical low temperature spectra of human liver is shown in Fig. 1. The spectra are qualitatively identical to those obtained with rat liver. From these spectra, the concentration of the cytochromes were calculated using the equations described above. The absorbances of the α and Soret bands of these cytochromes of rat liver homogenate have been shown to be a linear function of the concentration of the homogenate (*cf*, ref. 10).

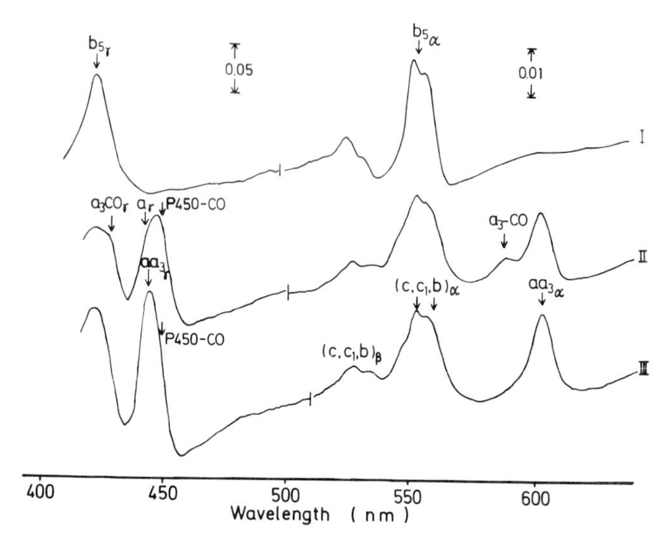

Fig. 1. Difference spectra in liver homogenates from a patient with chronic hepatitis. The measure and reference chambers of the split-beam spectrophotometer cuvet contained a mixture of 1.0 M sucrose, 1.0 mM EDTA, 0.02 M Tris-HCl buffer (pH 7.4), 5 μM rotenone and 15.0 mg sample , and bubbled with CO : O_2 mixed gas for 60 sec. Then the sample in the measure cuvet was treated with 0.1 mM NADH, and both chambers were immersed in liquid nitrogen. The first spectrum was taken at -196°C(I). The samples were thawed to room temperature, and solid Na-dithionite was added to the measure chamber and NADH was added to the reference chamber. 3 min later the samples were cooled to -196°C in complete darkness. The spectrum was taken from 650 to 430 nm (II). Then the samples were illuminated with a 200 W tungsten lamp at about 3 cm distant from the samples at -196°C and a spectrum was taken at this temperature (III).

III. RESULT

Cytochrome aa_3 content of the "normal liver" was 7.2 ± 2.4 n moles/g liver wet weight (mean ± S.D.). In patients with chronic hepatitis, it was 7.7 ± 2.9 n moles/g. A slight decrease (5.6 ± 1.8) was shown in patients with cirrhosis and a significant decrease (2.0 ± 1.0) in patients with primary hepatoma (Fig. 2). When the liver cytochrome aa_3 content was compared between drinkers and nondrinkers, the drinkers showed a slight increase but it was not significant ($0.10 < p < 0.20$).

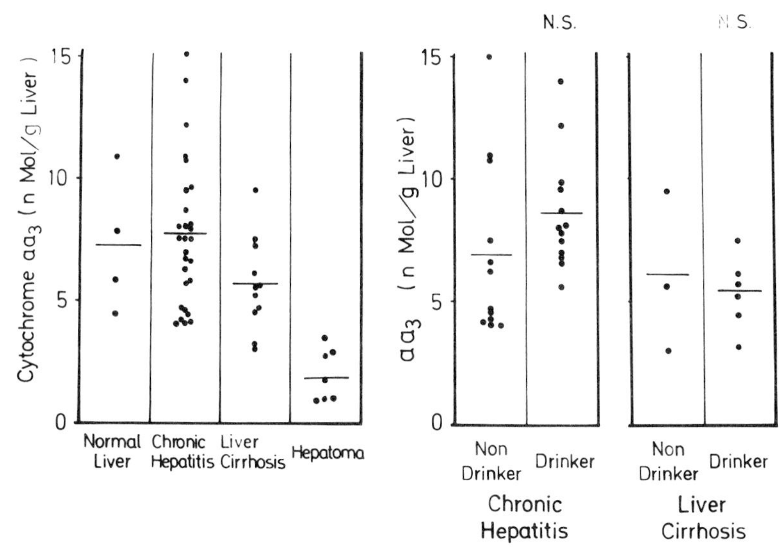

Fig. 2. (Left) Human liver cytochrome aa_3 content. (Right) Comparison of liver cytochrome aa_3 content between the drinkers and nondrinkers. Values were calculated from the low temperature difference spectra obtained under the condition as shown in Fig. 1. The mean ± S.D. (n moles/g liver wet weight) are given in the text.

No significant difference was observed in the contents of cytochrome P-450 among the livers of "normal" (13.6 ± 4.4 n moles/g liver), chronic hepatitis (13.6 ± 6.8) and "cirrhosis" (11.0 ± 5.2) groups (Fig. 3). However, a large variation in the cytochrome P-450 content was observed among the livers studied. The P-450 content decreased markedly in the primary hepatomas (0.3 ± 0.4), and no cytochrome P-450 was detected in four cases among seven, as was observed in experimental hepatomas in rats and mice (14,15). It is interesting to note that the cytochrome P-450 content was increased significantly in the drinkers' livers of the "chronic hepatitis" group ($p < 0.05$). On the other hand, in the case of "cirrhosis", no difference was observed between the two groups. Fig. 4 shows the relation of cytochrome P-450 content in livers to the daily alcohol intake. A significant increase of P-450 content was observed in patients who had habitually been taking alcohol in relatively small doses (40-120 g/day). On the other hand, heavy drinkers (> 120 g/day) showed no significant increase.

The content of cytochrome b_5 in livers of the "chronic hepatitis" group (7.0 ± 3.3 n moles/g liver) was slightly higher than that of the "normal liver" group (5.3 ± 1.5), while

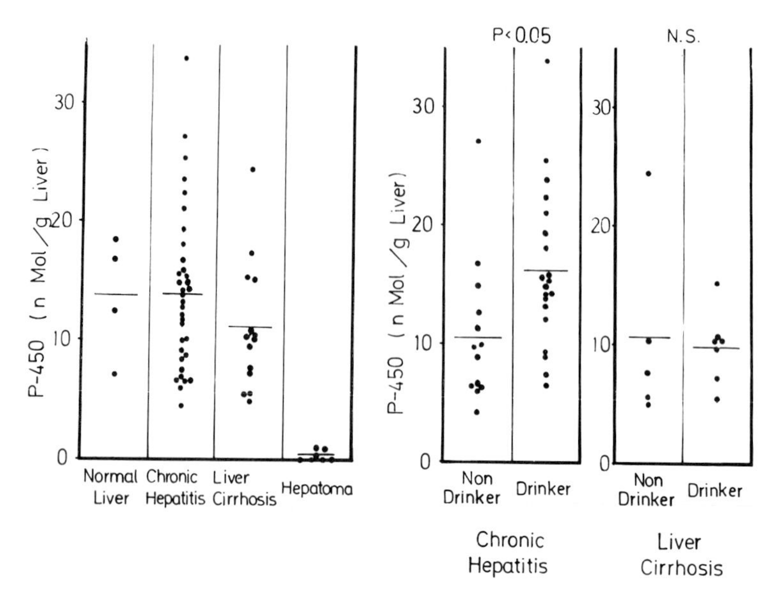

Fig. 3. (Left) Human liver cytochrome P-450 content. (Right) Comparison of liver cytochrome P-450 content between the drinkers and nondrinkers. Conditions as in Fig. 2.

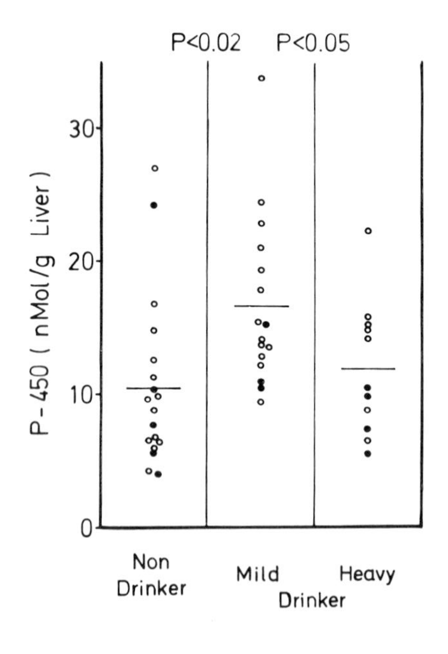

Fig. 4. Relation of liver cytochrome P-450 content and daily alcohol intake in patients with chronic hepatitis and cirrhosis. Daily intake in alcohol in mild drinkers was 40-120 g/day, and that of the heavy drinkers 120 g/day. Open circle, chronic hepatitis; closed circle, cirrhosis.

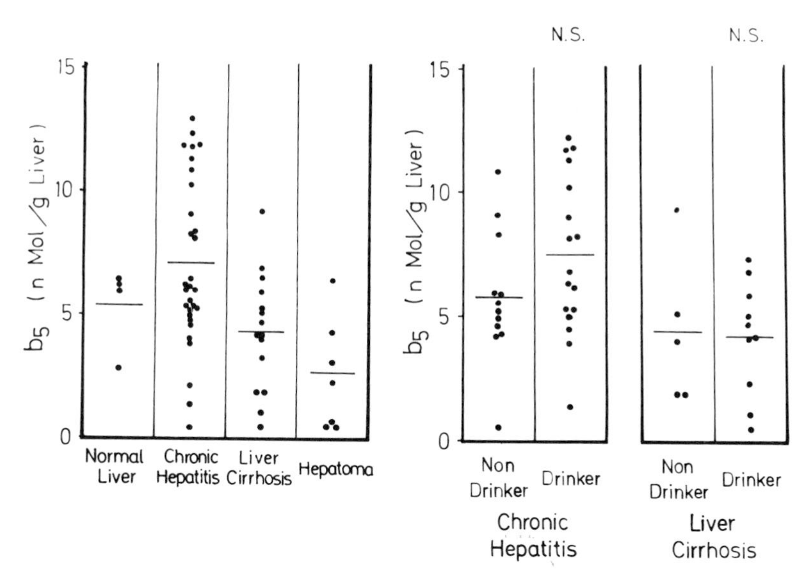

Fig. 5. (Left) Human liver cytochrome b_5 content. (Right) Comparison of liver cytochrome b_5 content between the drinkers and nondrinkers.

that of the "cirrhosis" group (4.2 ± 2.4) was slightly lower than that of the normals (Fig. 5). Cytochrome b_5 was detected in all cases of primary hepatoma (2.45 ± 2.07), in contrast to the almost complete disappearance of cytochrome P-450 in these tumors. In Fig. 6A, the contents of the microsomal cytochromes P-450 and b_5 in the livers of the nondrinker's patients are plotted. The regression line has been calculated, and

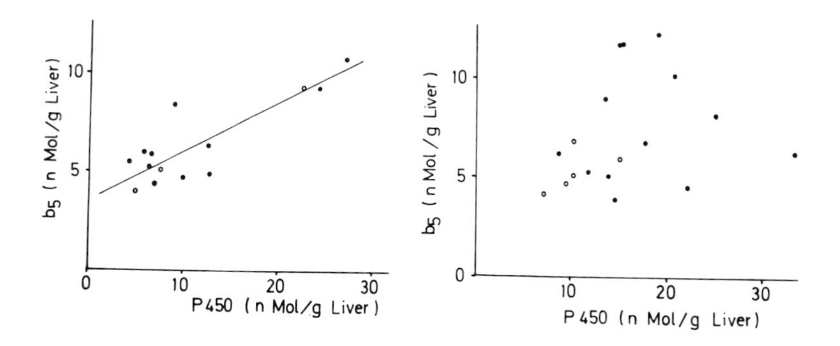

Fig. 6. Plot of the cytochrome P-450 content (abscissa) versus cytochrome b_5 content (ordinate) of the livers obtained from the nondrinkers (A, left), and drinkers (B, right). Chronic hepatitis, (•); cirrhosis, (○).

$Y = 0.245X + 3.637$ ($\gamma = .862$) was obtained. The regression coefficient is highly significant in the nondrinkers' livers ($p < 0.01$). In Fig. 6B, the contents of these cytochromes of the drinkers' livers were plotted in the same way as in Fig. 6A. The regression line was $Y = 0.119X + 5.228$ ($\gamma = 0.271$), and in this case is not significant. Thus the P-450/b_5 ratio in the whole livers showed a large variation in the drinkers, which is mainly due to the increased P-450 content and in some cases to the changes of the cytochrome b_5 contents.

IV. DISCUSSION

It is well established that alcoholics have profound mitochondrial changes in their livers (16). Ultrastructural changes of mitochondria were manifested in chronic ethanol consumption (e.g. 16-18). However, there is a controversy about the effect of ethanol on mitochondrial function. Rubin *et al.* (19) reported reduction in cytochromes a and b contents, and in succinate dehydrogenase activity. On the contrary, Videla and Israel (2) observed an increase in succinate dehydrogenase activity in total liver hemogenate of ethanol-fed rats. These mitochondrial activities are closely associated with the capacity of mitochondria to reoxidize the NADH generated via alcohol dehydrogenase in the cytoplasm, a pathway limited by the rate of NADH reoxidation (e.g. 1-3). The data of the current report, while extensive as possible, are perhaps premature for definitive conclusions. For example, in Fig. 2, right, we conclude that there are no significant changes in cytochrome aa_3 while in Fig. 3, right, we conclude that there is a change in P-450, but not in cytochrome b_5 (Fig. 5). The margins by which these conclusions are rendered significant or insignificant are rather small, and no doubt subject to change as further data is accumulated on wider populations of patients. Thus, we acknowledge that the size of the sample is too small, and the margins of significance and insignificance too tenuous for "hard" conclusions.

The present result with human livers shows that the daily alcohol intake for more than ten years induces practically no significant change in cytochrome aa_3 contents of livers both in cases of chronic hepatitis and cirrhosis. Thus, it would seem unlikely that the metabolic tolerance to alcohol in man results from the enhanced capacity of mitochondria to respire by virtue of the increased concentration of cytochromes in the mitochondrial respiratory chain. In making this statement of course it is appropriate to point out that cytochrome aa_3 itself is rarely rate limiting in the respiratory chain. In fact, it is the component least likely to be rate limiting. Respiration is regulated precisely by such factors as phosphate potential, ion gradients, and, from the electron transport point of view, dehydrogenase activity.

Interestingly, in contrast to cytochrome aa_3 content, the cytochrome P-450 content in livers was significantly increased in light drinkers with chronic hepatitis. Induction of cytochrome P-450 by ethanol-feeding in rats is well-known (9,23). The increase of cytochrome P-450 might be attributable to the direct involvement of this cytochrome in NADPH-dependent ethanol oxidation, and to the possible adaptive increase after chronic alcohol consumption. However, while drinkers had approximately the same degree of cirrhosis as the light drinkers, no increase in cytochrome P-450 was observed (Fig. 4). This is a rather strange result since one would expect that the greater the degree of alcoholism and liver damage, the greater degree would be induction of a system which might be associated with alcohol metabolism. These data are therefore at variance with those observed in rats (9,20) and indeed raise questions of the applicability of the induction of the P-450 system as an index of the functionality of the P-450 system in alcoholism metabolism.

After withdrawal of alcohol the blood clearance rate decreased with days up to 18th day (21) and the cytochrome P-450 content returned to the control value in rats (22). These findings might be related to the huge variation of P-450 content in drinkers. Our liver biopsy procedure was carried out 5 to 30 days after abstinence. The variation of cytochrome P-450 content in human livers might result in the variable drug metabolizing activities in normal subjects as well as in patients with liver cirrhosis (23).

The cytochrome P-450/b_5 ratio might be a convenient indicator of liver damage. This ratio is strikingly changed in the drinkers' livers not merely due to the increased content of cytochrome P-450, but to changes of cytochrome b_5 content as well. This might imply that a long term alcohol consumption results in the metabolic shift of the NADPH- as well as NADH-oxidase systems in human liver microsomes.

V. ACKNOWLEDGEMENTS

Our thanks are due to Drs. A. Takaoka and J. Kojima for their fruitful discussion. This work was supported by NIAAA grant AA-00292-06.

VI. REFERENCES

(1). Theorell, H., and Chance, B., Acta Chem. Scand. 5, 1127 (1951).
(2). Videla, L., and Israel, Y., Biochem. J. 118, 275 (1970).
(3). Israel, Y., Khanna, J., & Lin, J., Biochem. J. 120, 447 (1970).
(4). Rubin, E., & Cederbaum, A., in "Alcohol and Aldehyde Metabolizing Systems" (R. Thurman, T. Yonetani, J. Williamson and B. Chance, Eds.), p. 435. Academic Press, New York, 1974.
(5). Lieber, C., Hasumura, Y., Teschke, R., Matsuzaki, S., & Korsten, M., in "The Role of Acetaldehyde in the Actions of Ethanol" (K. Lindros & C. Eriksson, Eds.), p. 83. Finnish Foundation for Alcohol Studies, 1975.
(6). Hawkins, R., Kalant, H., & Khanna, J., Can. J. Pharmacol. 44, 241 (1966).
(7). Israel, Y., Videla, L., & Bernstein, J., Fed. Proc. 34, 2052 (1975).
(8). Teschke, R., Hasumura, Y., Joly, J., Ishii, H., & Lieber, C., Biochem. Biophys. Res. Commun. 49, 1187 (1972).
(9). Rubin, E., and Lieber, C., Science 162, 690 (1968).
(10). Sato, N., Hagihara, B., Kamada, T., & Abe, H., Analyt. Biochem. 74, 105 (1976).
(11). Vanneste, W., Biochim. Biophys. Acta 113, 175 (1966).
(12). Omura, T., and Sato, R., J. Biol. Chem. 239, 2370 (1964).
(13). Kajihara, T., & Hagihara, B., J. Biochem. 63, 453 (1968).
(14). Sato, N., and Hagihara, B., Cancer Res. 30, 2061 (1970).
(15). Hagihara, B., * Sato, N., Cancer Res. 33, 2947 (1973).
(16). Svoboda, D., & Manning, R., Amer. J. Path. 44, 645 (1964).
(17). Iseri, H., & Lieber, C., Amer. J. Path. 48, 535 (1966).
(18). Lane, B., & Lieber, C., Amer. J. Path. 49, 593 (1966).
(19). Rubin, E., Beattie, D., & Lieber, C., Lab. Invest. 23, 620 (1970).
(20). Joly, J., Ishii, H., Teschke, R., Hasumura, Y., and Lieber, C., Biochem. Pharmacol. 22, 1532 (1973).
(21). Ugarte, G., Pereda, T., Pino, M., & Iturriaga, H., Quart. J. Stud. Alc. 33, 698 (1972).
(22). Mezy, E., Biochem. Pharmacol. 21, 137 (1972).
(23). Levi, A., Scherlock, S., & Walker, D., Lancet 1, 1275 (1968).

A NOVEL ROUTE FOR THE METABOLISM OF ETHANOL: THE OXIDATION OF ETHANOL BY HYDROXYL FREE RADICALS

Gerald Cohen

The City University of New York

Hydroxyl radicals (·OH) are generated by certain enzymes, such as xanthine oxidase, or during the spontaneous autoxidation of certain agents, such as ascorbate, dialuric acid (reduced form of alloxan) or 6-hydroxydopamine. Ethanol is a good scavenger for ·OH; thiourea is a much more potent scavenger. These observations are relevant to the ability of thiourea to block the "artifactual" production of acetaldehyde from ethanol in tissue extracts. In studies in vitro, the production of acetaldehyde from ethanol during the xanthine-xanthine oxidase reaction or during the autoxidation of ascorbic acid was suppressed by superoxidase dismutase or by ·OH scavengers (n-butanol, thiourea). In studies with mice, ethanol and n-butanol and thiourea each prevented the toxic action of alloxan on the pancreas and, thereby, prevented the development of a diabetic state. A substituted thiourea and, to a lesser extent, ethanol and n-butanol, protected sympathetic nerves in vivo against destruction by 6-hydroxydopamine. These observations imply that ethanol can scavenge ·OH in vivo. It is not clear to what extent the scavenging of ·OH may represent a minor pathway for the metabolism of ethanol under normal physiologic conditions.

I. INTRODUCTION

Ethanol is a relatively potent scavenger of hydroxyl radicals (·OH) in solution (1). The product of the reaction is acetaldehyde (2).

Hydroxyl radicals can be produced in biologic systems from a number of sources. These include the interaction of ionizing radiation with water (2,3), the xanthine-xanthine oxidase reaction (4) and the rapid spontaneous autoxidation of a number of substances (5). The high reactivity of ·OH with cellular constituents has led to suggestions that ·OH may be responsible, in part, for certain forms of tissue damages, such as the destructive action of ionizing radiation (3), the diabetogenic action of alloxan (6) and the sympathectomy caused by injection of 6-hydroxydopamine (7). Sippel (8) has suggested that ·OH or other radicals arising from autoxidizing ascorbate may be responsible for the "artifactual" production of acetaldehyde from ethanol in tissue extracts.

This article has several purposes: 1. To further document a catalase-independent pathway for the production of acetaldehyde from ethanol during the oxidation of xanthine by purified xanthine oxidase; 2. To provide evidence for ·OH as an intermediate in the production of acetaldehyde by xanthine-xanthine oxidase and by autoxidizing ascorbate; 3. To provide indirect evidence for the scavenging of ·OH by ethanol in intoxicated experimental animals.

1. Superoxide-Dependent Production of Hydroxyl Radicals

In 1934, Haber and Weiss (9) suggested the following reaction to account for aspects of the kinetics of the decomposition of hydrogen peroxide by iron salts:

$$(1)\quad H_2O_2 + O_2^{\overline{\cdot}} \rightarrow \cdot OH + OH^- + O_2$$

In this reaction, the superoxide anion radical ($O_2^{\overline{\cdot}}$) reacts with hydrogen peroxide to form ·OH as one of the products. Thus, one condition for the production of ·OH in biological systems is the generation of superoxide radicals in the presence of H_2O_2.

Superoxide radicals are produced as intermediates in a number of enzyme-catalyzed oxidation reactions (10). Superoxide arises as the result of the transfer of one electron to oxygen:

$$(2)\quad O_2 + e^- \rightarrow O_2^{\overline{\cdot}}$$

Superoxide radicals dismute spontaneously to yield a reduced product (oxygen) and an oxidized product, hydrogen peroxide ($k < 10^2\ M^{-1}\ sec^{-1}$) (11):

$$(3)\quad O_2^{\overline{\cdot}} + O_2^{\overline{\cdot}} + 2H^+ \rightarrow O_2 + H_2O_2$$

Therefore, reaction mechanisms that yield superoxide as an intermediate also lead to the accumulation of H_2O_2. These conditions are conducive to formation of ·OH via reaction 1. The dismutation of superoxide radicals (reaction 3) can be catalyzed by superoxide dismutase ($k = 2.3 \times 10^9\ M^{-1}\ sec^{-1}$) (11).

Beauchamp and Fridovich (4) have provided good evidence for formation of ·OH during the xanthine-xanthine oxidase reaction.

They observed oxidation of ferrocytochrome c and the decomposition of methional to form ethylene gas during the oxidation of xanthine by xanthine oxidase. Both reactions were blocked by adding either catalase or superoxide dismutase, in accord with equation 1. In addition, the reactions were blocked by adding known hydroxyl radical scavengers such as ethanol, benzoate or mannitol. It was also observed that while H_2O_2 by itself was ineffective in producing ethylene, H_2O_2 added to the reaction mixture at zero time enhanced the production of ethylene by xanthine-xanthine oxidase. This latter observation was consistent with the elimination of a lag phase which was due to the need to accumulate sufficient H_2O_2 to promote reaction 1.

Reaction 1 can be catalyzed by heavy metals (12,13). Therefore, trace heavy metals present normally in tissues or trace metal contaminants in systems *in vitro* may promote the formation of hydroxyl radicals. Alternatively, metal chelates with EDTA can promote production of superoxide radicals in certain systems (14). The formation of $\cdot OH$ in biological systems has generally not been demonstrated directly, but is generally inferred from the results of experiments with radical scavengers, and with superoxide dismutase and catalase. The evidence in favor of reaction 1 (the Haber-Weiss reaction) in biologic systems was recently summarized (15).

2. Production of Acetaldehyde from Ethanol During the Xanthine Oxidase Reaction

The reaction system shown in Fig. 1 consisted of 0.05 M phosphate buffer, pH 7.4, containing 0.1 mM EDTA (disodium salt, Fisher Scientific, N.Y.) and 50 mM ethanol (Publicker Industries, Linfield, Pa.). Xanthine oxidase (0.63 units/mg; Sigma Chemical Co., St.Louis, Mo.) was added to 10 µg/ml and the reaction was initiated by the addition of 0.2 mM xanthine (Sigma). The concentration of ethanol (50 mM, 230 mg%) is in the range seen in the blood of experimental animals or human subjects during intoxication with ethanol. A steady production of acetaldehyde was observed between 10 and 30 minutes; however, a lag phase was evident during the first 10 minutes. No acetaldehyde was produced when either xanthine or xanthine oxidase were omitted.

The effect of adding various substances to the system is shown in Table I. H_2O_2 (0.1 mM, Fisher) added at zero time, promoted acetaldehyde production. Superoxide dismutase (Truett Labs., Dallas, Texas) diminished acetaldehyde production. These results were consistent with the production of $\cdot OH$ via equation 1. The potent hydroxyl radical scavengers, n-butanol (Baker) and thiourea (Sigma), inhibited acetaldehyde production in rough concordance with their faster rate constants (compared to

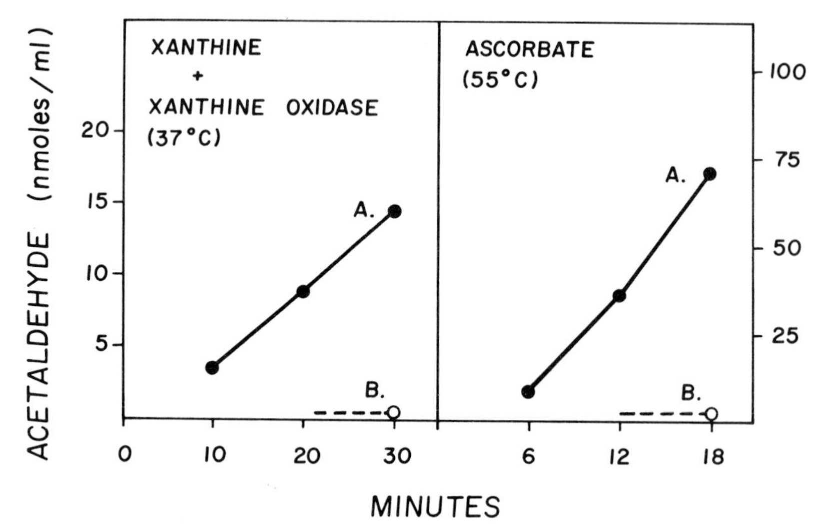

Fig.1. Production of acetaldehyde from ethanol by xanthine-xanthine oxidase (left) or by 1mM ascorbate (right). Curve A= complete system. Curve B = minus xanthine or xanthine oxidase or minus ascorbate. The system consisted of 0.05 mM phosphate buffer, pH 7.4, containing EDTA (0.1 mM) and ethanol (50 mM). Xanthine (0.2 mM) and xanthine-oxidase (10 μg/ml) were added to the system at 37°C. One ml aliquots of the xanthine oxidase system were removed and the reaction was stopped with 0.1 ml of 0.4 M perchloric acid; the system was assayed for acetaldehyde at 55°C by a head space method (18). The head space in the ascorbate system was assayed directly.

ethanol) for reaction with hydroxyl radicals. In separate experiments, benzoate, another good scavenger of ·OH ($k = 3.5 \times 10^{-9}\ M^{-1}\ sec^{-1}$ (1)), also suppressed acetaldehyde production by well over 50%. Urea, an agent that reacts poorly with ·OH did not suppress acetaldehyde production (Table 1). It should be noted that n-butanol is a relatively poor peroxidatic substrate for catalase (16).

Catalase would be expected to diminish the flux of hydroxyl radicals emanating from reaction 1 (4,15). However, catalase could not be used in these experiments because it generates acetaldehyde from ethanol in the presence of H_2O_2-generating systems such as xanthine-xanthine oxidase. As a precaution, azide (0.1 mM) was added in some experiments in order to evaluate the possible presence of catalase as a contaminant in the xanthine oxidase. Azide had no effect; therefore, the production of acetaldehyde cannot be ascribed to the peroxidatic activity of catalase.

TABLE 1

Effect of H_2O_2, superoxide dismutase (SOD) and hydroxyl radical scavengers on the generation of acetaldehyde by xanthine-xanthine oxidase or by ascorbate. The system was the same as that described in Figure 1. Rate constant estimates for the reactions of various scavengers with hydroxyl radicals have been taken from Anbar and Neta (1).

Additions	$k_{\cdot OH}(M^{-1}s^{-1})$	Acetaldehyde Production: Xanthine Oxidase	Acetaldehyde Production: Ascorbate
		Control Rate	
Complete System contains ethanol (50 mM)	1×10^9	0.3 µg/ml/15min.	1.4 µg/ml/12min.
		% Change ± SEM (N)	
H_2O_2 (0.1 mM)		+ 100 ± 12% (6)	+120 ± 15% (3)
Superoxide Dismutase (50 µg/ml)		- 97 ± 2% (4)	- 29 ± 4% (8)
n-Butanol (50 mM)	2×10^9	- 82 ± 4% (6)	- 69 ± 1% (5)
Thiourea (50 mM)	5×10^9	- 96 ± 1%(10)	- 98 ± 1% (5)
Urea (50 mM)	7×10^5	- 6 ± 4%(10)	0 ± 6% (3)

3. Production of Acetaldehyde from Ethanol During the Autoxidation of Ascorbic Acid

The autoxidation reaction of ascorbic acid (1 mM Fisher) was studied in the same reaction medium employed to study xanthine oxidase. However, the temperature was raised to 55°C which is generally used for the measurement of acetaldehyde in tissue extracts by "head space" methods (8,17, 18). The reaction was sampled serially to follow the appearance of acetaldehyde in the vapor above the heated sample. Fig.1 illustrates a progressive appearance of acetaldehyde with time. A portion of the initial lag phase can be ascribed to the time required for saturation of the head space with acetaldehyde. No acetaldehyde was observed when either ethanol or ascorbate were omitted.

Table 1 illustrates the effect of adding various agents. In general, results were similar to those observed in the xanthine-xanthine oxidase system. Specifically, b-butanol and thiourea (as well as benzoate) were inhibitory, while urea was not. H_2O_2

added at zero time stimulated acetaldehyde production; a similar phenomenon was noted earlier by Sippel (8).

Inhibition of acetaldehyde production by superoxide dismutase (Table 1) illustrates a role for superoxide radicals in the ·OH forming reaction. The fact that superoxide dismutase was much less effective in the ascorbate system requires some explanation. It may be that other reaction pathways exist for the producton of ·OH or that other reaction intermediates can react with ethanol. Alternatively, Rapp *et.al.* (19) have shown that catechol compounds can form complexes with superoxide dismutase; the similar ene-diol structure of ascorbate may do likewise, with some inactivation of the enzymatic activity.

4. Evidence of Hydroxyl Radical Scavenging by Ethanol *In Vivo*

The evidence for hydroxyl radical scavenging by ethanol *in vivo* derives from studies with alloxan and 6-hydroxydopamine. When alloxan is injected into experimental animals, the insulin-producing beta cells of the pancreas are selectively destroyed. The result is an experimental form of diabetes. 6-Hydroxydopamine, on the other hand, induces a selective destruction of adrenergic nerve terminals. The high tissue specificity of these agents lies in their selective accumulation by the pancreatic beta cells or adrenergic nerve terminals, respectively.

Alloxan, a quinoidal compound, can be readily reduced by ascorbate or other tissue reducing agents to dialuric acid, a phenolic compound. Both alloxan and dialuric acid are diabetogenic. There is evidence for the formation of dialurate from alloxan *in vivo* (20). *In vitro* studies (5) have shown that dialuric acid reacts rapdily and spontaneously with molecular oxygen to form superoxide radicals and hydrogen peroxide. During the autoxidation, hydroxyl radicals are also generated. Similarly, superoxide and hydroxyl radical formation have been observed during the spontaneous autoxidation of 6-hydroxydopamine.

Ethanol and other hydroxyl radical scavengers have proved effective *in vivo* as antagonists to both alloxan and 6-hydroxydopamine (6,7,21). In the results shown in Table 2, ethanol intoxicated mice failed to develop diabetes 72 hours after the injection of alloxan (i.v., 50 mg/kg). n-Butanol and thiourea were similaraly protective. Histochemical studies showed that the protective agents prevented the degranulation of the beta cells of the pancreas (6). n-Butanol proved to be the more effective alcohol at lower molar dose levels (6). This result was in keeping with the faster rate constant for reaction with hydroxyl radicals exhibited by n-butanol (Cf. Table 1).

TABLE 2
Prevention of alloxan diabetes by ethanol, n-butanol or thiourea. Male Swiss-Webster mice received i.v. alloxan (50 mg/kg) with and without pretreatment with i.p. ethanol (4 g/kg, 30 min), n-butanol (0.8 g/kg, 30 min) or thiourea (3 g/kg, 2 hours). Blood glucose was measured 72 hours later. Groups receiving drug + alloxan were $p < 0.01$ compared to the corresponding groups receiving alloxan alone.

	Blood Glucose in mg% ± S.E.M. (N)		
Drug	Control	Alloxan Alone	Drug + Alloxan
Ethanol[a]	152 ± 36 (10)	417 ± 23 (26)	195 ± 25 (23)
n-butanol[b]	120 ± 4 (15)	442 ± 19 (25)	133 ± 23 (14)
Thiourea[b]	124 ± 2 (61)	449 ± 9 (24)	147 ± 11 (20)

a. Data from Heikkila et.al. (20)
b. Data from Heikkila et.al. (6)

TABLE 3
Protection of the adrenergic nerve plexus of the left atrium against destruction by 6-hydroxydopamine (6-OHDA). Male Swiss-Webster mice were either untreated or pretreated with ethanol (4 g/kg), n-butanol (0.8 g/kg) or one of two thiourea derivatives (PTTU, 200 mg/kg or methimazole, 400 mg/kg). One hour later, each animal received i.v. 6-OHDA·HBr at a dose of either 4 mg/kg (ethanol, n-butanol, and methimazole experiments) or 7.5 mg/kg (PTTU experiment). The uptake and accumulation of tritium-labelled norepinephrine (^{3}H-NE) by the left atrium in vitro, measured 24 hours later, served as an index of the integrity of the sympatheticnerve plexus of the atrium. Groups treated with drug plus 6-OHDA were $p < 0.01$ compared to the corresponding groups treated with 6-OHDA alone. Data from Cohen et.al. (7).

	Accumulation of ^{3}H-NE as % Control + SEM (N)	
Experiment	6-OHDA Alone	Drug + 6-OHDA
PTTU	22 ± 2 (22)	97 ± 3 (22)
Methimazole	28 ± 3 (18)	63 ± 6 (21)
Ethanol	41 ± 4 (26)	59 ± 6 (26)
n-Butanol	33 ± 3 (26)	50 ± 4 (27)

In studies with 6-hydroxydopamine, the scavengers ethanol and n-butanol each provided partial protection against the destruction of adrenergic nerve terminals (Table 3), while two substituted thioureas provided partial to full protection. The thiourea derivatives, namely 1-phenyl-3-(2-thiazolyl)-2-thiourea (PTTU) and methimazole, were known to penetrate into adrenergic nerve terminals *in vivo*. (Thiourea itself was not protective (7)). PTTU was also an effective antagonist to the diabetogenic action of alloxan (21).

PTTU and methimazole were tested as ·OH scavengers in the ethylene-forming system described by Beauchamp and Fridovich (4). These thiourea derivatives proved to be as powerful as thiourea itself (7).

These studies have shown that ethanol and other ·OH scavengers provide protection against the specific cytotoxic actions of alloxan and 6-hydroxydopamine. As the generation of ·OH by these cytotoxic agents has been observed *in vitro*, the combined results indicated that ethanol and the other protective agents acted by scavenging deleterious hydroxyl radicals *in vivo*. Therefore, these studies constitute indirect evidence that ethanol can scavenge ·OH *in vivo*.

II. DISCUSSION AND CONCLUSIONS

Portions of these *in vitro* and *in vivo* experiments were prompted, in part, by the prior observations of Sippel (22), who showed that thiourea blocked the artifactual production of acetaldehyde from ethanol in tissue extracts. As thiourea is a powerful scavenger for ·OH (1), an implication was that ·OH may have been responsible for the oxidation of ethanol. Additional impetus for the experiments came from the observations of Beauchamp and Fridovich (4) who showed that ·OH was generated by xanthine oxidase and that ethanol served as a good scavenger in their system.

In the current studies, a superoxide-mediated (i.e. inhibited by superoxide dismutase, Cf. Table 1) production of acetaldehyde from ethanol was observed both during the xanthine-xanthine oxidase reaction and during the spontaneous autoxidation of ascorbate. Both systems were stimulated by the addition of excess H_2O_2 (0.1 mM) at zero time, but H_2O_2 alone (no xanthine oxidase or no ascorbate) was without effect. These observations were consistent with the production of the oxidizing radical, ·OH via reaction 1 in each of the two systems. Suppression of acetaldehyde production by known ·OH scavengers (n-butanol, thiourea, benzoate) but not by a substance known to react much more weakly than ethanol with ·OH (urea)supported the view that acetaldehyde arose from the reaction of ·OH with ethanol.

The superoxide-mediated production of ·OH was also observed during the spontaneous autoxidation of dialuric acid (reduced alloxan) or 6-hydroxydopamine. The cytotoxic actions of these agents on either the pancreas or peripheral sympathetic nerves, respectively, was blocked either completely or in part by ethanol, n-butanol and thiourea derivates. Thus, the available evidence points to effective scavenging of ·OH *in vivo* (pancreas, sympathetic nerves) by ethanol and the other protective agents.

These studies indicate that acetaldehyde can arise both *in vitro* and *in vivo* during enzymatic or non-enzymatic processes that lead to transient formation of hydroxyl radicals. It seems probable that the suppression by thiourea of artifactual production of acetaldehyde from ethanol in tissue extracts (17,22), results from the scavenging of hydroxyl radicals. Many enzymes that generate superoxide radicals have been identified (10), and can be looked on as potential sources of hydroxyl radicals. The activity of xanthine oxidase from rat liver is complicated by a change from dehydrogenase to oxidase activity during purification (23). It is not clear at the present time to what extent the scavenging of ·OH by alcohol in liver may represent a minor pathway for the production of acetaldehyde. This pathway may be of greater interest in tissues lacking significant alcohol dehydrogenase activity. It remains to be determined to what extent this pathway may contribute to tissue levels or circulating blood levels of acetaldehyde *in vivo*.

III. ACKNOWLEDGEMENTS

I thank Mr. Michael Zimbalist for carrying out the gas chromatographic assays for acetaldehyde.

IV. REFERENCES

(1). Anbar, M., and Neta, P., J.Appl.Radiat.Isotop. 18,493 (1967).
(2). Swallow, A.J., Biochem. J. 54, 253 (1953).
(3). Myers, L.S. jr., Fed. Proc. 32, 1882 (1973).
(4). Beauchamp, C., and Fridovich, I., J. Biol. Chem. 245,4641 (1970).
(5). Cohen, G., and Heikkila, R.E., J.Biol.Chem. 249,2447 (1974).
(6). Heikkila, R.E., Winston, B., Cohen , G., and Barden, H., Biochem.Pharmacol.25, 1)85 (1976).
(7). Cohen, G., Heikkila, R.E., Allis, B., Cabbat, F., Dembiec, D., MacNamee, D., Mytilineou, C., and Winston, B., J. Pharmacol.Exp.Therap., in press (1976).
(8). Sippel, H.W., Acta Chem. Scand. 27, 541 (1973).
(9). Haber, F., and Weiss, J., Proc.Royal Soc.Ser.A. 147,322 (1934).

(10). Bors, W., Saran, M., Lengfelder, E., Spottl, R., and Michel, C., Current Topics Radiat. Res. Quart. 9, 247 (1974).
(11). Fridovich, I., in :"Free Radicals in Biology" (W.A. Pryor, Ed.) Vol. I, 239, Acad. Press, New York (1976).
(12). Barb, W.G., Baxendale, J.H., George, P., and Hargrave, K. R., Nature 163, 692 (1949).
(13). Walling, C., Acts.Chem.Res. 8, 125 (1975).
(14). McCord, J.M., and Fridovich, I., J.Biol.Chem. 25,6049 (1969).
(15). Cohen, G., in:"Superoxide and Superoxide Dismutase" (A.M. Michelson, Ed.) Acad. Press, New York, in press.
(16). Tephley, T.R., Mannering, G.J., and Parks, R.E. jr. J. Pharmacol. Exp. Therap. 134, 77 (1961).
(17). Eriksson, C.J.P., Sippel, H.W., and Forsander, O.A., in: "The Role of Acetaldehyde in the Actions of Ethanol" (K. O. Lindros and C.J.P. Eriksson, Eds.) 9, Finnish Foundation for Alcohol Studies, (1975).
(18). Dembiec, D., McNamee, D., and Cohen, G., J.Pharmacol.Exp. Therap, 197,332 (1976)
(19). Rapp, U., Adams, W.C., and Miller, R.W., Can.J.Biochem. 51 158 (1973).
(20). Heikkila, R.E., Barden, H., and Cohen, G., J.Pharmacol. Exp. Ther. 190, 501 (1974).
(21). Cohen, G., Allis, B., Winston, B., Mytilineou, G., and Heikkila, R.E., Eur.J.Pharmacol. 33, 217 (1975).
(22). Sippel, H.W., Acta Chem. Scand. 26 , 3398 (1972).
(23). Waud, W.R., and Rajagopalan, K.V., Arch.Biochem.Biophys. 172, 354 (1976).

FOLATE DEFICIENCY AND METHANOL POISONING IN THE RAT

A.B. Makar and T.R. Tephly

The University of Iowa

Species differences in sensitivity to methanol toxicity is well known. Rodents do not develop metabolic acidosis or ocular toxicity after methanol treatment, whereas the monkey and man display these features. Since the administration of methanol results in formic acid accumulation in the monkey but not in the rat, and since both species metabolize formate primarily through a folate-dependent pathway, differences in the ability of the animal to metabolize formate could explain the relative sensitivity of the species to methanol. The rat metabolizes formate at rates approximately twice those observed in the monkey. Rats were placed on a folate-deficient diet and the maximal rate of formate oxidation was reduced to that of the monkey. Administration of methanol (4 g/kg, intraperitoneally) led to formate accumulation in the blood to levels comparable to those observed in the methanol-intoxicated monkey (18 mEq/l). This was accompanied by a marked decrease in blood pH (7.04). This represents the first demonstration of a metabolic acidosis in the rat after methanol and suggests that the species differences observed between the primate and the rat may be related to folate and its biological disposition.

II. INTRODUCTION

Unlike ethanol, which is rapidly metabolized in the animal organism to CO_2 and water, methyl alcohol displays a specific toxicity in man. It is slowly metabolized and produces metabolic acidoses, ocular toxicity and death. Until recently it has not been possible to completely characterize the toxicity except in man. But more recent studies show that the monkey may serve as an appropriate model for methanol toxicity in man since the syndrome in this animal has features common to those in man. McMartin *et al.*(1) and Clay *et al.*(2) have described an accumulation of formate in the blood of monkeys coincident with the development of metabolic acidosis. This is unlike the situation in ethanol ingestion where metabolic acidosis and

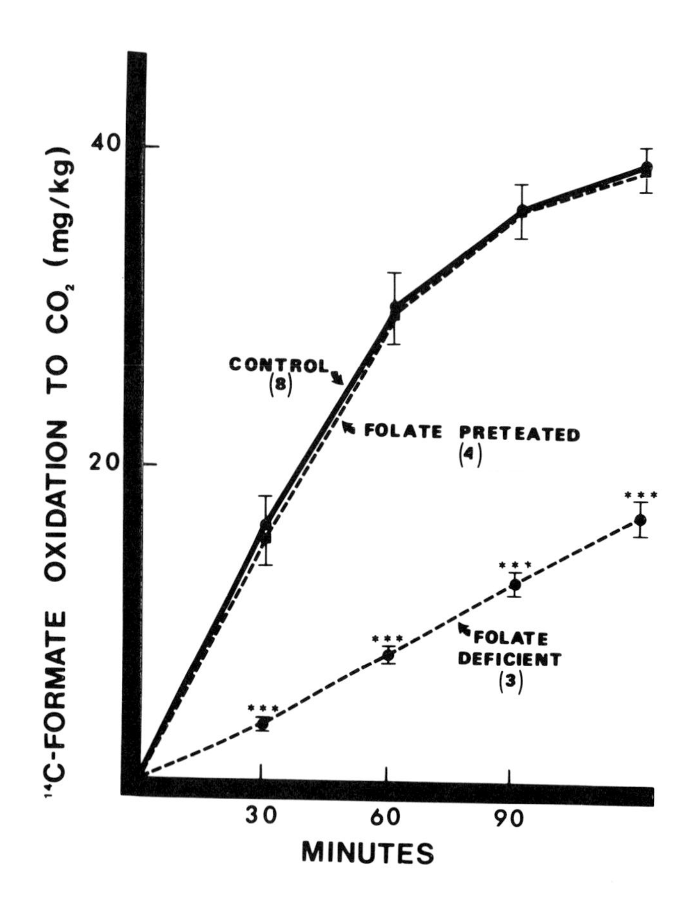

Fig. 1. In vivo rates of formate oxidation to CO_2 in control, folate-deficient and folate-pretreated rats. Procedures for production of folate deficiency and for folate pretreatment are described in the test. At zero time, rats were injected with ^{14}C-sodium formate (68 mg/kg intraperitoneally) and rates were determined as described by Palese and Tephly.

blindness do not develop. Although methanol is toxic in man and monkey it is not toxic to rat or other rodent species (3,4). An explanation for this may relate to formate disposition in the several species. Since formate accumulates in the monkey after methanol ingestion but does not accumulate in the rat it is possible that decreases in the ability to oxidize formate to CO_2 might lead to increases in sensitivity to methanol in a given species. Recent studies from our laboratory have shown

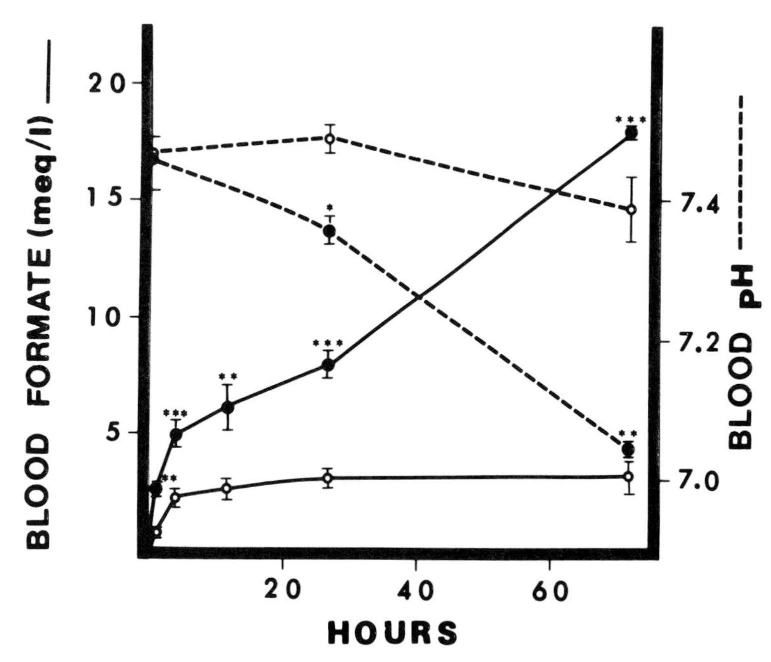

*Fig. 2. Blood formate levels and pH values of control (o) and folate-deficient rats (o). At zero time methanol was administered intraperitoneally at a dose of 4 g/kg. Each point represents the mean value obtained from 2-8 rats. Vertical bars represent ± S.E.M. *, **, and *** indicate statistically significant differences from corresponding control values at P levels of <0.05, <0.01, and <0.001, respectively. This work has been published previously (11).*

(5) that formate oxidation to CO_2 *in vivo* is dependent on a folate-related 1-carbon pool pathway. Others have held (6) that catalase played a major role in the metabolism of formate but studies using relatively selective inhibitors of the catalase pathway have been ineffective in decreasing formate oxidation in the rat (5) and in the monkey (7). The objective of the current work was to test whether folate deficiency and decreases in formate oxidation render the rat sensitive to methanol poisoning with the production of metabolic acidosis through increases in blood formic acid.

III. METHODS

Male Sprague-Dawley rats (about 200 g) were divided into groups fed either a folate-deficient diet or a control diet (Bio-Serve, Frenchtown, N.J.). A radioassay was employed to measure liver folate levels (8) using kits obtained from Diagnostic Biochemistry (San Diego, C.A.). Formate was determined by a method described previously (9) and formate meta-

TABLE 1
Hepatic Catalase Activity and Formate Oxidation in the Rat In Vivo

Treatment	Duration of Treatment	Hepatic Catalase Activity[c] (kat.f.units/g Liver)	*In vivo* [c] Formate Oxidation to CO_2 (mg/kg/hr)
Chow diet	-	863 ± 93	29.98 ± 2.12
Folate deficient diet	9 days	806 ± 88	31.63 ± 2.60
Folate deficient diet	10-12 weeks	590 ± 42	8.18 ± 0.40
Folate control diet	10-12 weeks	765 ± 57	28.81 ± 1.52
Chow diet + AT treatment[a]	-	46 ± 6	30.48 ± 2.29
Chow diet + MTX (1 mg/kg/day)[b]	9 days	228 ± 53	16.06 ± 1.10
Chow diet + MTX (0.5 mg/kg/day)	9 days	443 ± 71	23.01 ± 0.54
Folate deficient diet = MTX (1 mg/kg/day)	9 days	317 ± 66	9.67 ± 1.44
Folate deficient diet + MTX (0.5 mg/kg/day)	9 days	370 ± 29	18.89 ± 1.75

a. 3-Amino-1,2,4-triazole (AT) was ibjected 1 hour prior to the beginning of the experiment.

b. Methotrexate (MTX) was injected intraperitoneally once a day.

c. Values are the means obtained from 3 to 8 animals ± S.E.M.

A correlation coefficient of 0.37 between hepatic catalase activity and the rate of formate oxidation was obtained when data were submitted to linear regression analysis.

bolism was assessed as previously described (5). Blood pH was determined on samples obtained by cardiac puncture using a blood gas analyzer (Instrumentation Laboratories, Model No.713). In certain experiments folate was administered to animals receiving a diet adequate in folate in order to produce a folate hypervitaminosis. The treatment consisted of three i.p. injections of a preparation of sodium folate (Folvite R) obtained from Lederle Laboratories, American Company, Pearl River, N.Y. The dose was 50 mg/kg at 48, 24 and 1 hour prior to the injection of ^{14}C-sodium formate.

IV. RESULTS

Figure 1 shows results from experiments from rats fed a folate deficient diet for 10-12 weeks. The results are compared to those obtained in pair-fed animals on a control diet containing 2.25mg of sodium folate per kg of diet for a similar period of time. In another group, folate was injected into animals as described in Methods in order to produce a hypervitaminosis. There is marked decrease in formate oxidation in folate-deficient rats as has been reported previously (5). Injections of folate to animals on the control diet had no effect on formate oxidation to CO_2. The decrease in formate metabolism seen in rats or folate deficient diets brought the metabolic rate to those rates seen in the normal monkey.

Since the monkey metabolizes formate more slowly than the rat (7) it was of interest to determine whether the folate-deficient rat might be sensitive to methanol poisoning. Figure 2 compares blood formate after methanol levels and blood pH in control and folate-deficient rats. A marked elevation in formic acid levels and a marked decrease in blood pH was observed in folate-deficient rats whereas control rats did not accumulate formate in the blood and had no alteration of blood pH value. These results provide the first demonstration of metabolic acidosis in the rat after methanol treatment. The magnitude of changes seen in the folate-deficient rat are greater than those which have been reported for the rhesus and pigtail monkeys (1).

V. DISCUSSION

Results indicate the importance of the folate-dependent 1-carbon poll in formate metabolism and its role in methanol poisoning. It may be argued that folate-deficiency produces decreases in catalase activity and that decreases in catalase activity are responsible for the accumulation of formic acid in blood and the metabolic acidosis observed after methanol treatment in the folate-deficient rat. In order to demonstrate the relationship of catalase activity in the liver and the *in vivo* rate of formate oxidation to CO_2 in the rat results were obtained from animals which had been on a folate-deficient diet, rats fed control diets for 10-12 weeks, rats fed folate-deficient diets for 9 days and rats fed a Purina Chow diet which were treated daily with i.p. injections of methotrexate (1mg/kg) for 9 days. Other treatments of rats are included in this information in Table 1. There is a very poor correlation between hepatic catalase activity and the ability of the rat to metabolize formate to CO_2. For example, aminotriazole-treated rats which have 5-10% of control catalase activity showed no inhibition of formate oxidation . A correlation coefficient of 0.37 was obtained when information was submitted to linear regression analysis. Thus, we suggest that formate is metabolized to CO_2 via a folate-dependent pathway in the rat and that by reducing the rate of formate metabolism it is possible to sensitize the rat to methanol poisoning. Recent studies in this laboratory have shown that the monkey is exquisitely sensitive to methanol poisoning and that formate accumulates in the blood coincident with the production of metabolic acidosis. Other studies have shown (10) that low level chronic formic acidemia leads to ocular toxicity in the rhesus monkey and that formate infusion alone at normal pH, produces this toxicity. Thus, formate is a major determinant in methanol poisoning and further studies are underway to determine which catalyst is responsible for formate oxidation to CO_2 in the rat and monkey.

VI. ACKNOWLEDGEMENTS

This work was supported by NIH Grant 19420.

VII. REFERENCES

(1). McMartin, K., Makar, A.B., Martin-Amat, G., Palese, M., and Tephly, T.R., Biochem.Med. 13, 319 (1975).
(2). Clay, K.L., Murphy, R.C., and Watkins, W.D., Toxicol.Appl. Pharamcol. 34, 49 (1975).
(3). Roe, O., Pharmacol, Rev. 7, 399 (1955).
(4). Gilger, A.P., and Potts, A.M. Amer.J.Ophthalmol. 39, 63 (1955).
(5). Palese, M., and Tephly, T.R. J. Toxicol. Environ. Hlth. 1, 13 (1975).
(6). Friedmann, B., Nakada, H.I., and Weinhouse, S., J. Biol. Chem. 210, 413 (1954).
(7). McMartin, K., Martin-Amat, G., Makar, A.B., and Tephly, T.R., Alcohol and Aldehyde Metabolizing Systems (R.G. Thurman, T. Yonetani, J.R. Williamson, H. Drott, and B. Chance, Eds.) Vol, II Acad. Press, New York, In Press (1977).
(8). Longo, D.L., and Herbert, V., J.Lab.Clin.Med, 87, 138 (1976).
(9). Makar, A.B., McMartin, E.Ke., Palese, M., and Tephly, T.R., Biochem. Med. 13, 117 (1975).
(10). Martin-Amat, G., McMartin, K.E., Makar, A.B., Baumbach, G., Cancilla, P., Hayreh, M.M., Hayren, S.S., and Tephly, T.R. Alcohol and Aldehyde Metabolizing Systems, (R.G. Thurman, J.R. Williamson, J.R. Drott and B. Chance, Eds.) Vol. II Acad. Press, New York, In press, (1977).
(11). Makar, A.B., and Tephly, T.R., Nature 261, 715 (1976).

THE MONKEY AS A MODEL IN METHANOL POISONING

G. Martin-Amat, K.E. McMartin, A.B. Makar, G. Baumbach, P. Cancilla, M.M. Hayreh, S.S. Hayreh, and T.R. Tephly

The University of Iowa

Methanol poisoning in man is characterized by a mild central nervous system depression, metabolic acidosis and ocular toxixity followed by coma and death. Previous work from this laboratory has described the production of metabolic acidosis, coma and death in rhesus and pigtail monkeys without the demonstration of definitive ocular lesions. Since animals used in those studies died rapidly after methanol administration a prolonged and less intense state of intoxication was deemed necessary for the production and recognition of ocular toxicity. Thus, methanol was administered at a dose of 2 g/kg followed by subsequent doses of 0.5 g/kg until signs of ocular toxicity were observed: usually at 48 hours, or later, after the first dose of methanol. Ocular toxicity was characterized as optic disc edema with dilated pupils and a slow reaction of the pupillary reflex to light. A rapid intraarterial perfusion of appropriate fixatives was used in order to minimize autolysis of tissues. Histopathologic changes included intracellular swelling and mitochondrial disruption in the area of the optic disc but otherwise retinal histology was normal. Clinical symptoms appeared to be similar to those described in man and may provide a basis for our understanding of the mechanism of methanol toxicity with respect to the ocular lesions observed in man.

I. INTRODUCTION

A marked species difference in susceptibility to methanol poisoning is well known and has been the subject of numerous reviews (1.2.3.4). It has been shown recently that the rat does not accumulate formate in the blood after methanol administration whereas the monkey rapidly accumulates formic acid coincident with the production of a metabolic acidosis (5,6). Potts and coworkers (2,7) observed metabolic acidosis in the rhesus monkey and certain signs of ocular toxicity, but Cooper and Felig (8) were unable to confirm these results. Recently, Clay et.al. (6) and McMartin et.al. (5) have reproduced certain of the results reported by Gilger and Potts (2) with respect to the development of the metabolic acidosis in the monkey. They clearly demonstrated formic acid accumulation coincident with acidosis. However, animals usually died within 28 hours without any obvious clinical demonstration of ocular disfunction. It was reasoned that because of the rapid onset of metabolic acidosis and death that ocular toxicity might not have been apparent clinically. Thus, a different protocol for intoxication was devised and biochemical measurements were supplemented with clinical ophtalmological evaluations and histopathological studies.Methanol was given to rhesus monkeys in levels that yielded a chronic moderate state of metabolic acidosis and formic acidemia. We wish to report the development of an optic nerve lesion which develops after such treatment with methanol. In addition, recent studies show that similar lesions may be reproduced by the infusion of formate under conditions where blood pH was maintained at neutrality.

II. METHODS

Male rhesus monkeys weighing 2.5 - 4.5 kg were intoxicated by repeated oral adminstration of methanol as indicated in Figure 1. An initial dose of 2 g/kg was administered orally followed by subsequent doses of 0.5 g/kg. These doses were sufficient to maintain an arterial blood pH between 7.1 and 7.3 and arterial blood bicarbonate values of 10 mEq/l or above. Animals were prepared as described by McMartin *et al.* (5). Other procedures are described by Martin-Amat *et al.* (9). Preparations for ophtalmologic and histopathologic studies are described elsewhere (10,11).

A formate buffer 0.47 M, pH 6.0, was infused intravenously at various rates in order to maintain blood levels of formate between 20-30 mEq/l. The concentration of blood formate was measured periodically using the method described by Makar *et al.*(12). Clinical observations were made every 12 hours in formate-treated animals and, when toxicity was observed, rapid perfusion techniques were employed to obtain tissues for histopathological evaluation.

III. RESULTS

In Figure 1 results are shown which describe the rate of formate formation, bicarbonate depletion and methanol disappearance. Methanol was administered at zero time at a dose of 2 g/kg and 0.5 g/kg methanol was given at time periods indicated by arrows. Under these conditions, formate blood levels were maintained at about 7 mEq/l or higher; blood bicarbonate values of about 10 mEq/l were achieved and blood pH values between 7.1 and 7.3 were usually attained. Thus, biochemical evaluations were the major determinants for subsequent doses of methanol. It should be noted that increases in formate anion concentration corresponded to decreases in bicarbonate anion concentrations in arterial blood. It is our experience that decreases in blood bicarbonate provide the most reliable measure of the degree of metabolic acidosis after methanol. Obvious clinical ophtalmologic toxicity was seen at time periods indicated by stars in Figure 1.

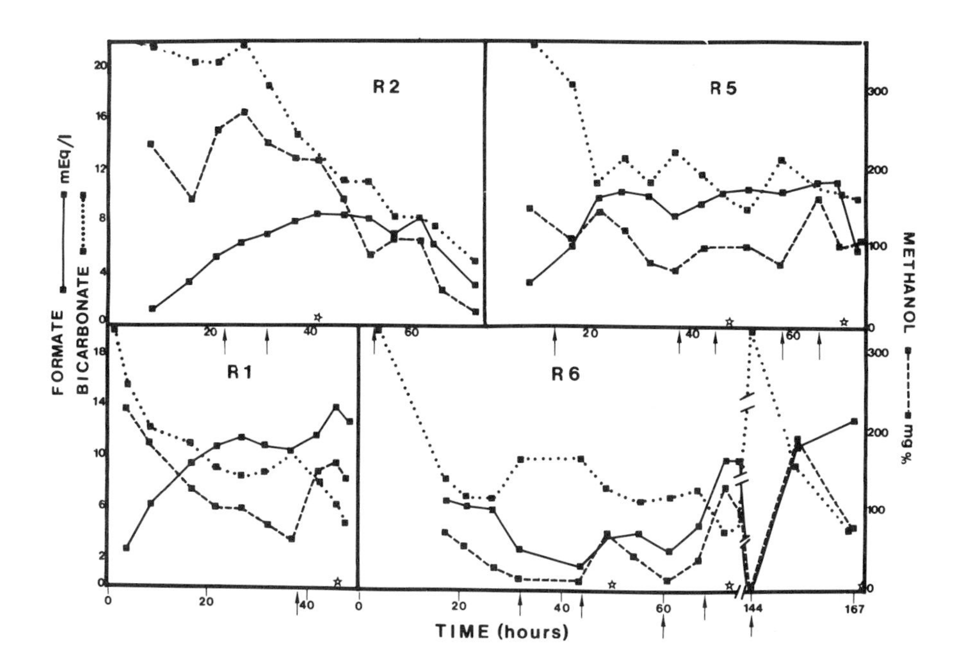

Fig.1. Formate accumulation and metabolic acidosis in four rhesus monkeys. Methanol (2 g/kg) was administered orally at zero time. Arrows indicate times at which methanol supplements were given (0.5 g/kg), except for R 6 at 44 and 72 hours when 1 g/kg was given. Stars indicate times at which ophtalmoscopic evaluations were performed. From G. Martin-Amat et.al. (9).

Table 1 shows data which describe optic disc changes and changes in the pupillary reflex in response to light. In general, marked optic disc edema was observed and, in certain cases, this clinical observation was progressive as indicated in animals R 2,R5 and R 6. The marked slowing of pupillary response to

TABLE 1

Clinical Observations in Methanol-Poisoned Monkeys

Animal	*Optic Disc Changes*	*Pupillar Reflex*	*CSF Pressure*[1]
R 1			
46 hrs.	Marked optic dis edema	Dilated,slow reaction to light	46 mm H_2O_2
R 2			
43 hrs.	Disc hyeremic (edema starting)	Dilated, slow reaction to light	
67 hrs.	Disc edema increased		
R 5			
48 hrs.	Disc blurring	Slow reaction Does not reach complete contraction	53 mm H_2O_2
70 hrs.	Disc edema increased		
R 6			
50 hrs.	Normal	Slow reaction Does not reach complete contraction	
171 hrs.	Disc hyeremic		

[1]*Normal CSF pressure is 100 mm* H_2O_2 *or less.*

light was a major observation and indicates functional alterations of the optic pathway. Since optic disc edema is also seen in cases of increased cerebrospinal fluid pressure, measurement of CSF pressure was taken in several animals in order to determine whether methanol was producing its effects via an effect on CSF pressure. It can be seen (Table 1) that CSF pressure is within normal limits in animals poisoned with methanol. Cerebrospinal fluid formate concentrations in animals R 1 and R 5 were at about the same level as the blood at the time the CSF was removed from the animal.

The funduscopic picture of a rhesus monkey (R 1) after 46 hours following the initial dose of methanol is shown in Figure 2B. The optic disc shows a marked degree of edema, a picture that is representative of other animals studied (Fig. 1), although in certain cases a marked progression was observed. This is described in detail in a report by Hayreh *et al.* (10).

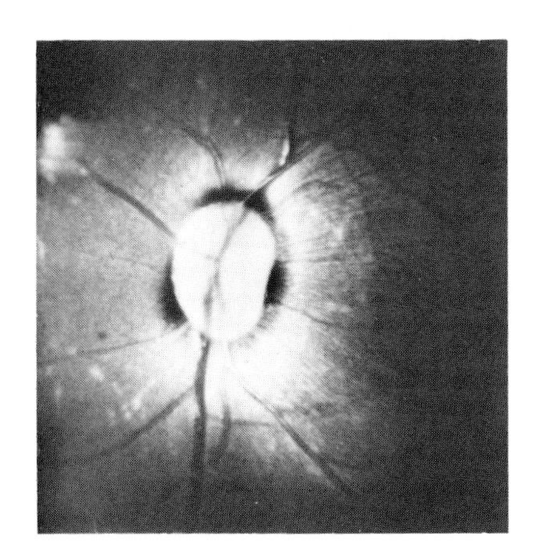

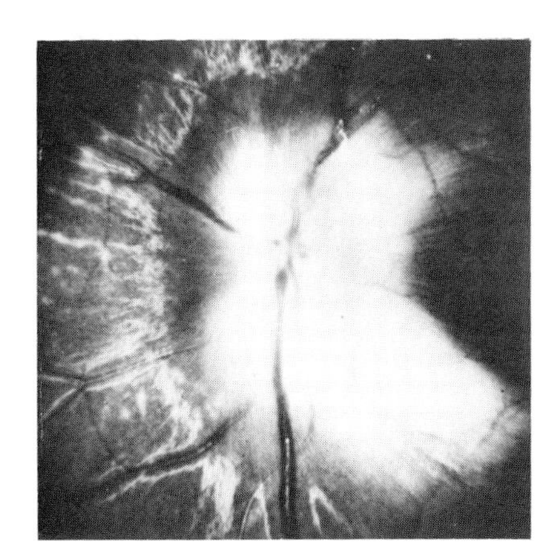

Fig. 2. (A) Fundus photography of control rhesus monkey (R 1) taken the week previous to the start of the experiment. (B) Fundus photography of rhesus monkey (R 1) 46 hours after the first dose of methanol, showing a marked degree of disc edema.

Figure 3B shows a cross section of the optic disc area of a control and methanol-treated monkey (R 6). Viewed at a light microscopic level one can appreciate marked intraaxonal swelling after methanol treatment. This alteration was confined to the area of the optic disc and anterior portion of the optic nerve. Sections of the posterior portion of the optic nerve revealed essentially normal histological patterns.

Figure 4B shows that the histological pattern of the retina under conditions which revealed marked changes in the area of the optic disc was essentially normal.

An example of ultrastructural changes seen in the optic disc or anterior portion of the optic nerve is seen in Figure 5B. A section taken from an untreated animal is shown for reference (5A). There is marked intraaxonal swelling, drastic mitochondrial destruction and neurotubular disruption to a marked degree.

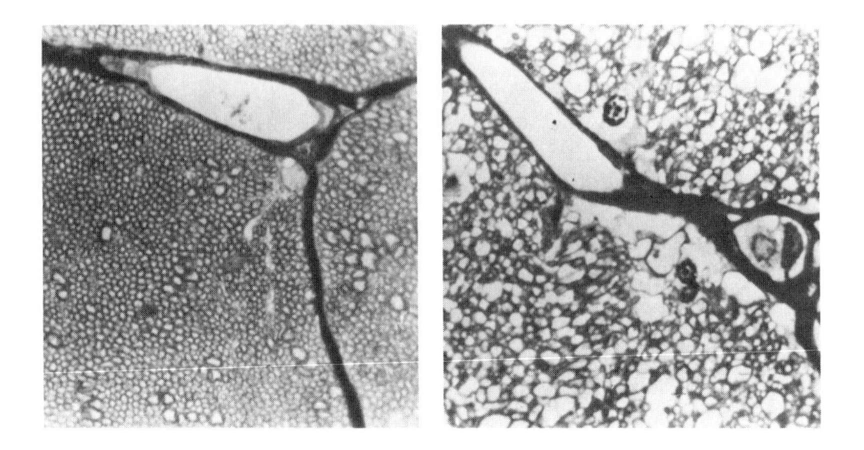

Fig. 3 (A) Light microscopic picture of the optic disc, in cross section, of a control animal, showing normal histology. (B) Light microscopy in the optic disc area, in cross section, of animal R6 (171 hours after the first dose of methanol) showing intraaxonal swelling (spongy degeneration), with high degree of disruption of the normal architecture.

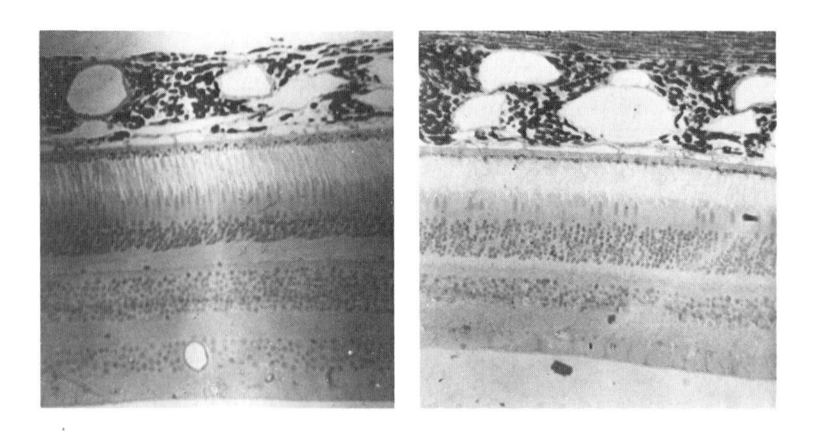

Fig. 4 (A) Light microscopic picture of the retina of a control animal. (B) Light microscopic picture of the retina of rhesus monkeys (R 1) sacrificed 48 hours after the first dose of methanol showing normal retina.

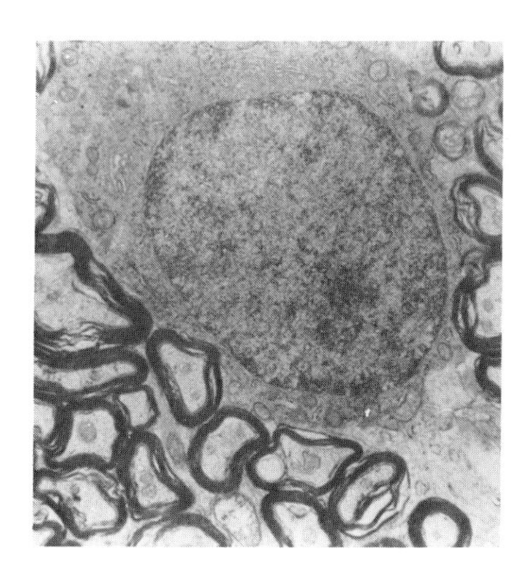

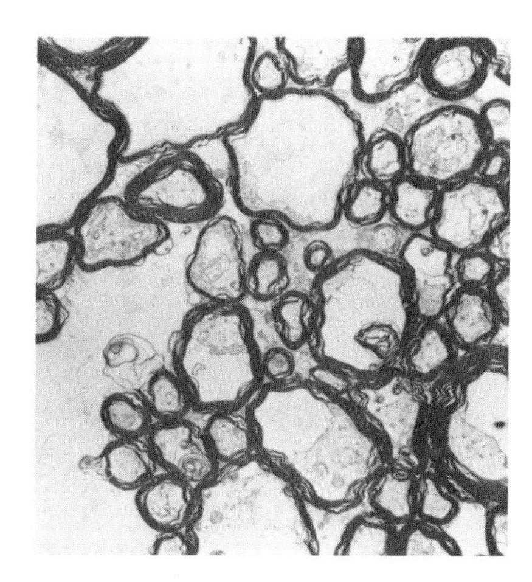

Fig. 5 (A) Electron microscopic cross section from the anterior portion of the optic nerve of a control monkey showing normal axons, mitochondria and neurotubules. (B) Electron microscopic cross section form the anterior portion of the optic nerve (R 5, 48 hours after first dose of methanol) showing mitochondrial neurotubular disruption with intraaxonal swelling.

Several monkeys were infused with formate buffer and maintained at normal blood pH values. Formate and bicarbonate levels in arterial blood were determined during the course of the experiment. Maximal concentrations of about 28 mEq/l of formate were achieved in these animals and marked optic disc edema was usually seen between 40 and 50 hours after the initiation of formate infusion. Optic disc edema was similar to that observed in the cases of methanol poisoning with high blood formate levels and metabolic acidosis (Figure 6). The edema observed after formate infusion where high blood formate levels achieved at normal blood pH and were even more marked than in monkeys intoxicated with methanol. Preliminary histopathological findings appear to be identical to those taken from methanol tissues of poisoned rhesus monkeys. Thus, it appears that formate infusion can reproduce the ocular toxicity seen when rhesus monkeys are treated chronically with methanol.

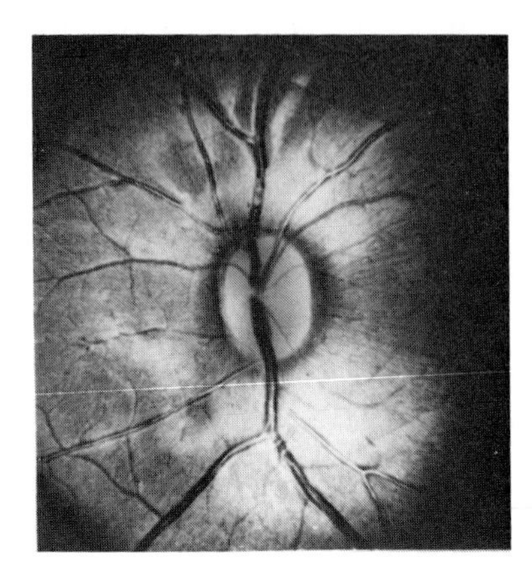
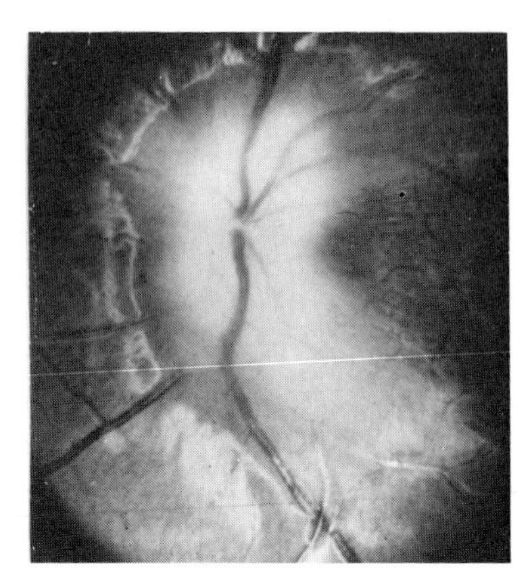

Fig. 6 (A) Fundus photography of rhesus monkey taken one week prior to start of the experiment. (B) Fundus photography from a rhesus monkey taken 53 hours after start of infusion with formate.

V. DISCUSSION

These studies show that it was possible to produce a marked ocular toxicity by methanol in the rhesus monkey when monkeys were treated with an initial high level of methanol followed by supplemental doses to maintain a moderate state of metabolic acidosis for prolonged periods of time. This treatment provided for an accumulation of formic acid in the blood of about 10 mEq/l. Major clinical observations were a marked optic disc edema which developed usually at about 40 hours after the initial dose of methanol. Dilated pupils and markedly reduced reaction to light were also seen in these animals but CSF pressures were within normal limits. Marked intraaxonal swelling in the area of the optic disc and anterior optic nerve can be observed when studies were performed at a light microscopic level whereas no discrete lesions were seen in the posterior portion of the optic nerve. It has been suggested by previous workers (1) that the retinal ganglion cell layer is affected in human cases of methanol poisoning. It should be pointed out that in most human cases of methanol poisoning sufficient time lapses occur between death and histopathological evaluation that autolytic changes compromise definitive interpretation. Early experiments in this laboratory showed retinal changes which were seen in both control and methanol-poisoned monkeys. Therefore, animals in the current study were perfused *in vivo* with fixative in order to minimize autolytic changes that occur after death. In our experiments the architecture of the retina appeared normal under light and electron microscopy. The clinical observations seen in monkeys resemble clinical observations in man reported by Benton and Calhoun (13). Ultrastructural changes seen in the area of the optic disc and anterior portions of the optic nerve were dramatic. Marked intraaxonal swelling with severe mitochondrial disruptions were observed. In methanol-intoxicated monkeys marked disruption of mitochondria and accumulation of mitochondria in the anterior portion of the optic nerve have been described by Baumbach *et al.* (11). These changes have been described as criteria for axoplasmic flow stasis (14,15).

That formate should account for certain aspects of optic nerve toxicity is suggested by observations presented here. When formate was infused in a manner which allowed for maintenance of normal blood pH ocular toxicity similar to that seen in methanol-poisoned monkeys resulted.

If formate is the major toxic agent it would explain why correction of metabolic acidosis by bicarbonate might not reverse ocular toxicity. This observation has been made by others (7). Thus, our working hypothesis is that formate is responsible for the toxicity seen in methanol-poisoned monkeys. Studies by McMartin *et al.* (16) have been performed in order to determine whether formalaldehyde appears in blood, body fluids or tissues in folate-deficient methanol-poisoned monkeys, animals which are exquisitely sensitive to methanol (16). No accumulation of formalaldehyde *per se* was observed in any blood, body fluid or tissue sample in folate-deficient monkeys injected with methanol (16). Although one cannot rule out the participation of formaldehyde in the ocular toxicity of methanol there is a strong association of formate with the ocular findings seen clinically and microscopically. Recent observations by Nicholls (17) indicate that formate is capable of inhibiting cytochrome oxidase at concentrations which appear to be within the limits observed in studies reported here. That formate should also promote disruption of mitochondria is consistent with these observations and have led us to formulate a mechanism of toxicity based on the inhibition of cytochrome oxidase by formate followed by marked alterations of axoplasmic flow and physiologic function of the optic nerve.

VI. ACKNOWLEDGEMENTS

This work was supported by NIH Grants GM 19420 and GM 12675.

V11. REFERENCES

(1). Roe, O., Pharmacol. Rev. 7, 399 (1955).
(2). Gilger, A.P. and Potts, A.M., Amer. J. Ophtalmol. 39, 63, (1955).
(3). Koivusalo, M., in International Encyclopedia of Pharmacology and Therapeutics, Vol.II, Section 20, p.456, (1970).
(4). Tephly, T.R., Watkins, W.D. and Goodman, J.J., Essays in Toxicology 5, 149 (1974).
(5). McMartin, K.E., Makar, A.B., Martin-Amat, G., Palese, M. and Tephly, T.R. Biochem. Med. 13, 319 (1975).
(6). Clay, K.L., Murphy, R.C. and Watkins, W.D., Toxicol. Appl. Pharmacol. 34, 49, (1975).
(7). Potts, A.M., Amer.J.Ophthalmol. 39 (#2), 86 (1955).
(8). Cooper, J.R., and Felig, P., Toxicol. Appl. Pharmacol. 3, 202 (1961).
(9). Martin-Amat, G., Tephly, T.R., McMartin, K.E., Makar, A.B., Hayreh, S.S., Hayreh, M.M., Baumbach, G. and Cancilla, P., Arch. Ophthalmol., In Press.
(10). Hayreh, M.M., Hayreh, S.S., Martin-Amat, G., Tephly, T.R., McMartin, K.E., Makar, A.B., Baumbach, G. and Cancilla, P. Arch. Ophthalmol, In Press.
(11). Baumbach, G., Cancilla, P., Martin-Amat, G., Tephly, T.R., McMartin, K.E., Makar, A.B., Hayreh, S.S. and Hayreh, M.M. Arch. Ophthalmol., Submitted.
(12). Makar, A.B., McMartin, K.E., Palese, M., and Tephly, T.R. Biochem. Med. 13, 117 (1975).
(13). Benton, C.D. and Calhoun, F.P., Amer.J.Ophthalmol. 36, 1677 (1953).
(14). Weiss, P. and Pillai, A., Proc.Natl.Acad.Sci. 59, 48 (1964).
(15). Kapella, K. and Mayer, D. J.Physiol. 19, 70 (1967).
(16). McMartin, K.E., Martin-Amat, G., Makar, A.B. and Tephly, T.R., in Alcohol and Aldehyde Metabolizing Systems (R.G. Thurman, J.R. Williamson, H. Drott and B. Chance, Eds.) Vol. II, Academic Press, New York 1977, In Press.
(17). Nicholls, P., Biochem. Biophys. Res. Commun. 67, 610 (1975).

METHANOL POISONING: ROLE OF FORMATE METABOLISM IN THE MONKEY

K.E. McMartin, G. Martin-Amat, A.B. Makar, and T.R. Tephly

The University of Iowa

The accumulation of formic acid is an important factor in the production of the metabolic acidosis after methanol administration to the monkey. Formic acid accumulation following methanol administration does not occur in the rat, which oxidizes formate at rates that are markedly higher than those seen in the untreated monkey. The objectives of this study were to determine the pathway responsible for formate oxidation in the monkey and to compare the disposition of formate in the monkey with that of the rat. This information was used to increase and decrease the rate of accumulation of formate in the monkey and to increase or decrease the sensitivity of the monkey to methanol poisoning. Results show that a folate-dependent pathway is the major route of formate metabolism in the monkey as has been shown previously in the rat. Folate administration to the monkey increases formate oxidation. 4-Methylpyrazole treatment, after the production of metabolic acidosis and formic acidemia in the monkey, reverses the acidotic state and formic acidemia. Although the folate-deficient monkey is the most sensitive animal towards methanol poisoning, formaldehyde did not appear in the blood, body fluids, or tissues. Since formaldehyde does not accumulate in the presence of high methanol and high formate, and since reversal of formate accumulation leads to reversal of the features of the methanol poisoning syndrome, it is suggested that formate rather than formaldehyde is the major determinant of methanol poisoning in monkeys and probably in man.

I. INTRODUCTION

The course of development of metabolic acidosis in methanol poisoned monkeys parallels the accumulation of formate and depletion of bicarbonate in the blood (1). In general, increases in blood formate can be accounted for by decreases in blood bicarbonate levels (1,2). Recent studies have shown that it is possible to produce ocular toxicity in the monkey when chronic administration of methanol is employed to produce a moderate metabolic acidosis and formic acidemia (3,4,5). Other studies have shown that pretreatment of animals with 4-methylpyrazole (4-MP) delays the rate of accumulation of blood formic acid, the metabolic acidosis and death (1).

The rat is not sensitive to methanol poisoning: no metabolic acidosis nor significant formate accumulation occurs (6). Since the accumulation of formate in the animal is involved in the generation of metabolic acidosis and represents an important aspect of methanol toxicity, information on the metabolic disposition of formate in sensitive and insensitive species has been studied.

Recently, Palese and Tephly (7) showed that formate is oxidized to CO_2 by a folate-dependent pathway in the rat. While it has often been held that formate oxidation to CO_2 is mediated by a catalase-peroxidative system, 3-amino-1,2,4-triazole (AT), a catalase inhibitor and ethanol, an alternative substrate for the catalase-peroxidative system, had no effect on formate oxidation to CO_2 *in vivo* in the rat. Only in folate-deficient rats was there a decrease in the rate of formate oxidation to CO_2 and, only in folate-deficient rats, could a role for catalase in formate oxidation be demonstrated (7).

II. RESULTS

The dose dependency of the rate of formate oxidation to CO_2 in monkeys and rats is shown in Figure 1. Over the dose range studied, the rat oxidized formate at rates that were markedly greater than the monkey. Makar and Tephly (6) have shown that, when rates of formate oxidation to CO_2 in rats are reduced to values similar to those seen in the monkey, the rat exhibits features of methanol poisoning following the administration of methanol. These include the accumulation of formic acid in the blood and marked decreases in blood pH values. Therefore, it was reasoned that the rate of formate oxidation to CO_2 is an important factor in the determination of species sensitivity to methanol poisoning.

The pathway involved in formate oxidation to CO_2 in the rat is dependent upon folate (7). Since the catalase system appears not to function in the rat and since it has been shown that methanol oxidation to CO_2 in the monkey is not affected by the catalase inhibitor, AT, it was not surprising to learn

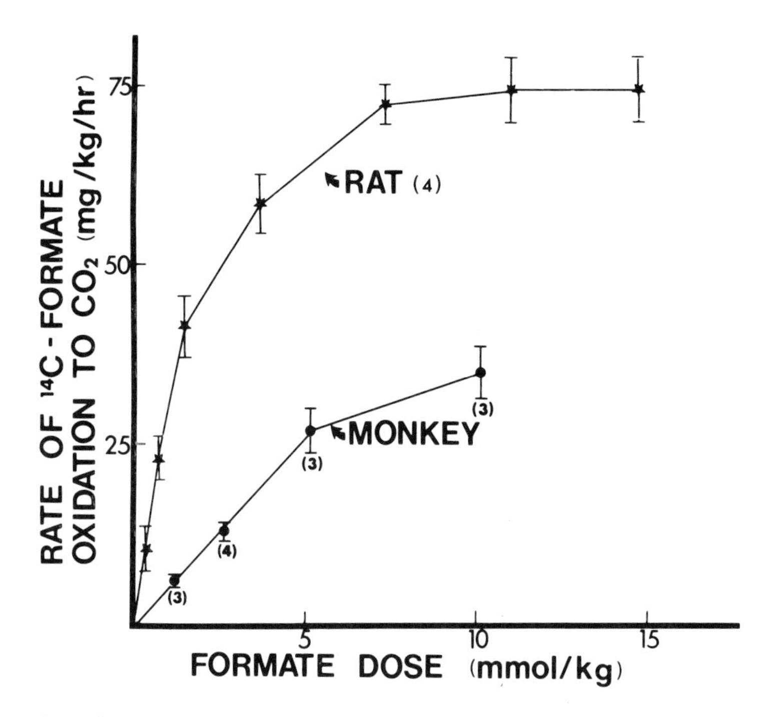

Fig. 1. Species differences in the rat of formate metabolism. Sodium [^{14}C]-formate was administered intravenously to the monkey. Rates of formate oxidation to $^{14}CO_2$ were determined during periods of time when the rate of $^{14}CO_2$ formation approximated linearity. Values for the rat are obtained from a paper by Palese and Tephly (7). Each point represents the mean value ± S.E.M. Numerals within parenthesis indicate the number of animals.

that AT had no effect on ^{14}C-formate oxidation to $^{14}CO_2$ in the monkey (8). Furthermore, AT pretreatment of monkeys sufficient to reduce hepatic catalase activity to 10% of control values had no effect on the rate of elimination of formate from the blood (8).

It was presumed that the pathway for formate oxidation to CO_2 in the monkey was folate-dependent. Studies to test this hypothesis were performed by placing male cynomolgus monkeys into 2 groups and presenting one with a folic acid-deficient diet and one with the folate-deficient diet which contained an additional 2.5 mg of folic acid per kg diet. Animals were maintained on these diets for about 14 weeks during which time hepatic folate levels and urinary formimino-L-glutamic acid (FIGlu) excretion in the urine were measured. Liver samples

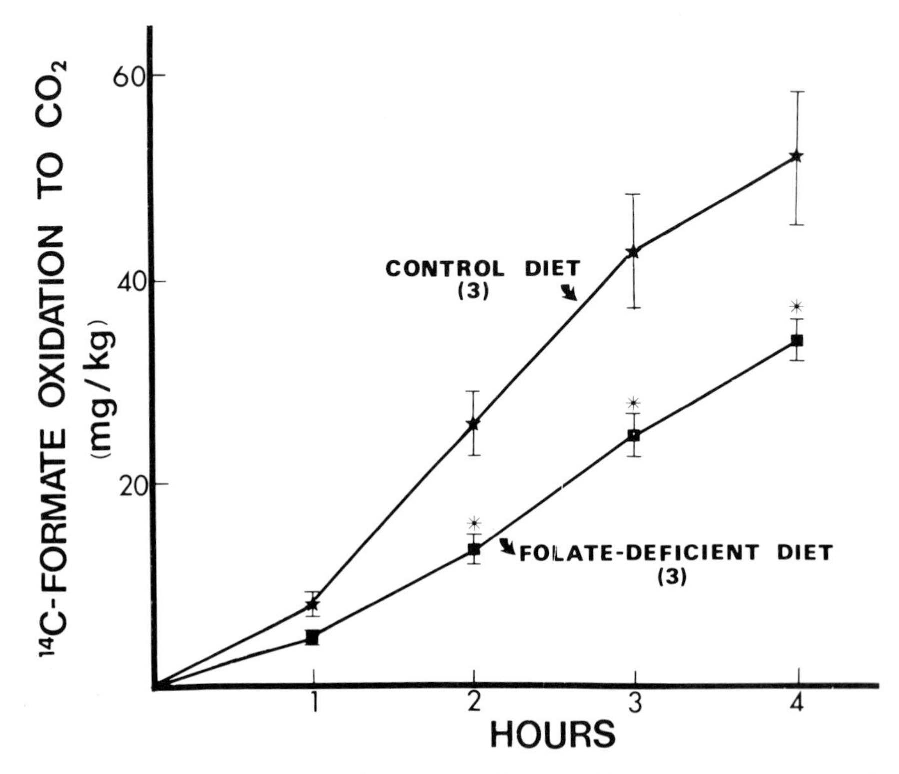

Fig. 2. Effect of folate-deficient diet on formate metabolism in the monkey. Sodium [^{14}C]-formate (2.5 mmol/kg) was administered intravenously to monkeys maintained on the control diet or on the folate-deficient diet. Each point represents the mean value ± S.E.M. Numerals within parenthesis indicate the number of animals. An asterisk (*) indicates significant difference from the control value ($P < 0.05$).

were obtained from each animal by percutaneous needle biopsy and prepared for folate analysis by a method based on that of Chanarin *et al.* (9). Measurements were made using a radioassay similar to that described by Longo and Herbert (10) using kits supplied by Diagnostic Biochemistry. Fourteen weeks after the initiation of these experiments liver folate values were reduced from about 8 μg/g of liver to about 1.5 μg/g of liver. At this time urinary FIGlu concentrations in the folate-deficient monkeys were increased about 10-fold. Thus, folate-deficient diets were effective in inducing folate-deficient states in these animals.

Results in Fig. 2 show that there is a marked decrease in the rate of ^{14}C-formate oxidation to $^{14}CO_2$ in monkeys that had been placed on a folate-deficient diet. A 50% decrease in the rate of formate oxidation is observed. Figure 3 shows that in these animals there is a 50% decrease in the rate of formate elimination from the blood and thus, a 2-fold increase in the half-life of formate disappearance from the blood (74 min) as compared to the value in control monkeys (39 min). Thus, a folate-dependent pathway is primarily responsible for formate oxidation in the monkey *in vivo*. However, unlike results obtained in the rat, administration of AT to folate-deficient monkeys produced no further decrease in the rate of oxidation of formate to CO_2. This indicates that the catalase-peroxidative system does not function in the oxidation of formate to CO_2 even in monkeys whose rate of formate oxidation has been decreased by folate deficiency.

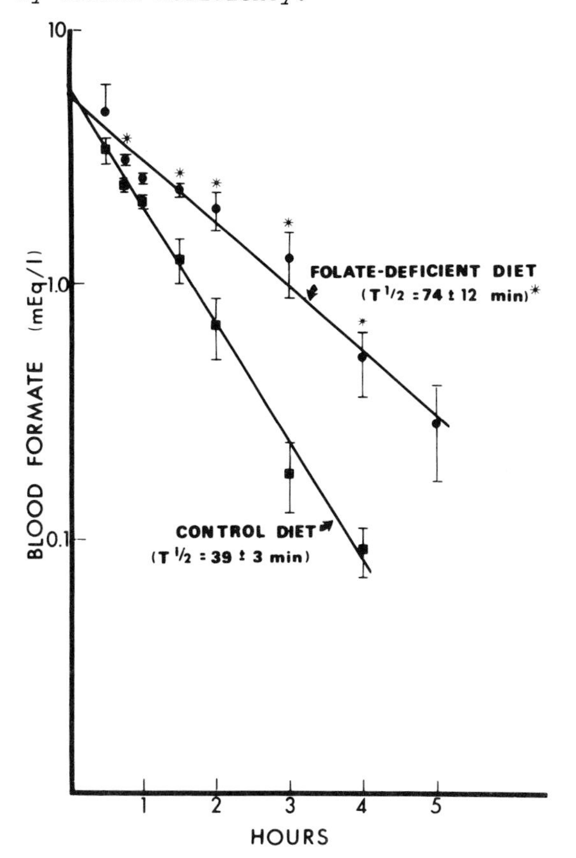

Fig. 3. Formate elimination from the blood of folate-deficient monkeys. Sodium [^{14}C]-formate (2.5 mmol/kg) was administered intravenously to monkeys maintained on the folate-deficient diet or on the control diet. Formate blood levels determined by an assay previously described (11). Each point represents the mean value ± S.E.M. for 3 animals. Solid lines represent regression analyses on the linear phase of the disappearance curves. The half-life represents the mean value ± S.E.M. for the values obtained from regression analysis of the disappearance curve for each animal. An asterisk (*) indicates significant difference from the control value ($P < 0.05$).

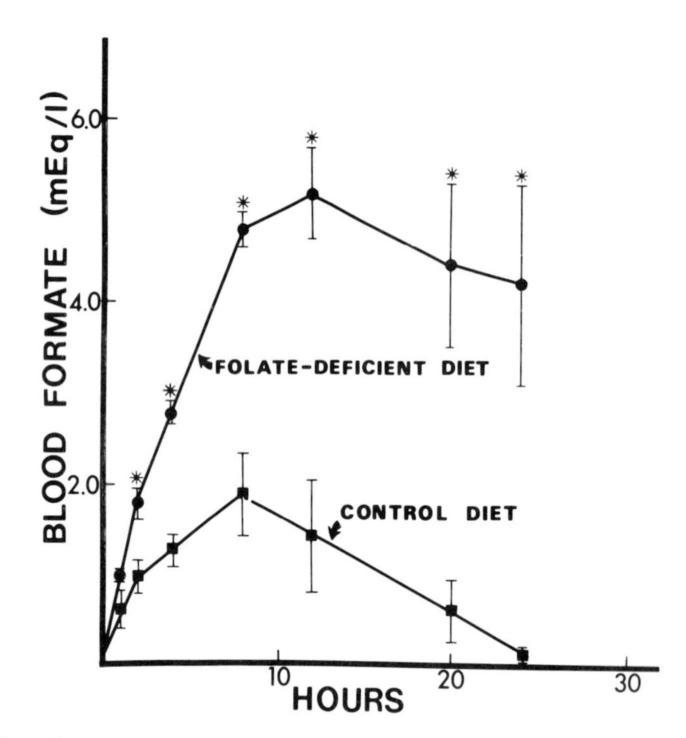

Fig. 4. Effect of folate deficiency on methanol toxicity in the monkey. Methanol (0.5 g/kg) was administered orally as a 20% (w/v) solution to monkeys maintained on the control diet or on the folate-deficient diet. Each value represents the mean value ± S.E.M. for 3 animals. An asterisk (*) indicates significant difference from the control value ($P < 0.05$).

Since folate-deficient monkeys display a reduced rate of formate oxidation, folate-deficient monkeys should be more susceptible to the effects of methanol than control animals. Results in Figure 4 show that after the administration of a low dose of methanol (0.5 g/kg, orally) there is a striking increase in the accumulation of formic acid in the blood of the folate-deficient monkeys. In these experiments, the rate of methanol disappearance from the blood of monkeys from each group was the same. Therefore, the increased accumulation of formate in the blood of folate-deficient monkeys can be accounted for by decreases in the rate of formate oxidation to CO_2 and not by increased methanol metabolism. Furthermore, these results suggest the major role of folate in

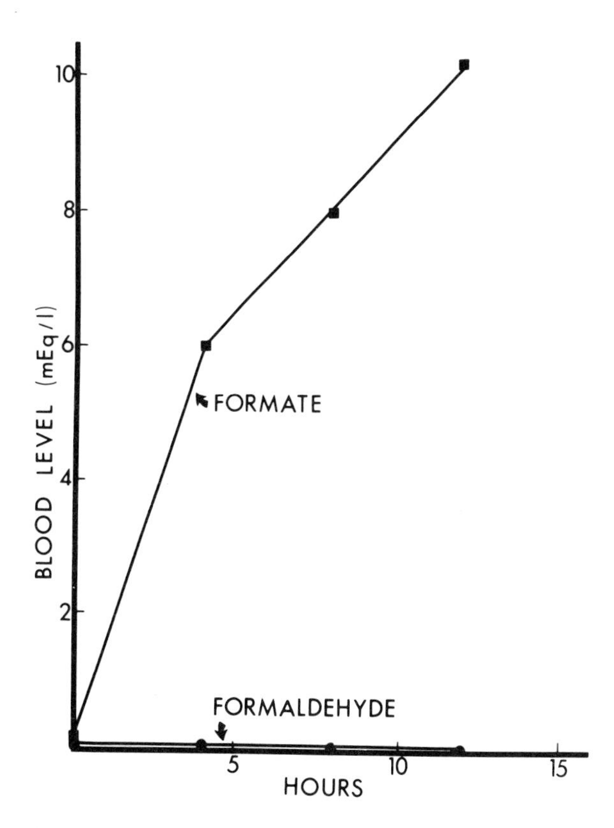

Fig. 5. Formaldehyde and formate blood levels in a methanol-poisoned folate-deficient monkey. Methanol (2.0 g/kg) was administered orally as a 20% (w/v) solution to a monkey maintained on the folate-deficient diet. Blood samples were obtained and analyzed for formaldehyde content by methods described in the text. Formate levels were determined as described previously (11).

formate oxidation and in regulating the susceptibility of the monkey to methanol poisoning. These results also indicate that the folate-deficient monkey is more sensitive to methanol poisoning than is the control monkey.

Since the folate-deficient monkey appears to be a sensitive animal model for methanol poisoning the folate-deficient monkey was employed in studies designed to test whether formaldehyde accumulates in methanol-poisoned animals and whether formaldehyde accumulation relates to methanol toxicity. Fig. 5 shows results obtained in a folate-deficient monkey which had been administered methanol (2 g/kg of body weight) orally at zero time. Over the course of 12 hours, blood and urine samples were obtained and analyzed for formaldehyde and formate. Blood samples were drawn from an arterial cannula into syringes containing semicarbazide. Tissue samples were obtained at 12 hours after methanol and homogenates were prepared with 1.15% KCl containing 112.5 μmol/ml of semicarbazide. Body fluid samples were diluted with a semicarbazide solution to a final concentration of 75 μmol/ml of semicarbazide. All samples were distilled from acid and analyzed for formaldehyde content using the sensitive and specific chromatropic acid technique described by MacFadyen (11). This method has been previously described by Tephly *et al.* (12). No formaldehyde accumulation occurred in blood or in urine samples determined over 12 hours following the administration of 2 g/kg of methanol. Furthermore, no formaldehyde was detected in the following body tissues (Detectability limits at about 50 nmol/g wet weight): liver, kidney, skeletal muscle, cerebrum, cerebellum, midbrain, hypothalamus, and optic nerve. Vitreous humor and cerebrospinal fluid were also studied and no formaldehyde was detected in these fluids (detectability limit at about 15 nmol/ml). Formate values for blood are indicated and they represent values which are about twice those seen in the normal monkey over the course of this experiment and 500 times the upper limit of potentially detectable formaldehyde. In the vitreous humor the formate concentration was 7.9 mEq/l 12 hours after methanol administration, a value that corresponds to that seen in the blood and cerebrospinal fluid at this time period. Therefore, in the presence of high methanol and formate values, formaldehyde concentrations were negligible in the folate-deficient monkey.

Figure 6 shows results which compare the rat and the monkey with respect to folate pretreatment and the rates of formate oxidation to CO_2. The rate of formate oxidation is enhanced by folate pretreatment in the monkey whereas folate pretreatment of the rat has no effect on the rate of formate oxidation. These results highlight another feature of the species difference seen in the animal's capacity to metabolize formate. Other differences of the monkey are the ability to accumulate formate and the lack of any role of catalase in methanol metabolism in the folate-deficient monkey.

A reduction in the rate of formate accumulation in the blood (Figure 7) can be observed in animals that have been administered methanol (3 g/kg) followed by the alcohol dehydrogenase inhibitor, 4-MP, once an acidotic state had been achieved (at about 10 hours). The administration of 4-MP to the methanol-intoxicated monkey led to a rapid reversal of the acidotic state; blood formate values rapidly decreased to negligible levels. At 40 hours following the administration of methanol,

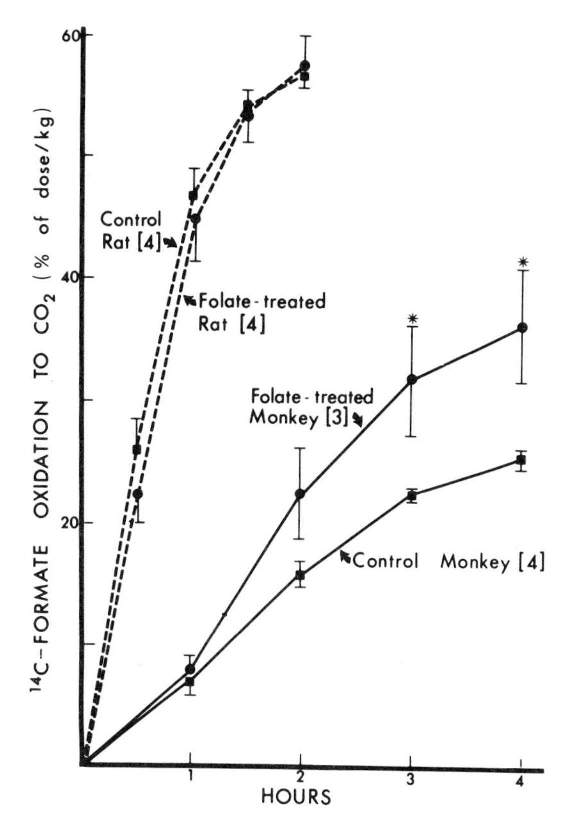

Fig. 6. Species difference in formate metabolism with folate pretreatment. Sodium folate (folic acid, 50 mg/kg in a 7.5% (w/v) solution of soidum bicarbonate) was injected intraperitoneally 48,24, and 1 hr prior to sodium [^{14}C]- formate (2.5 mmol/kg in monkey, 1 mmol/kg in rat) which was administered intravenously. Each point represents the mean value ± S.E.M. Numerals within parenthesis indicate the number of animals. An asterisk (*) indicates significant difference from the value obtained for animals treated with formate alone ($P < 0.05$).

the metabolism of methanol to formate and carbon dioxide was noted to increase and another dose of 4-MP was administered. This was again followed by a rapid reversal of the metabolic acidosis and formic acidemia. These results indicate that, by administering an inhibitor of alcohol dehydrogenase and decreasing methanol oxidation, it is possible to reverse methanol intoxication in the monkey. Previous results from this laboratory (1) have shown that pretreatment with 4-MP led to a protection of the animal against metabolic acidosis and formic acidemia after methanol administration.

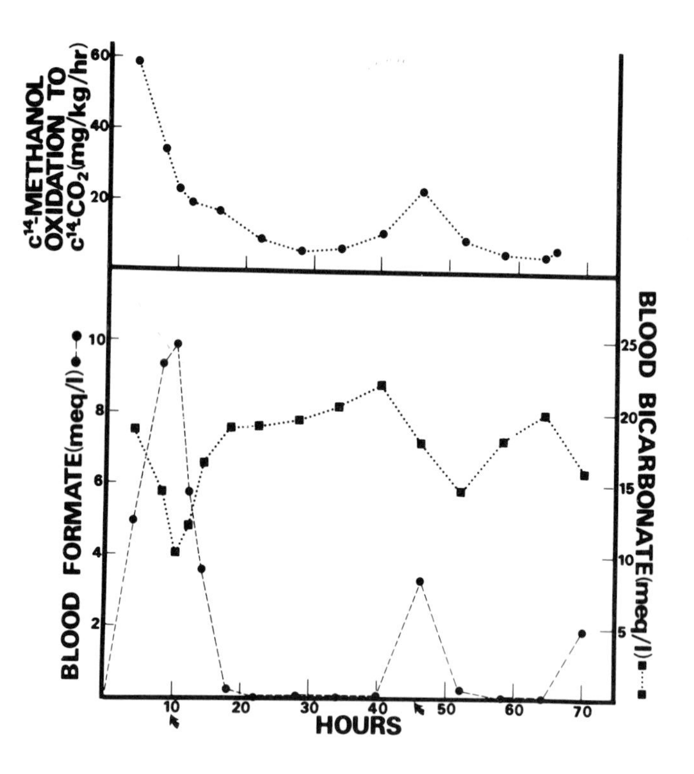

Fig. 7. Prevention of methanol toxicity in the monkey by repeated doses of 4-methylpyrazole (4-MP). [^{14}C]-methanol (3 g/kg) was administered orally at zero time. 4-MP (50 mg/kg) was administered at times indicated by the arrows. A final dose of 4-MP (5 mg/kg) at 81 hr was necessary to completely prevent the toxicity. Blood bicarbonate levels were calculated from the arterial blood pH and pCO_2 values. Blood formate levels were determined as previously described (11).

III. DISCUSSION

The sensitivity of the monkey to methanol poisoning appears to be related to the rate of formate oxidation in this species. The rat, which metabolizes formate to CO_2 faster than the monkey, is insensitive to methanol poisoning and is only rendered sensitive by reducing the rate at which it metabolizes formate to CO_2 (6). We have demonstrated that the monkey depends upon a folate-dependent system for oxidation of formate to CO_2. It was possible to decrease the rate of formate oxidation simply by inducing a folate-deficient state in the monkey. Folate-deficient monkeys oxidized formate to CO_2 at rates much lower than control animals and displayed increased sensitivity to methanol intoxication with a marked increase in capacity to accumulate formate in the blood after methanol administration.

Other workers have suggested that methanol exerts its toxicity through formaldehyde (13). In studies by Potts (14), no relationship between formate and methanol toxicity in monkeys was observed. He proposed that formaldehyde was responsible for the ocular toxicity seen after methanol administration, and concluded that formate accounted for only a small portion of organic acid excretion observed following methanol administration. Studies that have been performed in other laboratories as well as ours (1,2) demonstrate that bicarbonate depletion matches formate accumulation in the blood. In addition, we have shown that it is possible to produce ocular toxicity after administration of methanol and, more recently, we have reproduced most of the ocular toxicity seen after methanol by treating monkeys with formate (5). Thus, the association of formate with metabolic acidosis and blindness is formidable. On the other hand, in the animal model which is very sensitive to methanol poisoning, i.e., the folate deficient monkey, studies performed to detect formaldehyde in blood, urine, body fluids and tissues at a time when methanol poisoning was rampant showed that formaldehyde could not be detected by sensitive and specific methods. Thus, while one cannot rule out the participation of formaldehyde in some aspects of methanol poisoning, formate accumulation in methanol poisoning appears to represent a major factor. Also, the strong association of formate accumulation with methanol poisoning indicates that the best determinant of methanol poisoning in man should be the level of blood formate. It should be the guide by which treatment of methanol poisoning should be followed and it should assist in the regulation of treatment whichever modality is used.

IV. ACKNOWLEDGEMENTS

Supported in part by NIH grants GM 19420 and GM 12675.

V. REFERENCES

(1). McMartin, K.E., Makar, A.B., Martin-Amat, G., Palese, M., and Tephly, T.R., Biochem. Med. 13, 319 (1975).

(2). Clay, K.L., Murphy, R.C., and Watkins, W.D., Toxicol. Appl. Pharmacol. 34, 49 (1975).

(3). Martin-Amat, G., Tephly, T.R., McMartin, K.E., Makar, A.B., Hayreh, M.M., Hayreh, S.S., Baumbach, G., and Cancilla, P., Arch. Ophthalmol. In press (1976).

(4). Hayreh, M.M., Hayreh, S.S., Martin-Amat, G., Tephly, T. R., McMartin, K.E., Makar, A.B., Baumbach, G., and Cancilla, P., Arch. Ophthalmol. In press (1976).

(5). Martin-Amat, G., McMartin, K.E., Makar, A., Baumbach, G., Hayreh, M.M., Hayreh, S.S., and Tephly, T.R., in "Alcohol and Aldehyde Metabolizing Systems" (R.G. Thurman, J.R. Williamson, H. Drott, and B. Chance, Eds.), Vol. 2. Academic Press, New York, 1977. In press.

(6). Makar, A.B. and Tephly, T.R., Nature 261, 715 (1976).

(7). Palese, M. and Tephly, T.R., J. Toxicol. Environ. Hlth. 1,13 (1975).

(8). McMartin, K.E., Martin-Amat, G., Makar, A.B., and Tephly, T.R., J. Pharmacol. Exp. Ther. (Submitted).

(9). Chanarin, I., Hutchinson, M., McLean, A., and Moule, M., Brit. Med. J. 1,396 (1966).

(10). Longo, D.L. and Herbert, V., J. Lab. Clin. Med. 87, 138 (1976).

(11). Makar, A.B., McMartin, K.E., Palese, M., and Tephly, T. R., Biochem. Med. 13, 117 (1975).

(12). MacFadyen, D.A., J. Biol. Chem. 153, 107 (1945).

(13). Tephly, T.R., Parks, R.E., Jr., and Mannering, G.J., J. Pharmacol. Exp. Ther. 131, 147 (1961).

(14). Gilger, A.P., Potts, A.M., and Farkas, I.S., Amer. J. Ophthalmol. 42, 244 (1956).

(15). Potts, A.M., Amer. J. Ophthalmol. 39, 86 (1955).

MITOCHONDRIAL INJURY IN EXPERIMENTAL CHRONIC ALCOHOLISM

O.R. Koch*, A. Boveris* and A.O.M. Stoppani*

Universidad de Buenos Aires

Mitochondria of non-fatty livers from rats fed an alcoholic "super diet" for 4 months displayed enlargement and bizarre configuration. In vitro the isolated mitochondria showed reduced oxidation of succinate, malate-glutamate and 3-hydroxybutyrate by 50, 47 and 24%, respectively. Energy coupling remained unchanged in relation to mitochondria of control rats consuming regimens in which alcohol was isocalorically replaced by sucrose, fat or a basal diet. Mitochondria from the alcohol-fed animals exhibited a marked reduction in the content of cytochromes $a{+}a_3$ *(62%) and b (30%) and in the activity of cytochrome oxidase (36%). Succinate dehydrogenase and NADH dehydrogenase activities as well as the content of cytochromes c and* c_1 *and ubiquinone were not affected by ethanol feeding. Chronic alcohol intake inhibited* ^{14}C*-leucine incorporation into protein in liver mitochondria by 45% with respect to the control groups. Mitochondrial production of* H_2O_2 *in the presence of succinate and antimycin A was decreased by 56% in the alcohol group. The ATP-induced mitochondrial contraction after* Ca^{2+}*(+Pi) swelling was decreased by 33% in the alcohol-treated rats. These results clearly indicate that liver mitochondria are significantly damaged when ethanol at high levels is supplied for long periods of time to rats receiving a basal diet which does not induce a fatty liver. The data also suggest that chronic interference by ethanol with protein synthesis may lower the content of some components of the electron transport chain resulting in lowered respiratory activity.*

I. INTRODUCTION

Chronic ingestion of alcohol produces morphological

---

* Career Investigator from Consejo Nacional de Investigaciones Cientificas y Tecnicas, Argentina.

alterations of liver mitochondria, such as enlargement, distortion and appearance of megamitochondria, in humans (1) and in experimental animals (2). The functional significance of these alcohol-associated mitochondrial changes is still uncertain. Reported results on *in vitro* measured mitochondrial parameters are conflicting, probably reflecting differences in the *in vivo* experimental conditions. In our experimental model, animals were subjected to conditions in which the following requirements were fulfilled: (a) a high alcohol intake; (b) a true chronic period of alcohol feeding (4 months); and (c) the absence of fatty liver. With this model, it is possible to develop marked structural changes in liver mitochondria without other morphological alterations of the hepatocyte. In order to characterize the functional significance of the alcohol-associated mitochondrial pathology we have determined: (a) respiratory parameters involving oxygen uptake; (b) the content of some components of the respiratory chain; (c) the activity of the primary dehydrogenases; (d) some enzymatic activities related to specific cytochromes; (e) volume changes associated with ion transport; and (f) amino acid incorporation as a measure of mitochondrial protein synthesis.

II. MATERIALS AND METHODS

Groups of 6-10 male Wistar rats of approximately 100 g initial body weight were fed a solution of 32% ethanol-25% sucrose as drinking fluid (alcohol group) for a period of four months. In addition, the animals consumed a basal diet highly supplemented with lipotropes and vitamins ("super diet") as previously described (3). With this method, ethanol provided 35-40% of the total calories consumed by the rats. The control groups were offered water *ad libitum* and their solid diets were based on the final regimen (alcohol + basal diet) consumed by the alcohol group but in which alcohol was isocalorically replaced by sucrose (sucrose group), fat (fat group) or the whole basal diet (basal diet group). After four months the rats were killed and their livers were excised. Tissue samples were taken for light and electron microscopy studies by the same methods previously described (3). The remainder of the liver was homogenized in 0.25 M sucrose, 1 mM EDTA, 5 mM Tris-HCl buffer (pH 7.4) and mitochondria were isolated.

Mitochondrial oxygen uptake was measured with a vibrating platinum electrode in a reaction medium containing 0.24 M sucrose, 34 mM KCl, 5 mM $MgCl_2$, 1 mM EDTA, 5 mM phosphate, 9 mM Tris-HCl buffer (pH 7.4). Mitochondrial protein (1-2 mg/ml) and ADP (0.2-0.3 mM) were added successively. NADH dehydrogenase (4), succinate dehydrogenase (5), and cytochrome oxidase

(6) were measured in mitochondrial fragments obtained after osmotic shock (7). Total cytochromes were determined by differential spectrophotometry (dithionite reduced-oxidized) in a split beam spectrophotometer (8). Total ubiquinone (9), mitochondrial production of H_2O_2 (10), catalase (11), mitochondrial swelling and contraction (12), ^{14}C-leucine incorporation (13) and mitochondrial protein (14) were measured as described in the corresponding references. The values in the tables indicate mean values ± S.E.M. Differences between means were analyzed by the variance method.

III. RESULTS AND DISCUSSION

A. Light and Electron Microscopy

Very few fatty changes, mainly localized in central areas of the liver lobule, were found in the alcohol-treated rats. In sections stained with light green-chromotrope 2 R, numerous megamitochondria were observed in central and midzonal areas. Sections obtained from the different control groups showed fatty changes similar to those of the alcohol group, but the hepatocytes affected were those of periportal areas. No megamitochondria were observed in the control groups.

The ultrastructural alterations in the liver of rats from the alcohol group were essentially similar to those described previously (2,3,15,16). Mitochondria were enlarged and displayed bizarre shapes. Their cristae were shorter and diminished in number. The ultrastructural configuration of the hepatocytes from control groups was normal.

B. Respiratory Activity

The maximal physiological rate of oxygen uptake (state 3) with succinate, malate-glutamate and 3-hydroxybutyrate as substrates, of liver mitochondria from alcohol fed rats was significantly lower than the corresponding controls (Table 1). This decrease in the state 3 respiration was the cause of the lower respiratory control values also observed in this group. The ADP:O ratio was unaffected when succinate or malate-glutamate were the substrates and was diminished 12% with 3-hydroxybutyrate. A depression of the succinate-supported state 3 respiration was previously observed (17,18) in rats treated with an alcoholic liquid diet that produces fatty liver. Rats fed on normal solid diets plus 15-20% ethanol in the drinking fluid for periods ranging from a few weeks to several months,

TABLE 1
Oxidation and Phosphorylation in Liver Mitochondria in Alcohol-Treated and Control Rats

Group	*n*	*Maximal rate of respiration (ng-at O/min/ mg of protein)*	*ADP:O*	*Respiratory Control Ratio*[a]
		Substrate: succinate (10 mM)		
Alcohol	10	83 ± 9	1.34 ± 0.06	3.9 ± 0.3
Sucrose	7	176 ± 6[b]	1.48 ± 0.05	5.4 ± 0.5[c]
Fat	7	163 ± 11[b]	1.46 ± 0.05	5.2 ± 0.5[c]
Basal Diet	4	159 ± 16[b]	1.37 ± 0.06	4.7 ± 0.6
Substrate: malate (5 mM) - glutamate (5 mM) - malonate (2.5 mM)				
Alcohol	8	61 ± 5	2.22 ± 0.05	4.3 ± 0.6
Sucrose	6	118 ± 5[b]	2.32 ± 0.05	7.2 ± 0.8[d]
Fat	6	108 ± 7[b]	2.34 ± 0.04	7.2 ± 0.7[d]
Basal Diet	4	117 ± 12[b]	2.28 ± 0.07	7.3 ± 1.2
		Substrate: 3-hydroxybutyrate		
Alcohol	8	45 ± 3	1.89 ± 0.09	3.6 ± 0.4
Sucrose	6	61 ± 4[c]	2.28 ± 0.05[f]	4.3 ± 0.5
Fat	6	59 ± 4[c]	2.24 ± 0.07[f]	4.6 ± 0.6
Basal Diet	4	57 ± 7	2.05 ± 0.21	4.1 ± 0.7
		Substrate: 1-glycerophosphate (10 mM)		
Alcohol	4	28 2	1.70 0.08	2.5 0.3
Sucrose	4	19 2[d]	-----------	---------

[a] *Ratio of respiratory rate in the presence of ADP vs. the respiratory rate before ADP addition.*
[b] $p < 0.001$
[c] $p < 0.05$
[d] $p < 0.01$
[e] $p < 0.005$
[f] $p < 0.02$

and apparently having non-fatty liver, show normal (19,20,21), decreased (19,20,21) or even increased (20,21) respiration with succinate or NAD$^+$-linked substrates. These contradictory observations can be explained by the different models used by these authors who failed to induce an alcohol consumption of over 25% of the total caloric intake. Following treatment of rats with liquid diets for 3-4 weeks, Videla *et al.* (22) were unable to detect changes in the state 3 mitochondrial respiration with 3-hydroxybutyrate and succinate as substrates, probably due to the short time period of alcohol administration.

Chronic alcohol feeding increased l-glycerophosphate oxidation in liver mitochondria by 50%, in agreement with previous reports that l-glycerophosphate-supported oxygen uptake (20,21) and l-glycerophosphate dehydrogenase (23) activity was increased in ethanol fed animals. However, our results are at variance with Rubin *et al.* (24) and Cederbaum and Rubin (25) who failed to show increase in either oxygen uptake with l-glycerophosphate or glycerophosphate dehydrogenase activity, in all probability because the short period of alcohol intake that did not allow the fully adaptive physiological response to ethanol to occur.

C. Components of the Respiratory Chain

Cytochromes $\underline{a}+\underline{a}_3$ and $\underline{b}$ were markedly reduced (62% and 30%, respectively) in the alcohol-treated group, whereas cytochromes $\underline{c}$, $\underline{c}_1$ and ubiquinone did not show significant variations (Fig. 1). In accordance with the decrease of cytochrome $\underline{a}+\underline{a}_3$ heme, the activity of cytochrome oxidase was 36% decreased in the ethanol-fed rats (Fig. 2). A smaller decrease in cytochromes $\underline{a}+\underline{a}_3$ and $\underline{b}$ was reported previously in rats fed only 24 days on liquid diets containing alcohol (17). It is known that at least 3 months of alcohol consumption are necessary to fully develop mitochondrial abnormalities in the liver (2,3,15,16). Figure 2 also shows succinate and NADH dehydrogenase activities which were unchanged in the alcohol-treated animals.

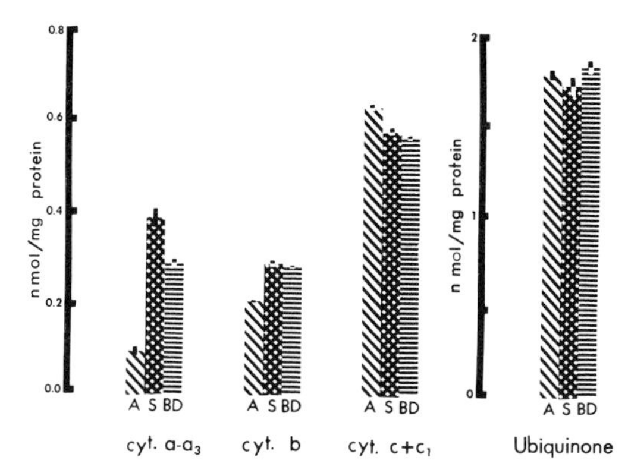

Fig. 1. Content of cytochromes and ubiquinone in liver mitochondria from alcohol-treated and control rats. A: control group; S: sucrose group; BD: basal diet group.

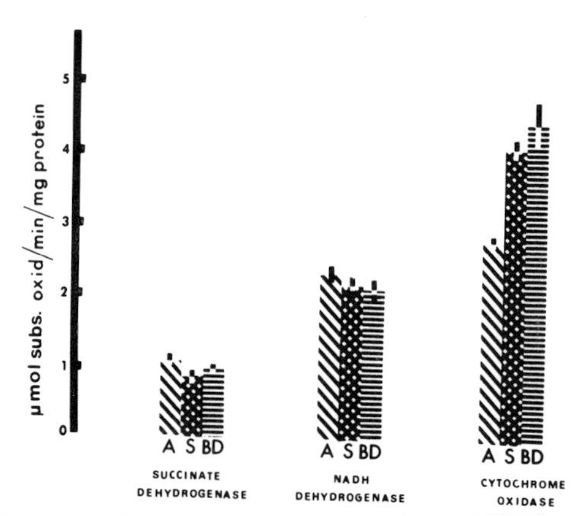

Fig. 2. Cytochrome oxidase, succinate dehydrogenase and NADH dehydrogenase activities in liver mitochondria from alcohol-treated and control rats. A: alcohol group; S: sucrose group; BD: basal diet group.

The mitochondrial production of H_2O_2 in the presence of succinate and antimycin A was decreased by 50% in alcohol-treated rats ($P < 0.01$; alcohol group: 0.35 ± 0.03; sucrose group: 0.78 ± 0.06; basal diet group: 0.76 ± 0.11; values expressed in nmole H_2O_2/min/mg of protein). This decrease is consistent with the lowered content of cytochrome b, assuming that some form of this cytochrome is involved in H_2O_2 production, particularly when we bear in mind the normal succinate dehydrogenase activity. The unchanged catalase activity in the mitochondrial fraction isolated from the rats of the alcohol-treated group rules out a possible increased catalase interference in the assay for H_2O_2 (alcohol group: 0.16 ± 0.02; sucrose group: 0.15 ± 0.02; basal diet group: 0.14 ± 0.02; values expressed in nmole catalase heme/mg of protein).

D. Mitochondrial Swelling and Contraction

Mitochondrial volumetric changes induced by Ca^{2+} in the presence of phosphate were measured to estimate mitochondrial ionic permeability. The swelling capacity after addition of Ca^{2+}, determined by the extent of swelling and by the time required for half-maximal swelling, was similar in the alcohol-treated and control groups (Table 2). On the other hand, chronic alcohol intake affected both the extent (34%) and the rate (32%) of contraction after ATP-Mg^{2+} addition to Ca^{2+}(Pi)-swollen mitochondria. This alteration may indicate a subtle damage of the system of ATP utilization that operates in the reversal of mitochondrial swelling.

TABLE 2

Mitochondrial Swelling in Alcohol-Treated and Control Rats

Group	n	Extent of Ca^{2+}-induced swelling ($A_{520} \times 10^3$)	Time for half-maximum swelling (sec)	Extent of ATP-induced contraction ($A_{520} \times 10^3$)	Initial rate of contraction ($A_{520} \times 10^3$/min)
Alcohol	10	159 ± 18	49 ± 6	66 ± 8	34 ± 5
Sucrose	7	173 ± 22	52 ± 4	113 ± 16[a]	59 ± 8[a]
Fat	7	132 ± 24	56 ± 5	100 ± 13[a]	59 ± 8[a]
Basal Diet	4	158 ± 18	40 ± 7	87 ± 16	44 ± 7

a. $p < 0.02$

E. Mitochondrial Protein Synthesis

Mitochondria isolated from the liver of chronically alcohol-fed rats showed a marked reduction in amino acid incorporation. They incorporated 25% of the labeled leucine as compared with the values corresponding to the mitochondria isolated from the sucrose group ($p < 0.01$; alcohol group: 1152 ± 170; sucrose group: 3876 ± 489; basal diet group: 1459 ± 142; values expressed as cpm/mg of protein). However, the values from the alcohol-treated group were not statistically different from those obtained from the animals fed the basal diet. In this connection, a similar though smaller difference in amino acid incorporation into mitochondrial protein between alcohol treated animals and controls was observed by Rubin *et al.* (26) in miniature swine. Again, the difference with the present investigation may be explained by the longer period of alcohol intake employed in this study.

IV. CONCLUSIONS

The fact that mitochondria isolated from ethanol-treated rats show significant decreases in several parameters of mitochondrial activity (Table 3) indicates that chronic ethanol feeding induces a multiple lesion of the mitochondrial inner membrane. The reduction of the functional parameters of liver mitochondria found in the present study largely exceeds those previously described by other authors. The longer feeding time (4 months) and the relatively high ethanol content of our diets may account for the significant ultrastructural and

TABLE 3
Functional Alterations of Liver Mitochondria in Chronic Experimental Alcoholism

Experimental Parameters	Percentage of control values
1) Respiratory rate (State 3)	
a) malate-glutamate	53
b) succinate	50
c) 3-hydroxybutyrate	76
d) 1-glycerophosphate	150
2) Respiratory control ratio	
a) malate-glutamate	61
b) succinate	69
c) 3-hydroxybutyrate	78
3) ADP:O ratio	
a) malate-glutamate	91 (n.s.)
b) succinate	94 (n.s.)
c) 3-hydroxybutyrate	88
4) Ca^{2+}-induced swelling and ATP-induced contraction	
a) extent of Ca^{2+}-induced swelling	104 (n.s.)
b) half-time for maximum swelling	102 (n.s.)
c) extent of ATP-induced contraction	66
d) rate of contraction	68
5) Enzyme activities	
a) NADH-dehydrogenase	104 (n.s.)
b) succinate dehydrogenase	119 (n.s.)
c) cytochrome oxidase	64
6) Cytochrome and ubiquinone content	
a) cytochrome $\underline{a}$ + $\underline{a}_3$	38
b) cytochrome $\underline{b}$	70
c) cytochromes $\underline{c}$ and $\underline{c}_1$	114 (n.s.)
d) ubiquinone	101 (n.s.)
7) Hydrogen peroxide production	44
8) ^{14}C-leucine mitochondrial incorporation	55

Percentage of control value was calculated from the ratio between the alcohol group and the mean value of the non-alcoholic control groups. (n.s.) not statistically different.

biochemical lesions. Chronic interference by ethanol (or products of ethanol oxidation) with the mechanism of mitochondrial protein synthesis may result in a lowered content of the components of the respiratory chain and the energy conserving mechanism (cytochromes $\underline{a}+\underline{a}_3$, cytochrome $\underline{b}$, F_1-ATPase) whose synthesis depends in part on the mitochondrial genome.

V. ACKNOWLEDGEMENTS

This work was supported by grants from Consejo Nacional de Investigaciones Cientificas y Tecnicas, Argentina.

VI. REFERENCES

(1). Porta, E.A., Bergman, B.J., and Stein, A.A., Amer. J. Path. 46, 657 (1965).
(2). Porta, E.A., Hartroft, W.S., and De la Iglesia, F.A., Lab. Invest. 14, 1437 (1965).
(3). Koch, O.R., Bedetti, C.D., Gamboni, M., Anzola Montero, G., and Stoppani, A.O.M., Exp. Mol. Path., in press (1976).
(4). Minakami, S., Ringler, R.L., and Singer, T.P., J. Biol. Chem. 237, 569 (1962).
(5). Arrigoni, O., and Singer, T.P., Nature 193, 1256 (1962).
(6). Smith, L., Arch. Biochem. Biophys. 50, 285 (1954).
(7). Parsons, D.F., Williams, G.R., and Chance, B., Ann. N.Y. Acad. Sci. 137, 643 (1966).
(8). Williams, J.N., Arch. Biochem. Biophys. 107, 643 (1964).
(9). Redfearn, E.R., Methods Enzymol. 10, 381 (1967).
(10). Boveris, A., Oshino, N., and Chance, B., Biochem. J. 128, 617 (1972).
(11). Chance, B., in "Methods of Biochemical Analysis" (D. Glick, Ed.), p. 412. Interscience Publisher, New York, 1954.
(12). Boveris, A., Stoppani, A.O.M., Arch. Biochem. Biophys. 142, 150 (1971).
(13). Sirotsky de Favelukes, S.O., Schwarcz de Tarlovsky, M., and Stoppani, A.O.M., Acta Physiol. Lat. Amer. 21, 30 (1971).
(14). Gornall, A.G., Bardawill, C.J., and David, M.M., J. Biol. Chem. 177, 751 (1949).
(15). Koch, O.R., Porta, E.A., Hartroft, W.S., Lab.Invest. 18, 379 (1968).
(16). Koch, O.R., Porta, E.A., and Hartroft, W.S., Lab. Invest. 21, 298 (1969).

(17). Rubin, E., Beattie, D.S., and Lieber, C.S., Lab. Invest. 23, 620 (1970).
(18). Cederbaum, A.I., Lieber, C.S., and Rubin, E., Arch. Biochem. Biophys. 165, 560 (1974).
(19). Sardesai, V.M., and Walt, A.J., in "Biochemical and Clinical Aspects of Alcohol Metabolism" (V.M. Sardesai, ed.), p. 117. C.C. Thomas, Sprigfield, 1969.
(20). Kiessling, K.H., Acta Pharmacol. Toxicol. 26, 245 (1968).
(21). Kiessling, K.H., and Pilstrom, L., Br. J. Nutr. 21, 547 (1967).
(22). Videla, L., Bernstein, J., and Israel, Y., Biochem. J. 134, 507 (1973).
(23). Israel, Y., Videla, L., MacDonald, A., and Bernstein, J., Biochem. J. 134, 523 (1973).
(24). Rubin, E., Beattie, D.S., Toth, A., and Lieber, C.S., Fed. Proc. 31, 131 (1972).
(25). Cederbaum, A.I., and Rubin, E., Fed. Proc. 34, 2045 (1975).
(26). Burke, J.P., Tumbleson, M.E., Hicklin, K.W., and Wilson, R.B., Proc. Soc. Exptl. Biol. Med. 148, 1051 (1975).

INDEX

H

I

L

M

N

P

S

T

U, X, Y

A
B 7
C 8
D 9
E 0
F 1
G 2
H 3
I 4
J 5